URBAN PLANNING IN A CHANGING WORLD

The Twentieth Century Experience

Studies in History, Planning and the Environment
Series editor Professor Anthony Sutcliffe

The Rise of Modern Urban Planning, 1800-1914 edited by Anthony Sutcliffe
Shaping an Urban World: Planning in the twentieth century edited by Gordon E. Cherry
Planning for Conservation: An international perspective edited by Roger Kain
Metropolis 1980-1940 edited by Anthony Sutcliffe
Arcadia for All: The legacy of a makeshift landscape by Dennis Hardy and Colin Ward
Planning and Urban Growth in Southern Europe edited by Martin Ward
Thomas Adams and the Modern Planning Movement: Britain, Canada and the United States by Michael Simpson
Holford: A study in architecture, planning and civic design by Gordon E. Cherry and Leith Penny
Goodnight Campers! The history of the British holiday camp by Colin Ward and Dennis Hardy
Model Housing: From the Great Exhibition to the Festival of Britain by S. Martin Gaskell
Two Centuries of American Planning edited by Daniel Schaffer
Planning and Urban Growth in the Nordic Countries edited by Thomas Hall
From Garden Cities to New Towns: Campaigning for town and country planning, 1899-1946 by Dennis Hardy
From New Towns to Green Politics: Campaigning for town and country planning 1946-1990 by Dennis Hardy
The Garden City: Past, present and future edited by Stephen V. Ward
The Place of Home: English domestic environments by Alison Ravetz with Richard Turkington
Prefabs: A history of the UK temporary housing programme by Brenda Vale
Planning the Great Metropolis: The 1929 Regional Plan of New York and Its Environs by David A. Johnson

Recent titles

Rural Change and Planning: England and Wales in the twentieth century by Gordon E. Cherry and Alan Rogers
Of Planting and Planning: The making of British colonial cities by Robert Home
Planning Europe's Capital Cities: Aspects of nineteenth-century urban development by Thomas Hall
Politics and Preservation: A policy history of the built heritage, 1882-1996 by John Delafons
Selling Places: The marketing and promotion of towns and cities, 1850-2000 by Stephen V. Ward
Changing Suburbs: Foundation, form and function edited by Richard Harris and Peter Larkham

Titles for 2000

The Australian Metropolis: A planning history edited by Stephen Hamnett and Robert Freestone
Utopian England: Community experiments 1900-1945 by Dennis Hardy
Urban Planning in a Changing World: The twentieth century experience edited by Robert Freestone
A History of British Development Control by Philip Booth
The Making and Remaking of England's Interwar Suburbs by Jeremy W.R. Whitehand

URBAN PLANNING IN A CHANGING WORLD:
The Twentieth Century Experience

edited by
ROBERT FREESTONE

E & FN SPON
Taylor & Francis Group

First published 2000 by E & FN Spon,
11 New Fetter Lane,
London EC4P 4EE

Simultaneously published in the USA and Canada
by Routledge
29 West 35th Street, New York, NY 10001

E & FN Spon is an imprint of the Taylor & Francis Group

Published with the assistance of the Australian Housing and Urban Research Institute

Typeset in Times and New Baskerville by NP Design and Print, Wallingford
Printed and bound in Great Britain by Biddles Ltd, Guildford and King's Lynn

This book was commissioned and edited by Alexandrine Press, Oxford

British Library Cataloguing in Publication Data
A catalogue record for this book is available from the British Library

Library of Congress Cataloging in Publication Data
Urban planning in a changing world : the twentieth century experience /
Robert Freestone
p. cm. — (Studies in history, planning, and the environment)
Includes bibliographical references and index
1. City planning—History—20th century. I. Freestone, Robert. II. Series.
HT166.U7387 2000
307.1'216'0904—dc21

99-087685

ISBN 0-419-24650-9

Contents

Contributors

Robert Bruegmann is Professor in the Department of Art History and School of Architecture at the University of Illinois at Chicago, Chicago, Illinois, United States.

Jeffry M. Diefendorf is Professor of History at the University of New Hampshire, Durham, New Hampshire, United States.

Raphaël Fischler is Assistant Professor in the School of Urban Planning at McGill University, Montreal, Quebec, Canada.

Robert Freestone is Associate Professor of Planning and Urban Development at the University of New South Wales, Sydney, Australia.

Brendan Gleeson is Senior Research Fellow in the Urban Frontiers Program at the University of Western Sydney-Macarthur, Sydney, Australia.

Sir Peter Hall is Bartlett Professor of Planning in University College London.

David Hamer, who died in May 1999, was Professor of History at Victoria University of Wellington, New Zealand.

Dennis Hardy is Pro Vice-Chancellor at Middlesex University, London, England.

Jon Lang is Professor of Architecture at the University of New South Wales, Sydney, Australia.

Nicholas Low is Associate Professor at the Faculty of Architecture, Building and Planning, the University of Melbourne, Melbourne, Australia.

Dirk Schubert is Senior Lecturer at the Technical University of Hamburg-Harburg, Hamburg, Germany.

Allen J. Scott is Professor in the Departments of Policy Studies and Geography at the University of California, Los Angeles, California, United States.

Gilbert A. Stelter is University Professor Emeritus, Department of History, University of Guelph, Canada.

Susan Thompson is Senior Lecturer in the Planning and Urban Development Program at the University of New South Wales, Sydney, Australia.

Maurits Van Rooijen is Director of International Education at the University of Westminster, London, England.

Stephen V. Ward is Professor of Planning History at Oxford Brookes University, Oxford, England.

CHAPTER 1

Learning from Planning's Histories

Robert Freestone

If the twentieth century is remembered as the city century, it will also prove memorable as the century of urban planning. According to the late Gordon Cherry, in its 'long-standing sense of common concern for environment and the setting it provides for individual life chances and community satisfactions', planning ranks as 'one of the significant movements' of the past one hundred years (Cherry, 1980, p. 20). The history of cities, urban development and urbanization has attracted a prodigious, multi-faceted literature dealing with everything from the morphology of town layouts to the experience of urbanism. A self-conscious history of modern planning, alive to linkages between and within the social sciences, humanities and design arts, is a more recent phenomenon of the last quarter century, although it could be said that the history of planning has always been part of planning.

PLANNING AND HISTORY

An immanent historicity characterized all the seminal early texts, particularly those in the English language. The first issues of the *Town Planning Review* – started in 1910 by the pioneering Department of Civic Design at the University of Liverpool – contained many articles dealing with an historical appreciation of old towns (Cherry, 1991). The core of Professor Stanley Adshead's opening article to the first issue was an 'historical retrospect' (Adshead, 1910). Raymond Unwin's classic *Town Planning in Practice* (1909) treats history as mostly contributing to the 'individuality of towns'. It introduces the neo-medievalism expounded by Viennese architect Camillo Sitte in *Der Städtebau nach seinen künstlerischen Grundsätzen* (1889) to help codify artistic, intimate techniques for the modern 'art of designing cities and suburbs'. (And there is a direct link to the present-day neo-traditionalism of Andreas Duany (1994) who welcomes the reprinting of Unwin's work nearly nine decades after its first appearance 'not as a memorial to its historical significance, but as a modern manual on technique'.)

Almost invariably, the first texts on town planning contained an obligatory historical preamble and this practice was carried into actual planning documents. Daniel Burnham and Edward Bennett's landmark *Plan of Chicago* (1909) included a preliminary review of 'city planning in ancient and modern times' as one of the inspirational underpinnings of a strategy for remaking Chicago as a global city renowned for its exemplification of modern planning goals: convenience, healthfulness, beauty and order.

History, as it was written in this manner, quickly became entrapped in a narrow ideological formula: the past was appropriated to legitimate the roots and emergent professionalization of what was a very new and practical discipline. The treatment was at best myopically focused, at worst idiosyncratic, and generally always historicist. The revisionist high points of the past – be they Pericles' Athens or Louis XIV's Paris – were selectively excavated to define a chronology of world's best practice. The mythology of the past was linked to the idealism of the present, with great male visionaries of the early twentieth century responsible for updating and carrying the glorious story forward into the modern era. Inevitably, it was not long before the hero-planner volumes appeared: Charles Moore (1921) on Daniel Burnham, Dugald MacFadyen (1933) on Ebenezer Howard, and landscape architect and civic artist Thomas Mawson (1927) on himself.

This was the model institutionalized by planning education, as the widespread bureaucratization of state intervention into urban development processes after World War Two demanded larger numbers of trained professionals. Most standard texts of the 1940s and 1950s stick, at most, to the inherited historiography. The first rite of passage for learning to plan anew was thus to see how it had been done before, although the contemporary relevance of the layout of fourth century Alexandria or thirteenth century Montpazier was not always apparent. The treatment of history became increasingly derivative and dull – 'dumbed down' to a tedious chronology of events and images estranged from any kind of political, social, or economic context. A couple of generations of planning students, at the least, had to endure this desultory historical prolegomenon before they broke through to the interesting, practical stuff. It was an exclusive, empirical archaeology divorced from practice. At the same time, any chance of an enlightened historical consciousness was undermined during the very period when planners were beginning to make decisive impacts on modern urban landscapes in tandem with architects, developers and government.

With planning machines comfortably bedded down in the postwar welfare state, the early signs appear of a social science flavoured approach to planning history. William Ashworth's (1954) account of the origins of British planning is a breakthrough contribution: conservative, atheoretical, and seeking historical continuities, but significantly embedded in social and economic context. Generally, however, urban planning remained a scholarly backwater, of little real interest to other academic disciplines. The dominant discourse was of planning as an unambiguously beneficent statist activity: comprehensive, technocratic, scientific, and still somehow magically reformist. More was better. Regardless of platitudes about community participation, this was strictly a professional activity for trained experts. The planning history of the day largely legitimated this conventional wisdom.

The literature of planning thus entered the 1960s in an innocent, dreamlike state, its antediluvian history increasingly remote from social reality, professional practice, and the new urban politics. This comfort zone was soon under siege. The postwar political consensus about the scope and direction of state intervention in the broad 'public interest' had been good to planning. The heady days immediately after World War Two saw planning elevated to one of the central planks of social reconstruction. Even when that could not be sustained, it quietly prospered from bi-partisan tolerance. But now a new world order was coming into play: the notion of the centralized, benevolent state was crumbling; economic policy was forced to look beyond the industrial Fordism which had been the touchstone of social progress in the first half of the twentieth century; there was dissent on the streets about unpopular wars and civil rights; and the environmental movement was awakening. The old verities, as has been said, were being blown away.

Foundational elements of modernist planning like new towns, comprehensive redevelopment, low density suburbs, mono-functional land use zoning, high rise public housing towers, and freeways came under greater scrutiny. All had had their downsides, a track record of economic disbenefits and socially redistributive outcomes belying the inclusive, visionary rhetoric at their birth. Urban communities did not seem to be inheriting the planning wisdom of the ages. All-knowing, all-seeing comprehensive master plans – along with allied scientistic methodologies (Taylor, 1998) – came under particular assault. Planning began to fragment into epithetical ventures around specialist concerns: transportation, the social environment, ecological issues. The fragmentation and uncertainty reflected planning's internalization of wider social tensions and crises (Harvey, 1978).

It was possible, in the face of this, to reaffirm the old values: the quest for urban salvation through all-encompassing city and metropolitan blueprints. Informed by interviews with some of the central participants in planning in the early twentieth century, Mel Scott's *American City Planning since 1890* (1969) did precisely this. It is an important account which unapologetically treats planning as a fundamental force in national history and breaks free of the tired, decontextualized descriptions which had dominated planning history. But encased in a narrow professionalism, and being sponsored by the American Institute of Planners, it not only failed to 'penetrate the rhetoric of reform that planners drew upon to legitimize their actions' but also denied the 'differential impacts' of planning on diverse communities (Sies and Silver, 1996, p. 8). While it pulls much together to remain an obligatory reference work to today, it leaves so much out by failing to situate the theory and practice of planning more critically within their fullest societal milieus.

The self-confidence of Scott's book contrasts achingly with other interpretations emerging from the 1960s. The contributions by Paul Davidoff (1965) and Robert Goodman (1972) are devastating professional auto-critiques with a radically different spin on top-down technocratic planning. As planning got political, it attracted wider interest. The really influential texts telling new, or re-interpreting old, history were being written by commentators outside the planning profession. The classic text is Jane Jacobs' dyspeptic view of mainstream planning ideology viewed from Greenwich Village, *The Death and Life of Great American Cities* (1962). At the other end of the political spectrum, more supportive of established structures and re-establishing the case for enlightened, neo-Fabian public policy, was the suburb-friendly argument of political scientist Hugh Stretton (1970). The main point here was that planning history was now beginning to attract wider interest and a re-invigoration from multidisciplinary contributions.

In the wake of this maelstrom, a new planning history emerges in the 1970s. Its appearance seems partly explained by a collective urge to make sense of a cathartic period from which planning emerged somewhat bruised and battered, its fortunes declining, and no longer riding the wave of the welfare state. This was also a time of generational change, with the baton of leadership in the profession passing from the die-hard physicalists of the early postwar period to progressive, social science types. Planning history provided a scholarly focus for academic planners, and was well placed to benefit from the rising interest in urban studies, urban history, local history, and heritage studies. The new planning history crystallized in an inter-disciplinary zone welcoming of practitioners, philosophers and teachers from all fields – 'a definitive specialism in its own right, notwithstanding its indeterminate boundaries' (Cherry, 1991, p. 40). An event which captures this moment is the formation in 1974 of what has become the International Planning History Society. National societies would follow in North America, Europe and South Africa.

Since the mid-1970s, a significant literature has been amassed. Its early growth is captured by Sutcliffe's (1981) herculean bibliography. Amid entries recording dogged institutional narratives, myth-building, and asocietal assemblages of facts and figures, are glimpses of a new critical literature. The better guides now are bibliographic essays reviewing national literature at the intersection of planning and history (for example, Burgess, 1996; Cherry, 1991; Freestone and Hutchings, 1993; Miller, 1998; Monclús, 1992). The nature of this literature is not easily typecast. The four main strands identified by Cherry (1981) are still evident: 'professional involvements' (the history of what planners have done); 'techniques and performance' (the history of town building and physical plans); 'social and institutional settings' (the history of planning events in their social context); and the biographical (post-hagiographic examination of broader themes through individual lives). Specific interests can be related to these broader approaches – institutions, legislation, methodologies, design, theory, post-colonial and cultural studies perspectives. The historiography of planning is now delivered through diverse modes, albeit with traditional narratives and case studies – indeed a propensity for virtual microscopic histology – predominating.

It would be erroneous to telescope all this into a homogeneous output varying only by subject matter and length. The tension identified in the early 1980s between the art historian and social historian (Birch, 1981) has largely dissolved with the reassertion of space in social theory and the erosion of physical fetishism in built environment studies. But different spins on common subject matter are evident depending on whether it is a historian, geographer, architect, political scientist or planner telling the story. The influences of Marx (Benevolo, 1967; Foglesong, 1986; Scott, 1980), Foucault (Boyer, 1983), and other postmodern theorists (Sandercock, 1998*b*) are now evident, usually making their initial mark on foreign-language literature before diffusing into anglophone discourse. Atheoreticism still abounds, but at least in a post-idealist vein quite distinct from earlier studies which saw planning disconnected from any political context (Hague, 1984).

This is not to cast the past quarter century as a golden era of reflexivity. Too many historical studies cast planning as an unproblematic endeavour, emphasize success over failure, and stress the 'supply side' (the neatness of theories and design intentions) rather than 'demand side' (the messiness of actual outcomes and lived experiences) of planning activity. Partly in response to these lacunae, a self styled, 'noir' history has lately been constructed. Yiftachel (1998) surveys the realities of planning as a tool of social control and oppression. This is darker, more sinister territory – planning's 'advancement', or at least technical facilitation, of regressive social goals involving exclusion, segregation, marginalization, and injustice. Also in extraordinary short supply, given planning's throughgoing practical and vocational aspirations, are detailed historically-informed and politically-contextualized policy evaluations. This genre is relatively weakly developed. The deterrents lie in the analytical complexities of such projects, the formidable methodological challenges of disentangling planning from other influences, and the necessary descent to the often untrendy realm of the local where planning and development control arguably have their greatest impacts on the urban landscape (cf. Hall *et al.*, 1973; McLoughlin, 1992).

Krueckeberg (1997) distills planning history's sins of omission and commission into three categories: exclusion (stories and values left out); unexamined contradictions (between theory and practice, philosophy and practice, social justice and narrow professionalism); and experimentalism (the trivialization of planning disaster). There is good and bad historical writing, and the flaws can be magnified if connections are made with current planning practice and urban policy (Abbott and Adler, 1989; Hutchings, 1998; Shaw and Robinson, 1998). While much of the literature on planning retains an unreconstructed present-mindedness, the contribution of an historical awareness will count for little if rooted in misunderstandings, misinterpretations and misreadings of the past. To avoid becoming 'the planners of history's mistakes', Krueckeberg counsels us to 'to look back more often' (p. 269).

Various texts have done that through the twentieth century. Even when forward-looking, they speak most directly to their own times. Five archetypal Anglo-American texts of different eras are those of Triggs (1909), Abercrombie (1933), Keeble (1952), Hall (1996), and Sandercock (1998*a*). The first of these, H. Inigo Triggs' *Town Planning*: *Past, Present and Possible*, is representative of the part-polemical, part-substantive treatises in the early decades of the emergence of modern town planning theory. Now an historical document in its own right, it was published in the extraordinary year 1909: the year in which, among other things, the British Parliament passed that nation's first Town Planning Act, Unwin's *Town Planning in Practice* appeared, the *Plan of Chicago* was completed, and the First American Conference on City Planning 'and the Problems of Congestion' was held in Washington DC. With the Haussmannization of Paris in living memory and Welwyn Garden City not even a glint in Howard's eye, it is ingrained with historicism and historic examples. Yet formal 'history' is confined to what becomes the obligatory early chapter on 'Types of Ancient and Modern Towns' dealing with town formation, siting, and layout.

Patrick Abercrombie's *Town and Country Planning* was first published in 1933, and reprinted and revised several times into the late 1950s. Abercrombie, the founding editor of *Town Planning Review*, was the great British patrician land use planning expert, and his book was probably the most widely-read imperial text of its day. It was explicitly synthetic, marrying town and country, city and region, past and future, in an accessible primer. His history, concentrating on physical town layout, was 'drawn from actual examples' – 'ancient', 'modern' and 'recent' antecedents. The historical early chapters were obviously selective with an evolutionary transition across the ages triumphing in consideration of contemporary theory and practice. Also written during the downtime of the Great Depression, Thomas Adams' *Recent Advances in Town Planning* (1932) promoted the same almost seamless segue from past to present to future.

Lewis Keeble's textbook, *Principles and Practice of Town and Country Planning*, first published in 1952, also went through countless editions into the 1960s. Keeble's concerns were ruthlessly pragmatic, factual, and atheoretical. His primary focus was on technical land use planning, and conveying the knowledge for getting things done within the new planning machinery put in place after World War Two. There was little pause to learn explicitly about the past, for better and worse. It was left to Professor R.A. Cordingley in his introduction to make the appropriate caveat: 'history, as such, here finds no place'. This almost de-historicized text, so influential in Britain and the Dominions, says a good deal about the perfunctory treatment of historical matters in the brave new world of the early postwar years.

By contrast, Peter Hall's *Urban and Regional Planning* (first published in 1974) was written in more reflective times. Intent on surveying the legacy of both the early utopian planners and the system technicians, the content was 'deliberately historical'. Hall's *Cities of Tomorrow,* first published in 1988, is the fuller account and widely acknowledged as a contemporary classic. It is an authoritative thematic-chronological overview of Western planning theory and practice from a liberal-rationalist, late modern perspective. It tidies up much of the received wisdom of planning history and updates it to the present-day through a grand paradigmatic evolution driven by the prescience and wisdom of a pantheon of visionaries: Ebenezer Howard, Lewis Mumford, Frank Lloyd Wright, J.F.C. Turner, Manuel Castells, and so on. One hundred years on from the birth of the modern planning movement, cities emerge as better places to live in. But this conclusion is heavily qualified by numerous degraded visions and what could-have-beens, the book ending gloomily with the recalcitrant urban problems and challenges in the 'deep pre-dawn chill' of the twenty first century recalling those a century ago (Hall, 1996, p. 422).

The fifth and most recent text is Leonie Sandercock's *Making the Invisible Visible* (1998*a*), an edited collection of diverse stories outside the mainstream of official reports, professional development, material progress and the triumph of modernist interventions. Representative of diffuse, revisionist perspectives (Muller, 1999), this is a postmodern potpourri with its theoretical basis in the writings of Michel Foucault, Jane Jacobs, Mike Davis, and, notably, urban anthropologist James Holston's concept of 'insurgent citizenship'. Critical of a planning history that concedes 'no fatal flaws', this book praises an alternative mix of bottom-up, subversive, softer plannings (community, social, cultural) sensitive to race, gender, and difference. The mainstream model that has evolved by century's end is typecast as reactionary and regulatory. Robert Goodman's (1972) 'soft cops' of the early 1970s become the 'spatial police' of the 1990s. As a series of exposés of the shortcomings of planning's real and imagined pasts (and hence its historians), the utopic alternative is sketched in the author's complementary *Toward Cosmopolis* (1998*b*). Sandercock's quest toward emancipatory, inclusive planning is not without its own critics (Sorensen and Auster, 1999), but this is a very different kind of planning history from that of Peter Hall who is assailed for writing tired, statist, 'male-stream' narratives. The contrast is even starker, of course, with Inigo Triggs' gentlemanly ride through the monumental landscape of civic art as the history that really mattered near the turn of the century.

THE HISTORY OF PLANNING

What emerges from all this historical activity about the origins, evolution and future of modern urban planning over the course of the twentieth century? With so many planning histories – products not just of their time, but their intent, method, theoretical position (or lack of), and their political and intellectual context – no one definitive story can emerge. But an overall picture can be constructed against which the contents of this volume can be set.

The history of urban planning is a complex and ongoing history of achievement, failure, resilience and challenge. It is a history of both continuities (quests for health, justice, efficiency, environment, amenity) and discontinuities (global technological, economic, and political shifts) (LeGates and Stout, 1998). In its modern guise, and integrating a mix of utopian, anarchist, authoritarian, and utilitarian reformist influences, it emerges in the late nineteenth century to address the calamitous by-products of industrial urbanization. Its evolution becomes inextricably linked with the processes of modernization and the experience of modernity (Swyngedouw, 1997). Its basic precepts are expressed through a rich array of modernisms: contrast, for example, what a dream team of Frederick Law Olmsted Sr, Joseph Stübben, Patrick Geddes, Eugène Hénard, Otto Wagner, Le Corbusier, Werner Hegemann, Clarence Stein, Robert Moses, Hans Blumenfeld, Victor Gruen, Thomas Sharp, and Ed Bacon variously stood for. Underlying these nuances were, of course, some common threads: the conceptual commitment to holistic templates; the emphasis on logic, order, science, and rationality; and the belief in enlightened public intervention for the greater social good. The development of planning thought is intertwined with the processes of professionalization which begin before World War One, experiencing its greatest periods of innovation in response to periodic social crises – economic downturns, global economic shifts, and bouts of reconstruction.

Planning's aspirations have always soared above the practicality of their implementation, as a constant stream of good city theories have been buffeted by broader economic, political and social constraints. At times, planning has been a deeply flawed discourse, co-opted as a technical device for social-spatial ordering by authoritarian regimes in east and west, and one whose outcomes even in more putatively enlightened political regimes have often been the reverse of those intended (destruction rather than enhancement of community, promotion of segregation rather than social mix, carbuncular aesthetics for enduring urban design). Nonetheless, there is also a raft of achievements in provision of open space and affordable housing and mass transit and basic public services, protection of natural and cultural resources, development of liveable, new communities, and the injection of sanity into myriad land use determinations. Historical hindsight can provide both a realistic and a sobering understanding of planning's practices which can be lost sight of by intellectual fashion. David Harvey, in surveying the 'heady brew of progressive bourgeois reformism' which marked the early twentieth century planning experience, writes:

> There are plenty of contemporary critics, of course, who, armed with their techniques of deconstruction and of Foucauldian analysis, might look back upon

> this period with jaundiced eye as a classic case of progressive reformism disguising capitalist plans for capital accumulation and speculative land development, a mask for concealing bourgeois guilt, paternalism, social control, surveillance, political manipulation, deliberate disempowerment of marginalised but restive masses, and the exclusion of anyone who was 'different'. But it is undeniable that the aggregate effect was to make cities work better, to improve the lot not only of urban elites but also of urban masses, to radically improve basic infrastructures (such as water and energy supply, housing, sewage, and air quality) as well as to liberate urban spaces for fresh rounds of organized capital accumulation in ways that lasted for much of the twentieth century. (Harvey, 1996, p. 406)

In common with Peter Hall, Harvey's conclusion is that a significant proportion of humanity on every continent finds itself living in better circumstances at the end of the twentieth century than at the beginning as a result of urban planning. Moreover, inherent in many of the best interventions over time has been a visionary notion of better cities or of better processes of resource allocation and conflict resolution – concerns that would be helpful to have back in many cities right now.

While the 'discourse of hygiene' (Sandercock, 1998*b*) may have been central to the urban planning project during the early twentieth century – applied to the city as 'a medicalizable object' (Foucault, 1980, p. 175) and extended to interrelated missions of social, moral and cultural cleansing – the agenda and scope of urban planning have expanded and changed dramatically over the last hundred years. Planning has had continually to reinvent itself: from physical design to technical rationality to systems modelling to urban management to joint private-public venturing to environmental monitoring. Always under capitalism there have been challenges to its scope, authenticity and social role, some of the greatest coming at the dawn of the new century. In recent decades, the terrains of planning have shifted markedly: from greenfields to brownfields; homogeneity to heterogeneity; exclusion to inclusion; control to deregulation; product to process; technocracy to democracy (and back again); from constraining to stimulating growth. The 1970s mark a clear break from what had evolved into the highpoint of planning as an enlightened modernist project of 'managed urbanism' (Swyngedouw, 1997, p. 104). It has even been argued that the skills and expertise that underpinned the early professionalization of the field are now increasingly irrelevant, if not defunct. 'Classical town planning' – the process of town design – is seen as having little purchase on contemporary environmental and land-use problems in Britain (Evans, 1997).

What is certain is that planning at the start of the new century is pulling in different directions. Over time, chameleon-like, its role has been constantly adapted to time, place, culture and circumstance, and not always for the good. These trends continue, driven by economic, environmental, cultural and technological forces.

The current turn towards planning as a 'hand-maiden to capitalism' through fast-tracking, place-marketing and legislative deregulatory activities is a cause of much contemporary soul-searching in its apparent preference for private above public interests. The 'totalizing sign of the commodity' rules (Swyngedouw, 1997, p. 105). Economic growth, employment creation, profitability, and global capital-friendliness have become the beacons of an entrepreneurial thrust facilitated, if not actively encouraged, by increasingly flexible, malleable planning systems. The real irony of the circularity of *fin de siècle* planning concerns that Peter Hall (1996) identifies is less the phenomenon of persistent 'trends' reasserting themselves so much as planning's need to contest again the view, fashionable under neo-liberalism, that it is the market which produces optimal social outcomes.

A second trajectory sees planning conceived more and more as a component of integrated environmental management (Atkinson *et al.*, 1999; Haughton and Hunter, 1994; Miller and De Roo, 1999). The historical moment is hard to ignore. According to Merrett (1995, p. 14), the rise of the environmental sustainability paradigm represents the third of post-medieval planning's 'three ages'. It follows on from the focus on public health, labour productivity and social unrest in the mid-nineteenth century and the early twentieth century preoccupation with rationalizing urban and regional land use and raising physical standards of provision with 'its tap root in the rise of the great European social democratic movements'. Planning for sustainability is seen as presenting the new agenda for the twenty-first century. Planning has given lip service to environmental concerns for many years, albeit usually diluting the ecological within a more liberal anthropomorphic definition. But since the Rio Earth Summit of 1992, the environmental imperative across all levels of government has proven irresistible (Blowers and Evans, 1997). Despite its status as one of the great synthesizing, compromising areas of public policy, the fundamental tension between planning as development and planning as environmental mitigation is proving increasingly problematic.

A third sea change lies in the realm of the cultural. There are now many plannings, with diverse actors – minority groups, non-governmental agencies, community organizations, special interest bodies, ephemeral action coalitions, and even more inchoate interests – all with legitimate and influential stakes in the generic task of seeking better urban outcomes. Planning is no longer the monopoly of the university-trained professional, but a more fluid, democratic set of arrangements expressed through a variety of forms and media within civil society. This is 'post-euclidean' planning. No longer grounded in 'the idea of professionals having privileged knowledge of what constitutes the public interest', the pressures of burgeoning, culturally polyglot metropolises, in particular, are calling forth a more dynamic, multifaceted, transformatively political planning acknowledging conflict, difference and citizens rights (Friedmann and Douglass, 1998, p. 3).

A fourth frontier for planning in the 1990s lies in coping with the demands of telecommunications and information society. Through the twentieth century, planning has had to respond to successive waves of technological innovation, particularly emergent forms of transportation and communication: trains, planes, automobiles. But the infobahn is radically different, since it challenges the time-honoured assumption of cities as physical containers for urban life in which key roles for planners are to improve mobility and allocate space for maximum efficiency. The wired city demands a different metric since instant real time eliminates the friction of distance (Graham, 1997). Interaction is despatialized, unleashing new forces of decentralization which break down traditional cartesian linkages (Mitchell, 1995). One outcome now unfurling is a more dynamic and complex urban geography at different scales than was the experience of the industrial cities of the twentieth century.

Whatever identifiable totality planning had at the beginning of the twentieth century appears to have been lost. It is no longer an autonomous, integrated policy and professional realm consensually tucked into the policy vectors of the welfare state. It is part of something much greater and more variegated, fanning into both broader patterns of governance and the unscripted machinations of civil society. It is being pulled in diverse, sometimes contradictory directions, adapting to roles far less heroic than those mapped out by the founders. Processes of reconciliation continue to define the field (environmental sustainability versus economic growth *vis à vis* equity; privatization versus the public realm; quick approvals versus community interrogation). It is as if planning is continually, relentlessly mediating its own future.

PURPOSE AND STRUCTURE OF THIS BOOK

The aim of this volume is to make some more sense of the gains, losses, continuities, disjunctures, and conundrums which surface when we take more than a superficial look at planning's past. The book's character might be first clarified by stating what it does not try to do. It does not set out to be a comprehensive, encyclopedic history of world planning, a mammoth league table of achievements and failures, or a systematic compendium of place, person or plan-specific case studies. The authors of this edited collection do not seek to make a collective argument, nor represent their individual contributions as definitive treatments. They were under no strict editorial instructions, with different theoretical perspectives and intellectual milieus clearly in evidence. The overall objective was thus not a singular, monolithic account of global planning history but, and in a way that has not really been attempted before, an integrated set of diverse, interpretative essays cutting across time, space and culture to explore a number of major themes in the twentieth century urban planning experience.

Conferences can be important stimuli to creative endeavour, and this has certainly been the case in planning history. The first three books in the Spon 'Studies in History, Planning, and the Environment' series were products of the first international planning history conference held in London in 1977 (Cherry, 1980; Kain, 1981; Sutcliffe, 1980). Sutcliffe's (1984) book on metropolitan development and then Ward's (1992) collation of perspectives on the garden city movement followed. The origins of the chapters in this book are in papers presented at a conference held at the University of New South Wales in Sydney, Australia in July 1998 on the theme 'The Twentieth Century Urban Planning Experience'. Nearly 200 papers were presented, with abbreviated versions compiled into printed proceedings (Freestone, 1998). Cumulatively, these contributions capture enduring concerns as well as charting different directions and venturing into new geographical domains.

The collection in this book, now substantially revised and developed into substantive essays, is a small but purposeful sample of what was on offer at the Sydney conference. The intention has been to assemble a cohesive set of authoritative, thematic and original essays which go beyond the traditional fare of empirical case studies and attempt to look critically and reflectively at 'big picture' topics which in themselves help define the scope and spirit of contemporary planning history. The contributors come from a variety of backgrounds, though predominantly European, American and Australian. They have diverse training in history, human geography, urban design, planning and architecture. They have approached their task using varied methodologies – from virtual bibliographic essays through taxonomy to discourse analysis.

The fourteen chapters are sifted into a running order of planning themes: from broad historical overviews to help set the scene through the realm of physical planning paradigms into discourses of urban reform and onto planning's incorporation of green and conservation issues with a final set of chapters dealing with more contemporary history and urban futures. In the opening stanza, Peter Hall provides a characteristically kaleidoscopic global narrative in the manner of a stripped down *Cities of Tomorrow*, telling a story which begins and ends with Ebenezer Howard. Events and names and places are gridded seamlessly within the framework of a fast-paced themed-chronological account spliced into five periods from a chrysalis period in which housing and planning were notably coupled to a present-day de-housed planning driven by the quest for sustainable urban development. Virtually all the more specialized dimensions of the twentieth century planning experience explored in subsequent chapters are touched upon here.

The internationalism scoped by Hall is examined more analytically by Stephen Ward. The notion of exporting planning was first conceptualized by King (1980) in a paper presented at the first international planning history conference. Sutcliffe (1981) developed this in a dazzling multi-national empirical survey exploding the staid nationalism of much traditional planning history. Ward trawls ever more widely for cross-cultural examples, but assembles them in a multi-dimensional matrix of categories derived from two fundamental processes of borrowing and imposition. The approach is a kind of theory-building allied to the established tradition of diffusion studies in the social and spatial sciences. Dennis Hardy's interest is also taxonomical, but focused specifically on utopian thought which has suffused planning's loftier aspirations worldwide over one hundred years (Fishman, 1982). He effectively takes up a challenge laid down by himself some years ago: 'the role of the planning historian might be to provide historical evidence and an understanding both of the strengths and of the weaknesses of different methods of anticipating the future' (Hardy, 1992, p. 500). The inevitable short-circuiting of aspirations towards the ideal city into dystopia (bad places) and subtopia (placeless spaces) is acknowledged, but the conclusion is buoyant. Utopias remain a noble quest, if constructed as realistic *quasi* utopias alive to multiple futures rather than fixed blueprints. The conclusion recalls Harvey's (1996, pp. 419-420) endorsement for 'a utopianism of process' over the 'dead spatiality' inherent in the 'utopianism of form'.

The literature on the history of physical planning is enormous. The next three chapters trek their way through this on more selective routes. Their most disarming implication – given the numerous well-known theoretical icons and showcase developments brought together here – is how planning over time has actually shaped a mere fraction of the built environment. The search for the 'Good Urban Plan' has been one of the central planks of planning theory (Yiftachel, 1989). Taking off from some of the same utopian schemes canvassed by Hardy, Jon Lang provides a synoptic account of this search for better cities through a century of 'empirical' and 'rational' models of urban design. This is applied design theory, concerned as much with the future as the past. While not unsympathetic to the modernist project (see also Gold, 1997), it ends with a plea for greater cultural sensitivity in implementation. The so-called 'city beautiful' movement of 1890s-1910s helped codify the agenda for modernist planning in the twentieth century. In his chapter, Gilbert Stelter advances the revisionist analysis of the movement initiated by Peterson (1976) and Wilson (1989). Drawing primarily from the Canadian experience, he further undermines two stereotypes of the city beautiful: its Americanism and its authoritarianism. The neighbourhood unit has been an 'article of faith' for the planning profession in the twentieth century (Reiner, 1963, p. 120), is continually being re-invented, and has had its history chronicled in different ways (e.g. Gillette, 1983; Kallus, and Law-Yone, 1997). Stephen Ward has already touched on the internationalism of neighbourhood theory in his chapter, and here Dirk Schubert provides a more intensive examination. With the neighbourhood unit construed as a major thematic strand in the history of community and site planning, this discussion adds value through exploring its longevity, commercialization, Germanization, and misapplication for political-ideological ends – the latter recalling the adaptation of central place theory to the spatial planning regime of National Socialism (Fehl, 1992).

Alerted in this way to the flexible applications of technical concepts, and linked by an interest in language, the next three chapters examine continuities and discontinuities across particular aspects of urban 'reform'. Raphaël Fischler explores a shifting concern over the twentieth century from the 'standard of living' to the 'quality of life'. The shift – intimately connected to a methodological revolution (from minimum standards to performance indicators, quantitative to qualitative measures) – is related to broader changes in the global economy and the role of the state in national welfare, plus material advances in living standards. Robert Bruegmann's analysis of the historical movement of planning concern from 'urban blight' in the early twentieth century to 'suburban sprawl' today covers a similar ideological terrain. Although the spatial objectives of these approaches are very different – one seeking decentralization, the other densification – their rhetorical expression is as remarkably convergent as professional and popular opinion is divergent. Jeffry Diefendorf describes how three cities in three different countries confronted the 'monster' of traffic. Differing outcomes driven by local circumstances belie remarkable similarities in planning theory and pick up an historical shift away from facilitating traffic flows to taming them (Newman and Kenworthy, 1999).

The focus of Diefendorf's chapter is the historic inner city, and in the first of two chapters which engage more directly with environmental issues, the late David Hamer tackles the historic environment. He defines the 1960s as the critical watershed. Before then, the forces of modernization usually set out to obliterate comprehensively traces of old urban ways; but from then on heritage conservation has continued to widen its concerns and techniques way beyond the 'scientific' identification of scattered noble edifices. While the roots of 'greenness' go back to the nineteenth century and earlier, explicit concern with the biophysical environment is more recent, with the importance of protecting natural ecosystems in turn appreciated ahead of conservation of built structures and settings. Maurits van Rooijen deals specifically with urban open space. Taking a pan-European perspective and digging deeper into history, he presents an historical typology of city types out of which emerges, in the context of sustainable urban development, the contemporary concept of 'the green city'. The ecological turn seems an irreversible tendency in all aspects of city planning (Platt, 1994). Apart from keeping token reminders of political power and national identity, the conservation of historic structures was not part of the early modernist project and few planning advocates were especially sympathetic, with notable exceptions such as Patrick Geddes.

All the contributions in this book usefully bring their stories and concerns into the realm of the present-day (if not beyond), but the final set of chapters is explicitly concerned with current trends and directions in planning in their historical context. Each addresses differently the increasing importance of locality, place and community as determinants of life styles, life chances, and growth fortunes. And all represent planning at a significant cross roads in its destiny, needing to re-configure itself in various permutations to meet the challenges of change. Susan Thompson's concept of the 'multi-layered' city focuses attention on the clamour of multiple publics seeking better ways for planning to better mediate difference and 'otherness'. Largely set in that new zone of planning defined by the 'nexus of the interplay of state, civil society and corporate economy' (Friedmann and Douglass, 1998, p. 3) and valuing cultural diversity (Beal, 1997), this provides a more practice-orientated interpretation than recent theoretical texts and draws on Australian local government's negotiation of multiculturalism. The last two chapters are more focused on urban economies, contemplating the implications for planning of small-government, pro-market 'neo-liberal' policies. Concerned mainly with post-Fordist mega-cities, Allen Scott's choice is for planning to be located within wider patterns of regional economic governance to address more proactively the dynamic uncertainties of global market forces. While reports of the 'death' of planning have been greatly exaggerated (Allmendinger and Thomas, 1998), Brendan Gleeson and Nicholas Low in the final chapter contemplate this scenario in jurisdictions (such as Australia) where a national political consensus privileging private investment, competition, entrepreneurialism, and deregulation threatens the historical core values of public planning. Their optimism resides not just in the renaissance of spatial planning in Europe but signs of an emergent counter-political movement seeking to 'revalue' traditional planning in meaningful economic, environmental and political terms.

CONCLUSION

Various themes and issues recur across these chapters: the utopian origins and aspirations of planning; the explanatory links between planning discourse and broader trends in society and governance; the export and imposition of planning concepts; the dictates of economic and political power; tensions between global forces and local identity; the commodification of planning ideas; the inexorable rise of environmentalism; and the need continually – and now more than ever – to revalidate planning as a social-democratic discourse for human advancement. The chapters are not able to package their contents with mutual exclusivity. Inevitably, there are topics which bob up several times – particularly the classic physical planning models like garden suburbs, neighbourhood units, the city beautiful movement, and the Haussmannization of Paris in the late nineteenth century which defined the project of modernist twentieth century planning. Other constructive patterns and textures will be discerned by the reader.

This book reaffirms many of the earlier themes established at the first international planning history conference over two decades ago (Cherry, 1980; Kain, 1981; Sutcliffe, 1980): the rise of planning as an international movement; its early descent to pragmatism and the drive to professionalism; the importance of the garden city movement; the development of the state of the art via accretions of knowledge; the chronic instability of the social climate for planning, and so on. But there are also differences, with new issues emerging such as the circularity and synchronicity of a century of urban reform. The earlier books appeared at a time of sustained critique of planning practice, particularly in Britain. Marxism was beginning to make its presence felt as an alternative to the dominant 'liberal-progressist' tone (Sutcliffe, 1980), but the 1977 conference did not pick up on latter-day postmodern and poststructuralist perspectives. The overwhelming impression is that the earlier focus on understanding the nineteenth and early twentieth century origins of modern planning has been supplanted by a liberation of the diversity of planning perspectives through the twentieth century and a concern with the destiny of planning into the twenty first.

History will likely judge this book as one of various titles on urban planning stimulated by the end of the twentieth century. Other books in this genre have styled themselves variously as 'exercises in retrospect and prospect' (LeGates and Stout, 1998, p. v), opportunities for 'retrospective pronouncements on a supposedly passing era and for speculative analysis of the prospects for the unfolding years to come' (Blowers and Evans, 1997, p. xi), and more forward-looking mappings of desirable future directions (Allmendinger and Chapman, 1999). The distinctiveness here is the collective 'drawing breath' which comes uniquely from planning history (Sutcliffe, 1981) – a reflective, if selective, stocktake. From this longitudinal perspective emerges a picture of planning tightroped in a balancing act in response to powerful, broader forces revolving around the competing demands of capital accumulation, environmental protection, technological change, democratic accountability, social justice, and community identity in a globalized world. At the close of the twentieth century, we can in fact perceive a flux of powerful forces akin to the turmoil that gave birth to modern planning at the end of the nineteenth century. The hope is that an historical appreciation of its history over the twentieth century might help us through the early challenges and uncertainties of the new millennium.

REFERENCES

Abbott, C. and Adler, S. (1989) Historical analysis as a planning tool. *Journal of the American Planning Association*, **55**, pp. 467-473.

Abercrombie, P. (1933) *Town and Country Planning*. London: Oxford University Press.

Adams, T. (1932) *Recent Advances in Town Planning*. London: J and A Churchill.

Adshead, S.D. (1910) An Introduction to Civic Design. *Town Planning Review*, **1**, pp. 3-17.

Allmendinger, P. and Thomas, H. (eds.) (1998) *Urban Planning and the British New Right*. London: Routledge.

Allmendinger, P. and Chapman, M. (eds.) (1999) *Planning Beyond 2000*. Chichester: John Wiley.

Ashworth, W.S. (1954) *The Genesis of Modern British Town Planning: A Study in the Economic and Social History of the Nineteenth and Twentieth Century*. London: Routledge and Kegan Paul.

Atkinson, A., Dávila, J.D., Fernandes, E. and Mattingly, M. (eds.) (1999) *The Challenge of Environmental Management in Urban Areas*. Aldershot: Ashgate.

Beal, J. (ed.) (1997) *A City for All: Valuing Difference and Working with Diversity*. London: Zed Books.

Benevolo, L. (1967) *The Origins of Modern Town Planning*, Cambridge, Mass: MIT Press.

Birch, E.L. (1981) Four perspectives on the history of urban planning. *Trends in History: Urban History*, **2**(1), pp. 79-92.

Blowers, A. and Evans, B. (eds.) (1997) *Town Planning into the 21st Century*. London: Routledge.

Boyer, M.C. (1983) *Dreaming the Rational City: The Myth of American City Planning*. Cambridge, Mass: MIT Press.

Burgess, P. (1996) Should planning history hit the road? An examination of the state of planning history in the United States. *Planning Perspectives*, **11**, pp. 201-224.

Cherry, G.E. (ed.) (1980) *Shaping an Urban World*. London: Mansell.

Cherry, G.E. (1981) Introduction, in Cherry, G.E. (ed.) *Pioneers in British Planning*. London: Architectural Press, pp. 1-18.

Cherry, G.E. (1991) Planning history: Recent developments in Britain. *Planning Perspectives*, **6**, pp. 33-45.

Davidoff, P. (1965) Advocacy and pluralism in planning. *Journal of the American Institute of Planning*, **31**, pp. 331-338.

Duany, A. (1994) Preface, Unwin, R. (1909) *Town Planning in Practice: An Introduction to the Art of Designing Cities and Suburbs*, facsimile edition 1994. New York: Princeton Architectural Press, p. v.

Evans, B. (1997) From town planning to environmental planning, in Blowers, A. and Evans, B. (eds.) *Town Planning into the 21st Century*. London: Routledge, pp. 1-14.

Fehl, G. (1992) The Nazi Garden City, in Ward, S. (ed.) *The Garden City: Past, Present and Future*. London: E & FN Spon, pp. 88-106.

Fishman, R. (1982) *Urban Utopias in the Twentieth Century: Ebenezer Howard, Frank Lloyd Wright, and Le Corbusier*. Cambridge, Mass: MIT Press.

Forlesong, R.E. (1986) *Planning the Capitalist City: The Colonial Era to the 1920s*. Princeton: Princeton University Press.

Foucault, M. (1980) *Knowledge/Power: Selected Interviews and Other Writings 1972-1977*. Brighton: Harvester Press.

Freestone, R. (ed.) (1998) *The Twentieth Century Urban Planning Experience: Proceedings of the 8th International Planning [History] Society Conference and 4th Australian Planning/Urban History Conference*. Sydney: University of New South Wales.

Freestone, R. and Hutchings, A. (1993) Planning history in Australia: The state of the art. *Planning Perspectives*, **8**, pp. 72-91.

Friedmann, J. and Douglass, M. (eds.) (1998) *Cities for Citizens: Planning and the Rise of Civil Society in a Global Age*. Chichester: John Wiley.

Gillette, H. (1983) The evolution of neighborhood planning: From the progressive era to the 1949 Housing Act. *Journal of Urban History*, **9**, pp. 421-445.

Gold, J.R. (1997) *The Experience of Modernism: Modern Architects and the Future City 1928-1953*. London: E & FN Spon.

Goodman, R. (1972) *After the Planners*. New York: Simon and Schuster.

Graham, S. (1997) Cities in the real-time age: The paradigm challenge of telecommunications to the conception and planning of urban space. *Environment and Planning A*, **29**, pp. 105-127.

Hall, P., Gracey, H., Drewett, R. and Thomas, R. (1973) *The Containment of Urban England.* London: George Allen and Unwin, two volumes.

Hall, P. (1974) *Urban and Regional Planning.* Harmondsworth: Penguin.

Hall, P. (1996) *Cities of Tomorrow: An Intellectual History of Urban Planing and Design in the Twentieth Century.* Updated edition. Oxford: Blackwell.

Hague, C. (1984) *The Development of Planning Thought: A Critical Perspective.* London: Hutchinson.

Hardy, D. (1992) Pragmatists and prophets: Planning history and the third millennium. *Environment and Planning B*, **19**, pp. 493-501.

Harvey, D. (1978) On planning the ideology of planning, in Burchell, R.W. and Sternlieb, G. (eds.) *Planning Theory in the 1980's: A Search for Future Directions.* New Jersey: The Centre for Urban Policy Research, Rutgers University, pp. 213-233.

Harvey, D. (1996) *Justice, Nature and the Geography of Difference.* Cambridge: Blackwell.

Haughton, G. and Hunter, C. (1994) *Sustainable Cities.* London: Jessica Kinglsey Publishers.

Hutchings, A. (1998) Planning history in practice. *Australian Planner*, **35**, pp. 122-126.

Jacobs, J. (1962) *The Death and Life of Great American Cities.* London: Jonathon Cape.

Kain, R. (ed.) (1981) *Planning for Conservation.* London: Mansell.

Kallus, R. and Law-Yone, H. (1997) Neighbourhood: The metamorphosis of an idea. *Journal of Architectural and Planning Research*, **14**, pp. 107-125.

Keeble, L. (1952) *Principles and Practice of Town and Country Planning.* London: The Estates Gazette.

King, A.D. (1980) Exporting planning: The colonial and neo-colonial experience, in Cherry, G.E. (ed.) *Shaping an Urban World.* London: Mansell, pp. 203-226.

Krueckeberg, D. (1997) Planning history's mistakes. *Planning Perspectives*, **12**, pp. 269-279.

LeGates, R. and Stout, F. (1998) Editors' Introduction, to *Early Urban Planning: Selected Essays.* Vol. 1. London: Routledge/Thoemmes Press, pp. v-xxxxii.

MacFadyen, D. (1933) *Sir Ebenezer Howard and the Town Planning Movement.* Manchester: Manchester University Press.

McLoughlin, J.B. (1992) *Shaping Melbourne's Future? Town Planning, the State and Civil Society.* Melbourne: Cambridge University Press.

Marcuse, P. (1980) Housing in early city planning. *Journal of Urban History*, **6**, pp. 153-176.

Mawson, T. (1927) *The Life and Work of an English Landscape Architect.* New York: Charles Scribner's Sons.

Merrett, S. (1995) Planning in the age of sustainability. *Scandanavian Housing and Planning Research*, **12**, pp. 5-16.

Miller, C.L. (1998) New Zealand's planning history: quo vadis? *Planning Perspectives*, **13**, pp. 257-274.

Miller, D. and De Roo, G. (eds.) (1999) *Integrating City Planning and Environmental Improvement: Practicable Strategies for Sustainable Urban Development.* Aldershot: Ashgate.

Mitchell, W.J. (1995) *City of Bits: Space, Place and the Infobahn.* Cambridge, Mass: MIT Press.

Monclús, F.J. (1992) Planning and history in Spain. *Planning Perspectives*, **7**, pp. 101-105.

Moore, C. (1921) *Daniel H. Burnham: Architect. Planner of Cities.* New York: Houghton Mifflin, two volumes.

Muller, J. (1999) Although God cannot alter the past, historians can: Reflections on the writing of planning histories. *Planning History*, **21**(2), pp. 11-19.

Newman, P. and Kenworthy, J. (1999) *Sustainability and Cities: Overcoming Automobile Dependence*. Washington DC: Island Press.

Peterson, J.A. (1976) The City Beautiful Movement: Forgotten origins and lost meanings. *Journal of Urban History*, **2**, pp. 415-34.

Platt, R. (ed.) (1994) *The Ecological City: Preserving and Restoring Urban Biodiversity*. Amherst: University of Massachusetts Press.

Reiner, T.A. (1963) *The Place of the Ideal Community in Urban Planning*. Philadelphia: University of Pennsylvania Press.

Sandercock, L. (ed.) (1998*a*) *Making the Invisible Visible: A Multicultural Planning History*. Los Angeles: University of California Press.

Sandercock, L. (1998*b*) *Towards Cosmopolis: Planning for Multicultural Cities*. Chichester: John Wiley.

Scott, A.J. (1980) *The Urban Land Nexus and the State*, London: Pion.

Scott, M. (1969) *American City Planning since 1890*. Berkeley: University of California Press.

Shaw, K. and Robinson, F. (1998) Learning from experience?: Reflections on two decades of British urban policy. *Town Planning Review*, **69**, pp. 49-63.

Sies, M.C. and Silver, C. (1996) The history of planning history, in Sies, M.C. and Silver, C. (eds.) *Planning the Twentieth-Century American City*. Baltimore: Johns Hopkins University Press, pp. 1-34.

Sitte, C. (1889) *Der Städtebau nach seinen künstlerischen Grundsätzen*. (City Planning According to Artistic Principles, translated by G.R. Collins and C.C. Collins, New York: Random House, 1965).

Sorensen, T. and Auster, M. (1999) Theory and practice in planning. *Australian Planner*, **36**, pp. 146-149.

Stretton, H. (1970) *Ideas for Australian Cities*. Melbourne: Georgian House.

Sutcliffe, A. (1981) Why planning history? *Built Environment*, **7**(2), pp. 64-67.

Sutcliffe, A. (ed.) (1980) *The Rise of Modern Urban Planning 1800-1914*. London: Mansell.

Sutcliffe, A. (1981) *Urban and Regional Planning: An annotated bibliography*. Mansell: London.

Sutcliffe, A. (ed.) (1984) *Metropolis 1890-1940*. London: Mansell.

Swyngedouw, E. (1997) The specter of the phoenix: Reflections on the contemporary urban condition, in Bosma, K. and Hellinga, H. (eds.) *Mastering the City 1: North-European City Planning 1900-2000*. Rotterdam/the Hague: NAI Publishers/EFL Publications, pp. 104-121.

Taylor, N. (1998) *Urban Planning Theory since 1945*. London: Sage.

Triggs, H.I. (1909) *Town Planning: Past, Present and Possible*. London: Methuen.

Unwin, R. (1909) *Town Planning in Practice: An Introduction to the Art of Designing Cities and Suburbs*, facsimile edition 1994. New York: Princeton Architectural Press.

Ward, S. (ed.) (1992) *The Garden City: Past, Present and Future*. London: E & FN Spon.

Wilson, W.H. (1989) *The City Beautiful Movement*. Baltimore: Johns Hopkins University Press.

Yiftachel, O. (1989) Towards a new typology of urban planning theories. *Environment and Planning B*, **16**, pp. 23-39.

Yiftachel, O. (1998) Planning and social control: Exploring the dark side. *Journal of Planning Literature*, **12**, pp, 395-406.

CHAPTER 2

The Centenary of Modern Planning

Peter Hall

The modern planning movement is just a little older than the twentieth century. It was in October 1898 that Ebenezer Howard published the original edition of *To-Morrow: A Peaceful Path to Real Reform*, to be republished four years later under the more familiar title *Garden Cities of To-Morrow* (Howard, 1898, 1946, 1985). And, though there were other important foundation stones of the modern planning movement – in France, in Germany, in the United States – surely without doubt this was the most significant.

It was not of course the birth of town planning, which had occurred millennia before that in the cities of the Middle East. But it was the effective birth of modern planning; a significantly distinct movement, distinguished by its social purpose. Classical town planning, which reached its apogee in the two centuries that began with Pope Sixtus V's reconstruction of Rome and ended in the building of Georgian London, Edinburgh and Dublin, was in essence a purely formal movement, an exercise in classical civic design that was an outgrowth of Renaissance architecture. Modern town planning, too, had its strong aesthetic side and even its aesthetic controversies: consider only Le Corbusier's tirade against Camillo Sitte (Le Corbusier, 1998), or the postmodernist reaction against the movement that Le Corbusier and Gropius represented. But, throughout the twentieth century, aesthetic considerations have been subsumed under social ones. Thus, for Le Corbusier, straight streets were preferable because they more easily accommodated his twentieth century 'machines for living in' and those other machines he worshipped, the machines we use to drive around. Equally, from Raymond Unwin to late twentieth century New Urbanists, others have contra-argued that the street pattern should be used to curb traffic, not to promote it.

So the story starts in 1898 and can conveniently be ended exactly a century later. Rather remarkably, we can consider the intervening century in the form of five snapshot pictures, taken almost exactly a quarter century apart. Of course, that is a huge simplification that could produce a historian's Procrustean bed: important developments occurred between those dates, and in particular much development – economic, social, political – was linear or cyclical. But the convention will serve to capture many of these underlying changes, and in particular some of the key themes: the birth of the garden city movement, its extension into regional planning, the attempt to create ideal modernist cities, the shift from top-down to bottom-up planning, and the attempt to forge a freewheeling entrepreneurial style of urban regeneration.

1898: The Rejection of the Palaeotechnic City

The first key date is of course 1898 itself. That year not only saw publication of Howard's tome, but of Kropotkin's influential *Fields, Factories, and Workshops* (Kropotkin, 1898, 1913). Howard later paid tribute to Kropotkin, then long resident in England: both books powerfully argued the case for planned decentralization from the congested Victorian industrial city, aided by the shift from what Patrick Geddes almost simultaneously called palaeotechnic to neotechnic urbanism, from the age of coal and steam to the age of electricity and the motor vehicle (Geddes, 1912, 1915; Kitchen, 1975; Mairet, 1957). Geddes was then conducting his early experiments in community self-help in the slums of Edinburgh, and inviting speakers to his summer schools. Immediately after reading Kropotkin, he could see the new technologies as the key to large-scale regional deconcentration of people from London and the other big cities to 'eutopia', communities in the open countryside, each in organic relationship with their surrounding agrarian regions.

These were theorists, utopians even; but Howard, in particular, was a doer as well as a thinker. He was instrumental in creating the Garden City Association in London, in June 1899, a mere eight months after publication of his book; the first garden city at Letchworth was already started by 1903. Even earlier than that, the London County Council (LCC) were building their first peripheral cottage housing estate at Totterdown Fields in Tooting, served by a new electric tramcar line that opened in 1903: the garden suburb was running neck and neck with the garden city.

It is significant that all were British, either by birth or adoption. The birth of modern planning took place in London and in Edinburgh. And the key individuals – Howard, Kropotkin, Geddes – met frequently in the first decade of the new century, at lectures and conferences and summer schools, commenting on each others' contributions. All were involved in the great London town planning conference of 1910, from which came the birth of an organized professional Town Planning Institute in 1914.

For all these people, and from those who came from mainland Europe for the 1910 conference, housing was the central question. The root problem was urban overcrowding and its inevitable consequences for public health. And on that score, it was difficult to say which city was worst. In London, it was typical to find one family of up to eight people in a single room, sharing with others a single water supply and water closet. Nearly one half of London families had to pay over one-quarter of their earnings for rent, and the poor could not escape, because of the casual nature of so much of their work (GB Royal Commission on Housing, 1885, I, pp. 7-9, 17-18). In Paris, the historic city's 2.45 million people, in 1891, lived at a density twice that of the LCC area; 14 per cent of the Paris poor, 330,000, lived in overcrowded dwellings; the poor were even worse housed than in London. Sellier calculated in 1911 that the total was still 216,000, with another 85,000 in the suburbs, living at two or more per room (Sellier and Bruggeman, 1927; Bastié, 1964). Berlin, which nearly doubled in twenty years, from 1.9 million in 1890 to 3.7 million in 1910, accommodated its growth in densely-packed five-storey 'rental barracks' around courtyards as narrow as 15 feet wide, the minimum necessary to bring in fire-fighting equipment. As late as 1916, no less than 79 per cent of all dwellings had only 1 or 2 heatable rooms (Eberstadt, 1917). In New York, the Tenement House Commission of 1894 estimated that nearly three in five of the city's population lived in tenement houses, so grossly over-built that on average nearly four-fifths of the ground was covered in buildings. In 1893, the 10th Ward, main destination of Jews from Eastern Europe, counted over 700 people to the acre, 30 per cent more than the most congested part of any European city. Part of the adjacent 11th Ward, had nearly 1000 to the acre, and was almost certainly the most crowded urban neighbourhood in the world. In 1908, half of East Side families slept at three or four to a room, nearly a quarter at five or more to a room, depending on a few communal taps. Fixed baths were non-existent (Ford, 1936; Howe, 1976; Scott, 1969).

The reformers sought an answer to these conditions – but also, in the case of both Howard and Kropotkin, to the problems of a depressed and depopulating countryside (figure 2.1). The answer would be central urban renewal at lower densities, accompanied by new garden cities and garden suburbs on green fields. These would be built either by public agencies such as the LCC, by limited-dividend philanthropic companies modelled on the Peabody and Waterlow Trusts which had built early tenement housing, by voluntary groups based on the principle of cooperation, or by philanthropic industrialists (in schemes as Port Sunlight, Bournville and New Earswick in England, or Margarethenhöhe in Germany – some of which actually predated and influenced Howard). The means to this planned dispersion would be the new technologies of electric power and low-cost public transport, above all the electric tramway.

The planning movement was thus an outgrowth of the housing reform movement, and remained firmly coupled to it. It was also a by-product of the intense discussion of land reform in late nineteenth century London, which mainly arose from the problem of landlordism in Ireland and the Highlands of Scotland. Howard's solution – purchase of land at depressed agricultural values followed by deployment of progressively rising values to create a local welfare state – was essentially a variant on land nationalization. Conversely, Howard believed that, once decentralization had done its work, ground rents in London would fall and the population of the County of London, then 4.5 million, would fall by four-fifths to less than one million (Beevers, 1987; Hall and Ward, 1998).

Within a decade or so after 1898, the first actual experiments were under way: at Letchworth and Hampstead Garden Suburb, and in early LCC estates like Totterdown Fields, White Hart Lane and Norbury; at Margarethenhöhe and Hellerau in Germany; and at Forest Hills Gardens (the first conscious embodiment of the neighbourhood unit principle) in New York City. Not long after that, from 1916, came the sixteen *cités-jardins* around Paris, designed and built by the *Office Public des Habitations à Bon Marché du Département de la Seine* under Henri Sellier. What is most notable about these schemes is that the great majority were garden suburbs, not garden cities. Indeed, only Letchworth fell into the select category of garden city, in that it aimed to decentralize people and their jobs well outside the range of the parent city. This was conscious; from 1912 Raymond Unwin was arguing that garden suburbs were a perfectly satisfactory solution. Henri Sellier well understood that his interpretation was not pure Howard, but Unwin's Hampstead variant; he took his architects to visit Unwin in England, in 1919, and used the Unwin text as a basis for design (Read, 1978; Swenarton, 1985).

Whether garden cities or garden suburbs, they shared a picturesque neo-vernacular style, which derived on the one hand from the Arts and Crafts movement of Ruskin, Webb, Morris, Lethaby, Unwin and Parker and their many continental allies in the craft workshop movement, and on the other hand from Camillo Sitte's immensely influential book *Der Städtebau*, published in 1889 (Sitte, 1965). Unwin spent long summer holidays in the small towns of France and Germany, seeking in his sketchbooks to capture the essence of their informal, organic design. His continental colleagues, such as Georg Metzendorf at Margarethenhöhe or Heinrich Tessenow at Hellerau outside Dresden, designed their garden suburbs in the same idiom (Unwin, 1911). Small gabled cottage-type homes, semi-detached or in short rows, cluster along curving streets or sometimes around greens or the dead ends of short streets; abundant tree growth, matured now over a century, contribute to a note of arcadian calm. And this quality is remarkably consistent across cities and countries: Unwin seems to be borrowing from German models in his shopping parade at Hampstead Garden Suburb, yet the same kind of grouping is repeated by Grosvenor Atterbury at Forest Hills Gardens.

1923: The Triumph of Vision and the Failure of Practice

The original impetus extended over at least the following quarter century, culminating in the vision of the Regional Planning Association of America (RPAA), which from 1923 effectively married the ideas of Howard, Kropotkin and Geddes into a central concept of decentralized communities in the distant countryside, powered by electricity and accessible through the private automobile (Sussman, 1976). Indeed, 1 May 1925, the date of publication of the RPAA's manifesto in the New York magazine *The Survey*, was a date almost as important to planning history as October 1898. But this grand vision was never realized. Through Alexander Bing's City Housing Corporation, the RPAA succeeded merely in building a small development, Sunnyside Gardens, in New York City, and the beginnings of a garden city, Radburn, just outside it, but the recession killed off the latter. Rexford Tugwell's plan for hundreds of Greenbelt cities was effectively truncated by a conservative Congress. And the Tennessee Valley Authority – essentially the realization of the RPAA's ideas on a huge spatial scale – became in essence a power generation and agricultural extension scheme, with just one minuscule new town, Norris in Tennessee, to recall the original vision. In Australia, Walter Burley Griffin's plan for the new federal capital of Canberra was continually stalled by disagreements and rivalries. Elsewhere, as in the United States, the prevailing ethos was to assist suburban development by private enterprise.

During the 1920s, in contrast, the significant developments were on the European mainland. Le Corbusier was publishing his Voisin Plan, proposing that central Paris should be razed to the ground and replaced by a new world of cruciform towers and multi-lane freeways. That vision, perhaps mercifully, was never realized. But Henri Sellier's team was at work on Suresnes, one of the first of the new Parisian *cités-jardins*. In Germany, Ernst May was appointed as Architect-Planner of Frankfurt am Main, there to develop his celebrated *Trabantensiedlungen* or satellite towns in the Nidda Valley. In Berlin, the city planner Martin Wagner was collaborating with other leading architects of the Federal Republic – Hugo Häring, Hans Scharoun, Walter Gropius – on a set of new housing developments which resolutely employed the Bauhaus principles of design for living. Vienna was at work on its great series of housing projects, most notably the Karl Marx Hof, which became so influential that it directly influenced similar developments in London (the Ossulston Street redevelopment) and elsewhere. In Amsterdam, H.P. Berlage was laying out the great Amsterdam South scheme.

All these European developments, though sometimes they confusingly adopted the garden city label, were quite different from the original British formulation. A few were pure inner-city renewal schemes; for the most part however they were peripherally-located, but even here they were essentially quite high-density schemes in the form of terraces, quite often with an admixture of apartment blocks. They were planned as an integral part of the city and were connected to it by good public transport. They stemmed from a very distinct continental style of urban apartment living, and although they were imitated in British slum clearance schemes, they did not become the norm there.

They also shared a new architectural style, derived from Le Corbusier and from the group under Walter Gropius that taught at the Bauhaus in Dessau from

1926: starkly functional, with a strong emphasis on clean lines and on rational planning to maximize exposure to sunlight. The predominant form was long, generally straight terraces of houses or apartments (*Zeilenbau*), typically based on walk-up access and so between three and five storeys in height, surrounded by mainly collective open space, and with predominantly straight or modestly curved streets. It was a deliberate reaction against the picturesque neo-vernacular style of the earliest garden cities and suburbs.

The reaction failed to travel across the English Channel. British housing schemes of the same era remained traditional in style, only occasionally swept by the modernist tides from Europe. One reason for that was that most followed the governmental pattern books that had been effectively drawn up by Raymond Unwin in 1919. The largest of all, the combination garden city-garden suburb of Wythenshawe outside Manchester, was designed by Barry Parker. What they did have in common with their continental counterparts was motivation and agency: they were essentially social housing schemes, built either by municipal agencies or by cooperative-type housing associations. Throughout this period and beyond, the link between housing and planning remained the key, as the American planner Catherine Bauer stressed in her influential book *Modern Housing* (Bauer, 1934). It could hardly be otherwise: everywhere, the slum legacy of the nineteenth century remained, and progress in removing it was all too slow.

America remained quite special in this regard from the turn of the century. It made no provision for public or cooperative housing on the European model, with the exception of a few emergency schemes for shipyard workers in World War One. Lawrence Veiller's immensely influential New York Tenement House Commission report of 1901 set the model: it relied on improved building codes and zoning regulations for privately-constructed housing, both in the cities and in the suburbs (Marcuse, 1980). With these in place, rising living standards plus increasing owner-occupation would deal with the problem. And indeed, after 1900, streetcar construction (plus subways or elevated lines in a few larger cities) had a major impact in allowing a second generation of inner-city tenement dwellers to escape into new suburban homes. Manhattan in 1910 had 2.3 million people, shrinking to 1.9 million in 1940; the Bronx's population shot from 89,000 in 1890 to 1.4 million in 1940; that of Queens increased from 100,000 to 1.3 million over the same period. In 1905, half of New York's population lived within four miles of City Hall, falling to only 30 per cent by 1925 (Jackson, 1985).

It was perhaps significant that London, the one European city to follow this American model of mass suburbanization in the 1920s and 1930s, injected American capital and operating expertise into its Underground system. Here, as in American cities, there was a minimal kind of planning for the new suburbs in the form of road layouts and segregated land uses and open space provision. But it was far from ambitious, and the resulting suburban sprawl was savagely criticized by architect-planners like Britain's Thomas Sharp and Clough Williams-Ellis, and by the RPAA's Lewis Mumford – most notably in the latter's celebrated criticism of Thomas Adams' New York Regional Plan of 1929-31 for the Russell Sage Foundation, which he berated for encouraging further commercial congestion at the centre and further sprawl at the periphery (Mumford, 1932). However, the results of these criticisms were significantly different. In Britain, architects and rural conservationists joined forces with the garden city movement to produce a united front, which proved effective when Neville Chamberlain, a politician who was a housing reformer and garden city enthusiast, became Prime Minister in 1937. Within months, he was instrumental in setting up a Royal Commission whose report of 1940 laid the essential foundations of the post-World War Two planning system. In the United States, the RPAA essentially broke up in the early 1930s, at just the point when it could have been expected to exercise maximum influence on the administration of Franklin Delano Roosevelt – a very similar politician to Chamberlain, who as Governor of New York State had enthusiastically espoused the cause of regional planning. Perhaps its demise provides part of the reason for Roosevelt's extraordinary failure in this field. Bauer at that time broke away from the RPAA and began her one-woman crusade for public housing in America, which bore fruit in the federal *Housing Act* of 1937 – nearly half a century later than equivalent legislation in Europe, and without much effect on the ground until after World War Two.

1948: The Great Postwar Rebuild

In this quarter-century chronology, the historic British *Town and Country Planning Act* of 1947 fell a few months early; but it was only one event in an unprecedented burst of planning activity that occurred immediately around and after World War Two, especially in Europe: in the United Kingdom, the publication in 1943 of Abercrombie and Forshaw's *County of London Plan* followed in 1945 by Abercrombie's *Greater London Plan 1944*, and then by the 1946 *New Towns Act* and the 1947 Act; in Copenhagen, the Fingerplan of 1948; in Stockholm, the Generalplan of 1952. This activity was overwhelmingly driven by the strong motivation to begin comprehensive postwar reconstruction of bomb damage and (in a few cases exclusively) of outworn slum housing. In effect, it represented a continuation or completion of the earlier movements after a long delay brought about by the great Depression and the war. Again, carried over from these earlier waves, there was the same emphasis on comprehensive schemes of urban renewal and construction of new communities by public agencies. The underlying assumption was that this was all part of a comprehensive programme to create a welfare state, administered by well-meaning public professionals – invariably, architect-planners – with little involvement either from the private developer, or from the ordinary citizen.

Perhaps in consequence, and perhaps because a generation of planners were learning from each other, there is a certain remarkable sameness about many of the resulting schemes: consider the earliest pedestrian town centres (Coventry, Rotterdam), the mixed high-rise/low-rise housing schemes (Berlin, Copenhagen, London, Stockholm) and the satellite or new town communities (London, Stockholm). Paris was a late arrival on the scene, making little effort to reconstruct or modernize its huge backlog of obsolescent housing until the 1960s; but then, with Paul Delouvrier's 1965 *Schéma Directeur* for the Paris region, ordered by Charles de Gaulle, the city began a massive job of catch-up. And again, the results were the same: major surgery in the inner suburbs, coupled with construction of five huge new cities outside them.

One point about these schemes, however, recalls the earlier episodes. The British, propelled by the enthusiasm and persuasiveness of Frederic Osborn, realized Ebenezer Howard's vision half a century late, building no less than twenty-eight new towns – eleven of them around London alone. Ironically, however, they did so through government-financed development corporations, quite contrary to his original prescription. What they did share with the original models at Letchworth and Welwyn Garden City was their location: well outside the major cities and their commuter rings (though sometimes absorbed into those rings later), separated by green belts or similar areas of open countryside, and designed to be both self-contained and socially-balanced communities. That meant a massive exercise in relocating not merely people but also jobs, made possible by the fact that in the long postwar boom, industry was expanding and therefore footloose. In mid-Hertfordshire north of London it even resulted in a fair approximation to Howard's original vision of Social City: Letchworth and Welwyn Garden City, together with two post-1946 new towns, Stevenage and Hatfield, correspond to his prescription for a polycentric garden city conurbation set against a rural background and linked by high-quality public transport.

No other nation attempted anything as ambitious. In Stockholm, Sven Markelius and Göran Sidenbladh consciously diverged from the British model by building satellite towns close to the city, from which residents could commute by a new transit system. In Copenhagen, the Fingerplan relied on the same mechanism. Paris in 1965 followed them, by placing the five new cities (*villes nouvelles*) at the edge of the agglomeration and connecting them to the city, and to each other, by new motorways and the new *Réseau Express Régional*. In other words, the model was precisely the same as Unwin had used at Hampstead, Sellier in Paris or May in Frankfurt: garden suburbs or satellites directly linked to the parent city. The pure garden city/new town remained almost exclusively a British device, though one can catch some shadow of it elsewhere, for instance the new town of Elizabeth outside Adelaide in Australia, based on a new car factory, or the attempts at commercial new town construction in America such as Reston in Virginia or Columbia in Maryland, which made some attempt to attract industry but ended by being absorbed in the Washington-Baltimore commuter belt.

Driven by the huge housing backlog in most European countries, and further by the postwar and subsequent baby booms, the resultant programme of construction lasted fully a quarter of a century, until the end of the 1960s or the start of the 1970s, and was indeed prolonged beyond that by a further wave of building beginning in the early 1960s which produced the British Mark Two New Towns such as Milton Keynes and Peterborough, or the Parisian new cities (*villes nouvelles*), which were realized mainly in the 1970s and 1980s. It produced many of the landmarks of the twentieth century planning movement, including the comprehensive reconstruction of London's East End, the Mark One and Mark Two British new towns, and the Stockholm satellite communities.

They had many remarkable similarities – though perhaps that is less remarkable when one considers the many common influences that entered into them. There was a very strong emphasis on the neighbourhood unit principle, in which housing was grouped around local shops and schools and other necessary services, and on the pedestrian scale of movement, reinforced in many cases by Radburn-style segregation of vehicles and pedestrians. There was also a stress on pedestrianized town centres which were focal points for public transport (buses in the British new towns, a metro system in the Swedish satellites). Housing was of uniform size and style, with limited variations: whether single-family housing in the British new towns, apartments in the Stockholm satellites, or a mixture of the two in the reconstruction of London, the main emphasis was on housing for families with children. Further, these developments represented a consistent ethos of growth-oriented comprehensive planning in which public planning took the lead and the private sector was reduced to a residual role. Indeed – especially in the 1960s, when the public housing programme reached a frenetic peak, and industrialized building methods were favoured – the resulting landscapes in Western European cities often came to be almost indistinguishable from their Eastern European socialist equivalents in Prague or Warsaw.

There were few parallels outside Europe. In the United States, despite further Federal housing legislation, large-scale American urban renewal, beginning in the 1950s, effectively uprooted urban ethnic communities without creating any effective means for overspill. In the worst instances, as in Chicago, the very poor were segregated in ghetto-style housing projects within the inner city (Bowly, 1978). Undoubtedly, that reflected the balkanized fragmentation of local government, which set suburb against city, coupled with underlying resistance to any large-scale movement of low-income ethnic minorities from the cities. Frederic Osborn gently chided Lewis Mumford for the RPAA's failure to emulate his own Town and Country Planning Association, which had scored such a triumph in the new towns programme (Mumford and Osborn, 1971); but it may have been that no such movement could have been effective in America. There, as the white middle class continued to spread out into ever more distant suburbs, following the model of the Levittown developments of the immediate postwar years, a potentially highly divisive pattern resulted: in city after city, concentrations of ethnic minority populations, often ridden by unemployment and social problems, ringed by white suburbia. Daniel Patrick Moynihan, an eminent urban sociologist turned politician, drew attention to the phenomenon in 1965 (Moynihan, 1965). Yet more money was poured into urban renewal, but the underlying problem was not being solved, because America could never bring itself to countenance large-scale planned urban dispersion from its urban slums. New towns, in this prescription, were essentially just a better kind of suburb, produced by commercial developers for buyers in the open market.

1973: THE GREAT SHIFT IN *ZEITGEIST*

Between 1967 and 1975, but especially in 1971-1975, a remarkable disjuncture occurred in these same countries, and indeed worldwide. It was dramatically illustrated by the almost simultaneous abandonment of the schemes to reconstruct London's Covent Garden and to build several hundred kilometres of urban motorway in London, in 1973; but there were many parallel events in other places. It was a change in *Zeitgeist* coinciding with the arrival of the postwar 'baby boomers' into active political and public life, and it was first marked by the remarkable public manifestations on university campuses, from Berkeley to Paris, in the middle and late 1960s. Essentially this generation, for the most part reared in postwar affluence, rejected many of the values of its parents: comprehensive reconstruction and construction, large-scale development and automobility were now seen as positively bad, and the prevailing slogan, borrowed from the influential environmental campaigner E.F. Schumacher, was *Small is Beautiful* (Schumacher, 1973). Protection of the environment now became a basic imperative, following the immensely influential 1972 Club of Rome report, *The Limits to Growth* (Meadows *et al.*, 1972).

Underlying this – fuelled by the civil rights movement and the war on poverty, the protests against the Vietnam War, and the campus free speech movement – was a general hostility to the advanced capitalist system and a desire for a return to simpler lifestyles, coupled with deep distrust about the ways in which the entire 'military-industrial complex' was managed by professional technocracies. There was now a widespread distrust of expert, top-down planning, and in particular a positive paranoia about the prevailing 'systems' approach, which was identified with the military machine. And all this was underlined by the riots that tore apart American cities, from Paterson, New Jersey, in 1964 to Watts, Los Angeles, in 1967. The reaction, embodied in highly influential contributions from planners like Richard Bolan and Paul Davidoff, was to turn the tables: planners must practice bottom-up planning by becoming advocate-planners in the service of local communities (Bolan, 1967; Davidoff, 1965). In particular, they would not try to set goals and objectives for their clients, but would help their communities set their own agendas, try to inform their client publics about possible alternatives, force public planning agencies to compete for public support, and help critics to develop alternative plans. The underlying assumption was that the planner did not have much power and did not deserve to have much either; indeed, one can date from this point a long decline in the status of the planning profession, from which it continues powerfully to suffer in the 1990s (Hall, 1996).

The immediate result was that public participation in planning for the first time became a major issue, particularly in the United States and the United Kingdom, where the influential Skeffington report appeared in 1969 (GB Department of the Environment, 1969). During the 1970s, in many countries, one result was a series of celebrated urban clashes in which local community groups clashed with established urban planning bureaucracies, sometimes triumphing (as at Covent Garden in London in 1973 or the celebrated Eldonian cooperative in Liverpool a few years later); sometimes engaging in protracted political battles (as in Berlin's Kreuzberg during the 1980s).

There was an interesting parallel in the developing world, where the British planner John Turner, heavily influenced by the kind of anarchist thinking that had permeated the origins of the modern planning movement, first proposed self-build site-and-service housing in Latin America – in effect, a legalization and planned organization of the urban occupation movements that had taken place on a large scale in these cities in the 1950s and 1960s, a result of movement off the land (Turner, 1967, 1976). Rather remarkably, it soon became orthodoxy in the World Bank, which saw it as a low-cost and effective alternative to bureaucratically-organized public housing schemes. By the 1980s, such schemes were proliferating throughout the fast-growing cities of the developing world. The problem that emerged, in the poorest such cities, was that it proved virtually impossible to maintain even minimal standards of site and service provision in advance of the occupation of the land; the resources were lacking (Alfaro, 1996).

These themes – bottom-up planning, environmental concern, adaptive urban change – continued to re-echo throughout the 1970s and beyond. They appeared to represent effectively the beginnings of a new political platform, and in Germany the Green Party had considerable success in local elections for some cities. During the 1980s, they existed rather anomalously side-by-side with right-wing movements stressing economic liberalism and freedom from control, which in the United Kingdom resulted in an attempt – ultimately futile, as it turned out – to scale down the planning system. Effectively it could be said that planning constituted one of the dimensions on which a new political division was being forged in advanced countries, replacing traditional class movements and interests.

On one issue both sides agreed: the need to regenerate decaying urban economies by injecting new activities into them. That resulted from a set of very fundamental economic changes, the outlines of which academic observers just began to perceive in the early 1970s. There was the globalization of the economy, in which transnational corporations could shift big parts of production from traditional manufacturing locations to new industrial countries and regions. Coupled with this, a locational divorce was opening up between command and control functions, with their attendant producer services, and production sites. Advanced economies were increasingly shifting towards knowledge-based industries, including both high-technology manufacturing and advanced services. Information technology was having an increasing impact on production, particularly through the ability to displace more routine service activities from major global cities to smaller lower-cost cities. Parallel to this, business travel increasingly depended on inter-city transportation hubs, including international airports and high-speed inter-city train stations. Finally, it became evident in the 1980s and 1990s that cities were acquiring an important enhanced role as cultural, creative and tourist centres (Castells, 1989; GB Government Office for London, 1996; Hall and Landry, 1997; Landry and Bianchini, 1995; Sassen, 1991).

In combination, these changes drastically transformed the economies of major cities. London lost 289,000 manufacturing jobs but gained 168,000 jobs in banking, financial and business services between 1981 and 1991. Overall, though it was successful in creating wealth, it suffered a net loss of 304,000 jobs during the decade. In New York, the core of the region (New York City plus Newark) lost 155,000 jobs while the suburban ring gained 1,882,000 (GB Government Office for London, 1996).

The urban economies, then, must be revived; on that, there was general agreement. But the two sides disagreed both on objectives and on mechanisms. During the 1970s, left-wing local councils emphasized the revival of the traditional manufacturing and goods-handling economy and the use of democratic mechanisms which clearly gave the predominant weight to existing interests. Right-wingers proposed the creation of new urban economic bases, especially in the knowledge-intensive service industries, through private real estate development supported by public infrastructure, using mechanisms which bypassed elected local councils. During the 1970s, there were years of dithering and delay in European cities, but both mechanisms eventually produced results. In the 1980s, the city of Rotterdam successfully redeveloped its waterfront, Salford in Greater Manchester did the same, while in New York, London and many other British cities, Urban Development Corporations (UDCs) achieved very similar impacts. In Britain, the early attempts to establish UDCs – above all, in London Docklands – proved fiercely contentious, being fought through the courts, and local councils pursued policies of non-cooperation for years after, to no avail. The evidence eventually showed that, whatever the mechanism, the old industries, once lost, would not come back. Even successful cities saw their manufacturing base erode and their more routine service activities relocate to smaller cities in the wider suburban rings around them. The key to success was the ability to develop employment in the new service industries, and this proved quite difficult because these too were beginning to show increases in productivity because of the injection of information technology.

The impact on planning was however quite clear: whereas in the long postwar boom the main emphasis had been on accommodating demographic and economic growth in new developments, now it was on urban regeneration. Plans for new towns were scaled back or abandoned, and resources shifted back into the cities, as in the United Kingdom through the Inner Urban Areas Act of 1978. Urban regeneration sought not only to rebuild the economic base, but also to create new housing in the cities in order to try to reverse long-continued population decline. Because deindustrialization and port closures had left large tracts of urban land derelict, it eventually proved possible to create large-scale schemes; in some cases, city populations actually began to rise again for the first time in decades.

1998: The Search for Sustainable Urban Development

One hundred years after Howard wrote, people in the advanced capitalist countries live in a very different world from his: one in which the great majority have achieved relative affluence, albeit with some reduction in economic security compared with the long boom years of the 1950s and 1960s, and in which as a result they have a considerable stake in their own homes and their own local environments. The majority, typically two-thirds in many advanced countries (though lower in some, such as Germany), own their own homes, at least on mortgage; they also tend to live in suburbs even if they still work in central cities, and many both live and work in suburbs, increasingly bypassing cities in the course of their everyday lives.

One key result is that, for the first time since the beginning of the modern planning movement, housing and planning have become decoupled: the so-called housing question has shrunk to the problem of how to provide for an unfortunate low-income minority, as in fact has been the case from the start in the United States and to a considerable degree in Australia (Hall, 1998). In some countries, such as the United Kingdom, large-scale municipal housebuilding has effectively come to an end; social housing is provided by housing associations, and is often scattered widely in and among owner-occupied housing. At the same time, the tenant mix has changed: the deserving poor, effectively the higher-income blue-collar workers and their families, who were the main concern of nineteenth century philanthropists and early twentieth century municipalities, have disappeared as blue-collar workers have ceased to be significant and as their grandchildren have been absorbed into the middle-class mainstream. Their place has been taken by a variety of new low-income groups, including poor lone parents, refugees, ethnic minorities suffering from high unemployment and poor social integration, older people on basic pensions, and people suffering from a variety of medical or mental problems. Because of this change, social housing now tends to carry a stigma long familiar in the United States or Australia, but previously unfamiliar in European welfare states. In particular, some large remaining areas of urban public housing were widely perceived as seats of multiple social problems, with high rates of long-term structural unemployment, high rates of family breakup, and high rates of petty crime (GB Social Exclusion Unit, 1998).

Because of this fundamental shift, the role of planning has been weakened: providing social housing is no longer seen as a key government priority (whether that is true in fact or not), and its occupants are no longer seen as particularly deserving by the mainstream majority. But planning is denigrated in other ways and for different reasons: critics, ranging from Prince Charles to social policy makers, generally accuse it of having destroyed traditional urban communities and of creating soulless, mechanistic urban environments (Charles, 1989). The fact, that in many of the most notorious cases planners had virtually no power and no role, is conveniently avoided; mistakes by architects and engineers are all too readily heaped at the doors of the 'planners'. Planning was widely perceived as a routine, unimaginative, bureaucratic regulatory activity, generally disliked but regarded as a necessary evil. Part of the reason was that, in the great majority of cases, it did not seem to play a very positive role: the great majority of development, whether for new jobs or new homes, came through private initiatives, to which the planning system was essentially reactive. That view downplayed the role of the planner in drawing up development plans and in shaping the system, but it seems to have been a widespread perception. At the same time, planning became more and more highly politicized, because planning decisions were seen as powerfully affecting people's quality of everyday life; 'Nimbyism' was a prevailing movement in the more affluent suburban and exurban areas around the cities.

This directly stemmed from another major trend: the continued movement of people and jobs from the cities. However, this is not universal; perhaps to a greater degree than ever before, there is now a divergence between the advanced industrial (or post-industrial) nations. In the United States and many parts of Europe, the continued out-migration of people and economic activity, from the cities to the suburbs and the countryside, has led to the emergence of areas of concentrated multiple deprivation, particularly of ethnic minorities, in the cities. These in turn have become a negative element encouraging further out-migration by the affluent majority, including the middle-class members of the minorities (Wilson, 1987, 1996). Coupled with the rapid growth in household numbers in these countries – a product less of population growth than of household fission through more young people leaving home, divorce and separation, and longer periods of widowhood – this brings a central dilemma for strategic planning: how far is it desirable and possible to repopulate and reanimate the cities for affluent households, thus promoting the objective of sustainable development? There is widespread enthusiasm for such policies from a wide and powerful coalition of interests: from environmentalists, from the city administrations themselves, from a new generation of architect-planners interested in creating new styles of urban living, and from defenders of the countryside, including many of the recent migrants from the cities (Hall and Ward, 1998; Rogers, 1997).

The problem is that, powerful as these interests may be at the turn of the century, they are trying to reverse a strong tide of people who show every sign

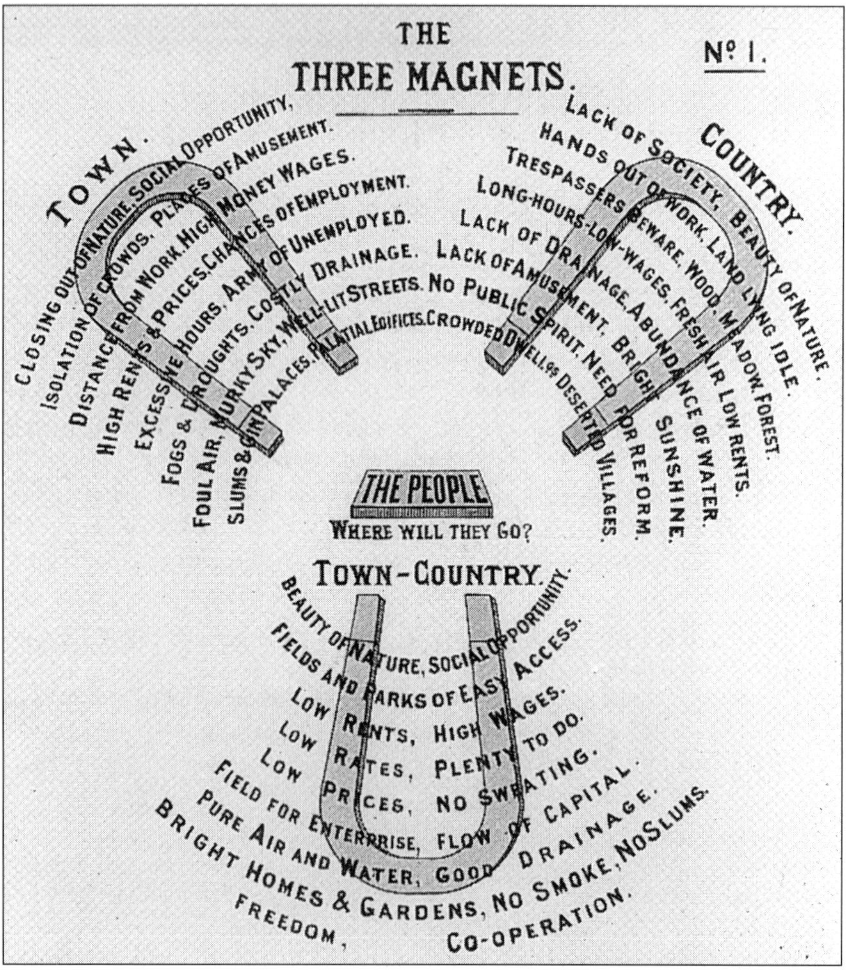

Figure 2.1. The Three Magnets, 1898. Howard's original diagram. The town offered economic and social opportunity, but a degraded environment. The country, after years of agricultural depression, in contrast offered an unpolluted environment but few jobs, poor social opportunities and overall poverty. But the third magnet, town-country, could offer all the advantages of town and country without the concomitant disadvantages. (*Source*: Howard, 1898)

that they want to escape the cities – above all, young parents with children, who have very negative perceptions of urban life and urban education. Proponents of urban revival point to the remarkable fact that the demographic projections show a large majority of single-person households. However, the question remains as to whether all of them, in their turn, will develop a taste for urban living. There are undoubtedly signs of a return to the city – in New York and a host of other major American cities, in London and one or two provincial British examples – but they are so far scattered and small-scale. Evidence suggests that, predictably,

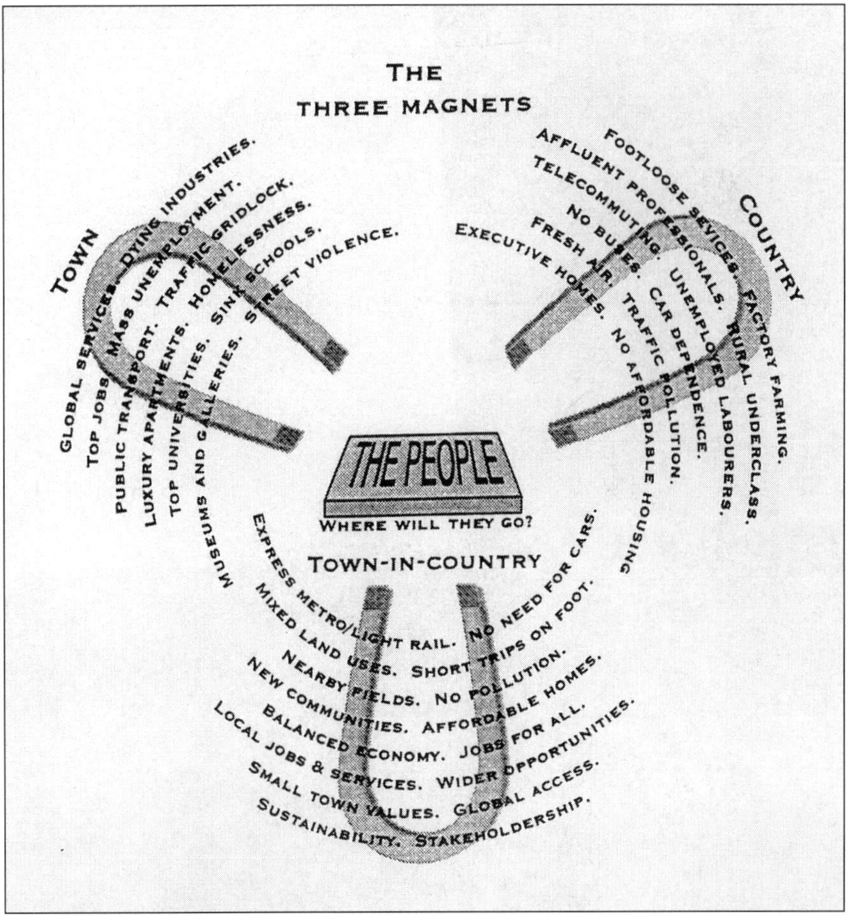

Figure 2.2. The Three Magnets, 1998. Howard's diagram still encapsulates the contrasts, but the polarities are reversed. Today, the deindustrialized town lacks economic opportunities for many workers, marginalized in social exclusion. The country has a buoyant economy because of electricity and highways and telecommunications, but is often socially exclusive. Oddly, town-in-country has happened through decentralization, but seldom according to Howard's prescription. (*Source*: Hall and Ward, 1998)

affluent single-person households want space and current developments, including the tendency for more work to be done at home, can only fortify that demand. Unless a significant share of the projected growth can be housed in the cities, the prospect is one of large-scale greenfield developments outside them; and this will bring forth further large-scale Nimby counter-movements.

These issues have less resonance in North America and in Australia, with their lower densities and their long traditions of unfettered suburbanization. But there are international trends shared by all the advanced nations. The most important, surely, is the new stress on sustainable development, marked in 1998 by the publication in English of a remarkable German work for the Club of Rome, *Factor Four: Doubling Wealth – Halving Resource Use* (Weizsäcker *et al.*, 1998). For planners, sustainability is widely interpreted to mean urban consolidation and compaction. Related to this is the attempt by many architect-planners to assert a more compact and traditional style of development – the 'new urbanism' in the United States, represented by such celebrated cases as Seaside in Florida, Kentlands in Maryland, and Laguna West in California (Calthorpe, 1993; Katz, 1994; Mohney and Easterling, 1991). Also there is the increasing emphasis in every country on preservation and conservation of historic buildings and entire urban districts, which over thirty years has spread from a few cities and countries to embrace countries and cities formerly immune, as in Pacific Asia.

Conclusion

Thus, against a totally transformed economic and social backcloth, the issues seem strangely similar to those of a century before; not least because rising agricultural productivity has again rendered substantial areas of countryside superfluous for farming, so that there is maximum pressure to preserve countryside against development at just the point when there is a minimal economic case for it. There is however an obvious and vital difference. When Ebenezer Howard drew his celebrated diagram of the Three Magnets, in 1898, the city offered economic and social opportunity while the countryside offered neither. Now, with the out-movement of people and activity, the roles are reversed. What in fact has happened is that twentieth century technology, in the shape of electricity and the motor car and the telephone and the internet, have made Howard's Third Magnet, Town-Country, as effective and attractive as he argued it would. But it has very seldom been realized in the planned forms, and via the planning mechanisms, which he proposed (Hall and Ward, 1998). The challenge he presented is as topical and as urgent now as then (figure 2.2).

Has planning taken a century to go around in a circle, then? Hardly, for in the process it has powerfully helped to shape better urban environments. That influence is not to be found merely in the classic achievements – Letchworth, Römerstadt, Radburn, Greenbelt, Harlow, Vällingby and the rest of the Pantheon

– but in their indirect influence on thousands of imitations, many reasonably successful, some few partial failures. City folk live immeasurably better lives than a century ago, and much of that they owe to what Keynes once memorably called science and compound interest. But they also owe something to better planned environments, which economic growth made possible, but which in turn have contributed directly to economic efficiency and indirectly to a better housed and more contented workforce. At the turn of the century, when it is fashionable in some quarters to attack planning as a brake on economic progress, that lesson from history is worth remembering and repeating.

REFERENCES

Alfaro, R. (1996) *Linkages Between Municipalities and Utilities: An Experience in Overcoming Urban Poverty.* Urban Environmental Sanitation Working Papers Series, UNDP-World Bank.

Bastié, J. (1964) *La Croissance de le Banlieue Parisienne.* Paris: Presses Universitaires de France.

Bauer, C. (1934) *Modern Housing*. Boston: Houghton Mifflin.

Beevers, R. (1987) *The Garden City Utopia: A Critical Biography of Ebenezer Howard.* London: Macmillan.

Bolan, R.S. (1967) Emerging views of planning. *Journal of the American Institute of Planners*, **33**, pp. 233-245.

Bowly, D., Jr. (1978) *The Poorhouse: Subsidized Housing in Chicago, 1895-1976.* Carbondale: Southern Illinois University Press..

Calthorpe, P. (1993) *The Next American Metropolis: Ecology, Community, and the American Dream.* Princeton: Princeton Architectural Press.

Castells, M. (1989) *The Informational City: Information Technology, Economic Restructuring and the Urban-Regional Process.* Oxford: Blackwell.

Charles, Prince of Wales (1989) *A Vision of Britain: A Personal View of Architecture.* London: Doubleday.

Davidoff, P. (1965) Advocacy and pluralism in planning. *Journal of the American Institute of Planners*, **31**, pp. 186-197.

Eberstadt, R. (1917) *Handbuch des Wohnungswesens und der Wohnungsfrage.* Jena: Gustav Fischer.

Ford, J. (1936) *Slums and Housing, with Special Reference to New York City: History, Conditions, Policy*, 2 volumes. Cambridge, Mass: Harvard University Press.

GB Department of the Environment (1969) *People and Planning: Report of the Committee on Public Participation in Planning.* (Chairman: A.M. Skeffington). London: HMSO.

GB Government Office for London (1996) *Four World Cities: A Comparative Analysis of London, Paris, New York and Tokyo.* London: Llewelyn Davies Planning.

GB Royal Commission on the Housing of the Working Classes (1885) Vol. I. *First Report*, Vol. II. *Minutes of Evidence and Appendices.* London: Eyre and Spotiswoode (*B.P.P.*, 1884-85, 30).

GB Social Exclusion Unit (1998) *Bringing Britain Together: A National Strategy for Neighbourhood Renewal.* London: Stationery Office.

Geddes, P. (1912) The twofold aspect of the industrial age: palaeotechnic and neotechnic. *Town Planning Review*, **31**, pp. 176-187.

Geddes, P. (1915) *Cities in Evolution.* London: Williams and Norgate.
Hall, P. (1996) *Cities of Tomorrow: An Intellectual History of Urban Planning and Design in the Twentieth Century.* Updated edition. Oxford: Blackwell.
Hall, P. (1998) *Cities in Civilization: Culture, Technology and Urban Order.* London: Weidenfeld and Nicolson.
Hall, P. and Landry, C. (1997) *Innovative and Sustainable Cities.* Dublin: European Foundation for the Improvement of Living and Working Conditions.
Hall, P. and Ward, C. (1998) *Sociable Cities: The Legacy of Ebenezer Howard.* London: John Wiley.
Howard, E. (1898) *To-Morrow: A Peaceful Path to Real Reform.* London: Swan Sonnenschein.
Howard, E. (1946) *Garden Cities of To-Morrow.* London: Faber and Faber.
Howard, E. (1985) *Garden Cities of To-Morrow.* New Illustrated Edition, with an introduction by R. Thomas. Builth Wells: Attic Books.
Howe, I. (1976) *The Immigrant Jews of New York: 1881 to the Present.* London: Routledge and Kegan Paul.
Jackson, K.T. (1985) *Crabgrass Frontier: The Suburbanization of the United States.* New York: Oxford University Press.
Katz, P. (1994) *The New Urbanism: Toward an Architecture of Community.* New York: McGraw-Hill.
Kitchen, P. (1975) *A Most Unsettling Person: An Introduction to the Ideas and Life of Patrick Geddes.* London: Victor Gollancz.
Kropotkin, P. (1898) *Fields, Factories, and Workshops.* London: Hutchinson.
Kropotkin, P. (1913) *Fields, Factories, and Workshops: or Industry combined with Agriculture and Brain work with Manual Work.* New, revised and enlarged edition. New York: G.P. Putnam's Sons.
Landry, C. and Bianchini, F. (1995) *The Creative City.* London: Demos/Comedia.
Le Corbusier (1998) *Essential Le Corbusier: L'Esprit Nouveau Articles.* London: Architectural Press.
Mairet, P. (1957) *Pioneer of Sociology: The Life and Letters of Patrick Geddes.* London: Lund Humphries.
Marcuse, P. (1980) Housing policy and city planning: the puzzling split in the United States, 1893-1931, in Cherry, G.E. (ed.) *Shaping an Urban World.* London: Mansell, pp. 23-58.
Meadows, D. and D., Randers, J. and Behrens, W.W. (1972) *The Limits to Growth.* New York: Universe Books.
Mohney, D. and Easterling, K. (eds.) (1991) *Seaside: Making a Town in America.* New York: Princeton Architectural Press.
Moynihan, D.P. (1965) *The Negro Family: The Case for National Action.* Washington, DC: US Department of Labor, Office of Policy Planning and Research.
Mumford, L. (1932) The Plan of New York. *New Republic,* **71**, pp. 121-126; 146-154.
Mumford, L. and Osborn, F.J. (ed. by Hughes, M.R.) (1971) *The Letters of Lewis Mumford and Frederic J. Osborn: A Transatlantic Dialogue 1938-70.* Bath: Adams and Dart.
Read, J. (1978) The garden city and the growth of Paris. *Architectural Review*, **113**, pp. 345-352.
Rogers, R. (1997) *Cities for a Small Planet.* London: Faber and Faber.
Sassen, S. (1991) *The Global City: New York, London, Tokyo.* Princeton: Princeton University Press.
Schumacher, E.F. (1973) *Small is Beautiful: A Study of Economics as if People Mattered.* London: Blond and Briggs.

Scott, M. (1969) *American City Planning since 1890.* Berkeley: University of California Press.

Sellier, H. and Bruggeman, A. (1927) *Le Problème du Logement: Son Influence sur les Conditions de l'Habitation et l'Aménagement des Villes.* Paris: Presses Universitaires de France.

Sitte, C. (1965) *City Planning according to Artistic Principles.* Translated from the German by G.R. Collins and C.C. Collins. New York: Random House.

Sussman, C. (ed.) (1976) *Planning the Fourth Migration: The Neglected Vision of the Regional Planning Association of America.* Cambridge, Mass: MIT Press.

Swenarton, M. (1985) Sellier and Unwin. *Planning History Bulletin*, **7**(2), pp. 50-57.

Turner, J.F.C. (1967) Barriers and channels for housing development in modernizing countries. *Journal of the American Institute of Planners*, **33**, pp. 167-181.

Turner, J.F.C. (1976) *Housing by People: Towards Autonomy in Building Environments.* London: Marion Boyars.

Unwin, R. (1911) *Town Planning in Practice: An Introduction to the Art of Designing Cities and Suburbs.* London: T. Fisher Unwin.

Weizsäcker, E.U. von, Lovins, A.B. and Lovins, L.H. (1998) *Factor Four: Doubling Wealth - Halving Resource Use: The New Report to the Club of Rome.* London: Earthscan.

Wilson, W.J. (1987) *The Truly Disadvantaged: The Inner City, the Underclass, and Public Policy.* Chicago: University of Chicago Press.

Wilson, W.J. (1996) *When Work Disappears: The World of the New Urban Poor.* New York: Knopf.

CHAPTER 3

Re-examining the International Diffusion of Planning

Stephen V. Ward

An intriguing footnote in the life of Ildefons Cerdà, author of the 1859 plan for Barcelona's *Eixample* (Extension), is that Georges-Eugène Haussmann, simultaneously engaged on the remodelling of Paris, is reported to have offered to buy Cerdà's plans and studies (Estapé, 1996, p. 55). Cerdà apparently refused, saying that he had drawn them up for Catalonia. If true, and the evidence rests on family recollection rather than a documentary source, this episode suggests that there was some degree of international linkage at a very early stage in the development of modern planning. In turn, Cerdà's refusal to sell may also partially explain why his remarkable innovatory work including, in 1867, the *Teoría General de la Urbanización* (General Theory of Urbanization), effectively the first modern theoretical work on urban planning, had a negligible international impact.

Within a few decades, however, there was abundant evidence that much more effective international flows of ideas and practices were becoming well established (Albers, 1997; Collins and Collins, 1965; Sutcliffe, 1981). Many cities, in Europe and beyond, sought to replicate the great new boulevards which Haussmann had driven through the old Paris at the behest of his Emperor. The emerging practices of *Stadterweiterungen* (town extension) in Germany, associated particularly with Reinhard Baumeister and Joseph Stübben, were also beginning to attract interest in Britain and the United States. So too was the notion of a more organic approach to town design, developed by Camillo Sitte and, to a lesser extent, Charles Buls from Austria and Belgium respectively. In England, Ebenezer Howard contributed a conceptually rich vision of the garden city in 1898, soon to be given a tangible form at Letchworth. Other seminal works followed in the new century, from Eugène Hénard and Tony Garnier in France, Charles Mulford Robinson and Daniel Burnham in the United States, Raymond Unwin in England, and Patrick Geddes in Scotland. Reflecting the nationalities of the authors of most of these works, four main innovatory planning traditions – Germany (*Städtebau*), Britain (town planning), France (*urbanisme*) and the United States (city planning) – were clearly apparent by 1914.

The key point, however, is that virtually all these works and the ideas they contained were also part of an emergent international discourse of planning. Written in German, French or English, many were translated within a few years of their publication into at least one other of the three languages. Some also appeared in other languages as well. The ideas they contained were interpreted, albeit selectively, into specific national contexts by a host of intermediaries throughout the world. Individual planners began to work in other countries, taking with them their own national planning conceptions and, in some cases, receiving others.

These early years set a pattern for the twentieth century. Despite the disruptive effects of major wars, internationalism has remained a powerful theme in urban planning. The dominant centres of innovation and patterns of emulation have not remained static, however. The balance of exchange between the main innovatory nations has shifted. Everywhere the influence of American ideas and practices has strengthened, paralleling the United States' rise to world dominance (Cody, 1998). The creation of the Soviet Union in 1917 unleashed a periodically innovative planning tradition, influential to some extent in Western Europe in the 1930s and 1940s and more directly so in Eastern Europe in the third quarter of the century.

Yet planning innovation has not been the exclusive prerogative of world powers. New innovators such as the Netherlands and Scandinavia emerged, to some extent replacing Germany as its influence waned in the Nazi years and their aftermath (e.g. Hall, 1991). Moreover, the decline of formal colonialism since 1945 has been paralleled by a weakening of the direct influence of British and French planning. A neo-colonialism built around foreign aid and international institutions such as the United Nations or the World Bank has, however, perpetuated a mainly (but not entirely) one way flow of planning ideas and practices from the developed to the developing world (e.g. Armstrong, 1987; Okpala, 1990; Sanyal, 1990). The European Union has also tried to play an important role in producing convergence in the planning repertoires of its member states (Williams, 1996).

INTERNATIONAL DIFFUSION AS A HISTORICAL THEME

Not surprisingly, the international character of the twentieth century planning movement has been a strong theme in historical writing. In particular, historians have been obliged to consider how and why planning ideas and practices came to spread between different countries. The best known works in English dealing with this theme are Anthony Sutcliffe's *Towards the Planned City* (1981) and Peter Hall's *Cities of Tomorrow* (1988, revised 1996). In fact, there is a significant volume of other work from many countries dealing to some extent with diffusion. The main concern has been to show how their principal subject, usually a particular city, country or group of countries, encountered key planning ideas or practices from elsewhere. This is apparent, for example, in Freestone (1989), Home (1997), King (1976, 1990), Smets (1977), Watanabe (1980, 1992), Wynn (1984), Yeoh (1996) and Yerolympos (1996).

Another approach to diffusion is that found in Buder (1991), Gold (1997) and Hardy (1991). These works have explored how the ideas and practices of particular planning traditions (namely the garden city and modern movements) have spread to different countries. In other cases the emphasis has been much more the exchange of planning ideas and practices, implying more complex diffusional flows between particular places. We can find this expressed in Albers (1997), Bosma and Hellinga (1997), Hall, P. (1996), Hall, T. (1997), Schubert (1997) and Sutcliffe (1981). It has also been the subject of a major international conference on planning history held in Tokyo in 1988 (TIPHC, 1988).

But, whatever the exact treatment of diffusion, we can identify three major concerns recurring throughout all these studies:

1. The mechanisms of diffusion. For example, key personalities, reformist or professional milieux, intergovernmental actions.

2. The extent to which ideas and practices are changed in their diffusion. How are they applied in specific national settings and why are differences apparent?

3. The fundamental causation of diffusion. For example, how much does it mirror the larger economic, political or cultural contexts of international relations? How far is the 'text' of planning's international diffusion more autonomous or reliant on chance actions?

No universal answers to these questions have emerged from historical studies and planning historians have adopted a variety of perspectives. We can illustrate this by reference to three planning historians whose work has been particularly influential. Thus Peter Hall's narrative, though acknowledging the role of structure and context, is largely one of great men with big ideas that spread because of the potency of message and the charisma and energy of those who carried it. By contrast, Anthony King's explanation rests much more on the global hegemony of Western imperialism, exporting its conception of planning in both the colonial and post-colonial eras. Between these two, Sutcliffe's account of the emergence of urban planning in the great powers of the late nineteenth century acknowledges both the larger impersonal forces and charismatic visionaries. Yet he stresses the critical intermediary role of the reformist and expert milieux, where imported ideas were distilled into locally relevant practices.

It will be immediately clear, however, that these different perspectives are not merely products of the interpretative stances of the historians concerned. They reflect real differences in the experiences of different countries. India's encounters with external planning models, the basis of King's work, were objectively different to those of Western Europe or the United States. In contrast to the pattern of promiscuous borrowing of external models that was typical amongst the latter group, planning in imperial India was externally imposed from one source. Significantly, when he deals with New Delhi, Nairobi or Lusaka, Hall readily acknowledges the central importance of this imperial domination. Here, at least, the ideas of the great men were triumphant because they were imposed by colonial power.

A wider review of writing about countries that were neither the major Western world powers nor their colonies increases this sense of real variety in experiences of diffusion. Thus the planning histories of southern European countries or the smaller countries of north western Europe, of Japan or the self-governing white Dominions of the British Empire typically show a high dependence on externally developed planning models. Yet such models were certainly not imposed in the classic imperial manner, without opportunity for indigenous political discretion. We can go further and note a growing awareness of the many subtleties of colonial planning, between imperial powers, between different parts of their Empires, and over time (e.g. Home, 1997; Wright, 1991). Also, it is clear that post-colonial experiences of planning imported as part of foreign aid, though replicating many features of the colonial era, have differed in some important respects.

These cases underline the general point that diffusion needs to be understood as highly variable, rather than as a single, uniform process. All the major dimensions of diffusion – the agencies and mechanisms by which it occurs, the extent to which ideas and practices are changed and, though it is more a matter of interpretation, its fundamental causation – have shown great diversity. Without denying that there will always be unique features in every episode of diffusion, it is possible (and indeed valuable) to generalize the different varieties.

A TYPOLOGY OF DIFFUSION

The remainder of the chapter elaborates on this general point, drawing widely on relevant historical writing to develop a typology for the diffusion of planning. It consists of a series of 'ideal types' of episodes of diffusion, highlighting salient features and giving some examples that reflect these features. Table 3.1 summarizes the different types. They fall into two distinct groups – 'borrowing' and 'imposition' – each having three types. This grouping thus marks a fundamental distinction between those episodes of diffusion where the 'importing' country has the greater role in shaping and controlling the diffusion process (borrowing) and those where the 'exporting' country is the main determining force (imposition).

As this rather implies, the essential basis of the typology is that of context,

Table 3.1 Typology of diffusion.

Type	*Indigenous Role*	*External Role*	*Typical Mechanisms*	*Level of Diffusion*	*Key Actors*	*Potential for Distinctiveness*	*Characteristic Examples*
Synthetic borrowing	Very high	Very low	Indigenous planning movements plus wide external contacts	Theory and practice	Indigenous	Very high	Major countries of of Western Europe & USA
Selective borrowing	High	Low	External contact with innovative planning traditions	Practice and some theory	Indigenous	High	Smaller countries of Western Europe
Undiluted borrowing	Medium	Medium	Indigenous deference to innovative external planning traditions	Practice with little or no theory	External with some indigenous	Fairly Low	Dominions of British Empire, Japan, & some European examples
Negotiated imposition	Low	High	Dependence on external planning tradition(s)	Practice	External with some indigenous	Low	Aid-dependent countries (e.g. Africa)
Contested imposition	Very low	Very high	High dependence on one external planning tradition	Practice	External	Low	'Enlightened' colonial planning
Authoritarian imposition	None	Total	Total dependence on one external planning tradition	Practice	External	None	Newly subjugated territories

specifically the power relationship between the countries originating and receiving planning models, is always of critical importance. By power relationship is meant simply the degree of domination, however expressed, of the one by the other. The individual types are, of course, rather generalized and certainly do not capture the subtleties of actual diffusion episodes. They merely represent the principal gradations of borrowing and imposition during the twentieth century. Thus the great Western powers, equivalents and rivals rather than deferential to each other, borrowed in creative rather than slavish fashion. Where the sense of deference was greater, then the borrowing became progressively less selective.

As deference graded into dependence, in colonial or post-colonial situations, then the power balance and the diffusional type shifted to imposition. Yet there are degrees of dependence and thus of imposition. In most cases receiving countries have been able to negotiate or contest the process of external imposition, thereby moderating its nature. However, in extreme cases, few in number in this century, dominance has been so complete as to eliminate any hint of indigenous modification. In all such situations the power balance, whether measured in governmental, economic or cultural terms, was very uneven as between the dominated colony and dominant imperial power. Among the rival great Western powers in the early twentieth century, however, the power relationship was more nearly equal.

DIFFUSION BY BORROWING

Synthetic Borrowing

Most familiar is the type of diffusion which has occurred between the main innovative planning traditions. As we have already noted of the early years, the patterns have typically been very open. Each of these innovatory countries drew on several external planning models, while the other innovators for their part borrowed back from them. To a greater extent than in other types of diffusion, the trade in ideas was closer to a state of balance. Before 1914, for example, Britain borrowed heavily from German town extension, zoning and organic approaches to urban design. In turn, the Germans (having already borrowed British public health innovations) looked admiringly on British housing design and, above all, the garden city. The United States, for its part, borrowed German zoning, the British garden city, and the French approach to grand urban design. It gave back to Europe the notion of the city-wide master plan and the grand approach to urban landscape design.

One of the key points about this form of diffusion is that the borrowed external models have typically been filtered through highly developed indigenous reformist movements and professional expertise. This filtering process has tended to deconstruct the models, breaking them down into component elements, and integrating them with planning ideas and practices that are already present. This deconstruction has occurred both consciously and unconsciously, through misunderstanding or partial understanding. In either case, though, the outcome has been that the diffused models were almost never transferred unaltered. Indigenous ideas and ideas already received from other sources were combined with newly-received models to create something distinctive and new. The overall effect was a process of synthetic innovation, with the further possibility that the resultant innovations might themselves be diffused elsewhere.

The history of the neighbourhood unit provides a classic example of how this process operated. The starting point was Britain, with Ebenezer Howard's indicative but barely elaborated concept of the 5000 population ward within his larger formulation of the garden city. It was, however, Clarence Perry in the United States who fashioned it into a workable physical model, sticking to Howard's 5000 population, in the 1920s. These ideas were further elaborated to reflect the growing importance of automobile traffic by Henry Wright and Clarence Stein. Their efforts culminated in the Radburn layout in the late 1920s.

Both ideas then spread back to Western Europe, where they were further overlain with new aspects and meanings. For the moment, Radburn principles were not applied, although there were signs of some German interest, especially in Hermann Jansen's road safety residential plans of the late 1920s (Hass-Klau, 1992). In Britain, however, planners were becoming more concerned with the neighbourhood as a device to promote social cohesion. Increasingly more ambitious objectives, involving a social class mix, were gradually added. Particularly influential was Barry Parker's plan for the Wythenshawe satellite town in Manchester, part of the garden city mainstream. Meanwhile in the Soviet Union there were some innovations that echoed Perry's ideas, though without apparent knowledge of Western developments (Tetlow, 1959). British modernists, in the shape of the MARS (Modern Architectural Research) group, began to experiment with Soviet-influenced neighbourhood ideas in the later 1930s (Gold, 1997). These two strands came together in the 1940s. Neighbourhoods, now with a notional population of up to 10,000, occupied a central place in Britain's wartime and early postwar plans, particularly in the first new towns.

American and British thinking had by then begun to influence planners in Sweden during the early 1940s. As a neutral country, Sweden suffered less wartime privation than other parts of Europe. This allowed it to assume an especially important role in the empirical elaboration of neighbourhood ideas in the immediate postwar years. Accordingly, Swedish experience became extremely important in the physical design of neighbourhoods. By 1947, variants of Radburn layouts were being planned, a few years before their first British use (Parsons, 1992). By 1950, Swedish experiences had also shown that 10,000 was far too low a population for an effective social unit in an affluent society (Sidenbladh, 1964).

Meanwhile, Dutch planners secretly replanning the devastated city of Rotterdam during the Nazi occupation had already come to similar conclusions but for quite different reasons (Lock, 1947). Aware to some extent of pre-war thinking on neighbourhood units but without knowledge of the important Anglo-American developments of the 1940s, the Rotterdam planners had already by 1945 proposed a socially mixed neighbourhood unit of 20,000. This reflected some very specific features of Dutch society, relating to the church's extensive role in social provision. Although planners from other countries, ever fascinated by Rotterdam's reconstruction, soon became aware of this variant of the neighbourhood, it did not seem as generalizable as the Swedish experience.

The most extraordinary aspect of the diffusion of the neighbourhood unit at this time involved its deployment in wartime Nazi Germany. Thus planners in Hamburg particularly made extensive use of a concept called the 'local group as neighbourhood cell' (Schubert, 1995). This reflected the idiosyncrasies of Nazi ideology, yet it also leaned on the Anglo-American concept of the neighbourhood. In part, this connection reflected pre-war links. Yet there was also a keen awareness of wartime developments in London and other cities. Via neutral Stockholm, the German intelligence services had secured copies of Patrick Abercrombie's plans for London, making them available to Hamburg's chief planner. The similarities ensured that, stripped of their Nazi overtones, they could therefore be perpetuated into the postwar years.

After 1945 all versions of the idea became the subject of even more international cross-fertilization, with Radburn principles being widely adopted and adapted. However enough has been said to show how this synthetic process gave an innovative dimension to diffusion, in circumstances where planning models came into countries which already had highly developed planning traditions.

Selective Borrowing

Where innovatory synthesis of imported and existing ideas and practices has been lacking, diffusion has often taken the form of a rather simpler process of borrowing. A characteristic feature of this non-innovatory borrowing has been a relatively shallow engagement within the importing country with the theoretical and conceptual bases of the borrowed model. This has limited the possibilities of deconstructing the ideas and reassembling them, with other ideas, to make something different. Instead planners in the receiving country have tended, rather atheoretically, to emulate specific aspects of external planning practice in a simple and direct manner.

This is not to say, however, that the borrowing has necessarily been slavish or uncritical. There has often been some degree of selection. Parts of the borrowed model may be discarded if they seem less appropriate. The main point, though, is that the importing country has added nothing significant to what is imported. In turn this offers little that is sufficiently distinctive for other countries to borrow. Yet countries which habitually borrow selectively may sometimes play an important intermediary role, facilitating the movement of innovations between more distinctive (and possibly competitive) planning traditions.

In fact, this type of diffusion episode can be found throughout the century in many different countries. It is, perhaps, most characteristic of the development of planning in the smaller and less powerful Western and Central European countries. These countries would have enjoyed fairly good access to more than one of the major innovative planning traditions. Typically they would also have had reformist movements pressing for planning and substantial indigenous professional expertise. Together these were capable of exercising some discretion over what was borrowed from external planning models. Yet the critical mass needed to innovate in more thoroughgoing fashion was absent.

A good example is Belgium. This small country was not entirely devoid of genuine innovation of international significance. Yet its planning tradition depended heavily on external sources of theory and practice, mainly French, German and British (Smets, 1977). Thus Parisian Haussmannism was emulated in the later nineteenth century (Hall, 1997). The early twentieth century brought growing awareness of the British garden city tradition, implemented through a combination of French-style social housing organizations and the British co-partnership model. More generally, Belgium (along with Switzerland) apparently played significant parts in moving British and German urban reformist ideas into the Francophone world (Claude, 1989). Another case is Norway, where traditional dependence on Swedish design in the early twentieth century was leavened, though not supplanted, by other influences, especially from Britain, the United States, France and the Netherlands (Lorange and Myhre, 1991). Again, however, no significant innovations arose from these borrowings.

Yet it would be incorrect to imply that selective borrowing has occurred only in smaller countries. Although there has always been a strong tendency in the major innovative planning traditions to use imported models in a more adaptive fashion, episodes of selective borrowing can be found everywhere. A recent example would be the rapid adoption throughout the developed world of the American approach to waterfront redevelopment. Pioneered in cities such as Baltimore and Boston in the 1970s, the model had by the 1990s appeared throughout Europe and beyond, with varying (but often very close) correspondence to the originals (Breen and Rigby, 1996).

Overall, however, examples of this type do not negate the main point: if synthetic innovation has been the dominant means of receiving externally generated ideas and practices, the cumulative result will be a national planning tradition that soon becomes distinct. If selective borrowing has been usual, particularly from more than one source, then differences will certainly arise, but more slowly. The cumulative result will also appear as an altogether more derivative planning tradition.

Undiluted Borrowing

This derivative quality has been even more marked where external ideas and practices have been received without conscious selectivity – where the borrowing, in other words, is undiluted. In such cases, the tendency has been to receive not just individual ideas or innovations but substantial packages of planning practice. As this implies, such borrowing has been rather uncritical and frequently with only very limited awareness of the full range of alternative external planning models that are available. In turn, this reflects a rather underdeveloped indigenous planning movement and, quite often, a high reliance on foreign planners to supply leadership. There is a real difference here from previous types, where imported ideas and practices were filtered through indigenous planning movements (and in some cases, intermediate countries).

This diffusional type has been characteristic of countries which exhibit a more general deference to ideas arising in those countries from which they borrow. The relationship between the two parties is therefore markedly more uneven than in the previous two types. Yet we should not exaggerate the aspect of external dependence. This type should be still understood, very definitely, as *borrowing*, clearly implying that the power to make decisions remains in the importing country.

The clearest examples of undiluted borrowing have undoubtedly been the white settled Dominions of the British Empire, whose early encounters with twentieth century planning came largely through the prism of British experience (Ward, 1997). These were self governing by the time modern planning thought and practice developed. Yet they had relatively small populations, underdeveloped reform movements and limited professional resources. When combined with more general ethnic and cultural affinities, these factors created a strong initial dependence on planning models from the imperial homeland. At varying rates, this was then overlain with what, initially at least, was an almost equally uncritical admiration for ideas and practices from the United States.

Canada borrowed uncritically from British planning in the first two decades of the century, to the extent of adopting planning legislation and founding a British-style professional body for planning (Simpson, 1983). Then, in the 1920s, American influences, often copied in an equally direct way, became dominant, coinciding exactly with the American replacement of Britain as the main foreign investor in Canada (Ward, 1999). After 1945, there was a resurgence of British planning (though not economic) influences. Yet external ideas were by then being received in a more critical and selective fashion. Canadian awareness of other European planning traditions, particularly the French, also increased.

There were many similarities between Canada and Australia. In the latter, however, British connections were dominant for much longer (Freestone, 1989; 1997). In part, this reflected the persistence of Australia's economic and cultural ties with Britain. Nor was the American model as conveniently located as it was for Canadian planners, who often found it easier to consult American planners than fellow Canadians. Thus the Australian planning system developed very much in the British image, with extensive British professional leadership. Although there was early awareness of American planning models (most strongly apparent in the chosen plan for the new federal capital at Canberra), it was not until the 1960s that they even began to match the extent of British influence.

Yet we should not see undiluted borrowing as a manifestation only of late imperialism. It could also arise in quite different circumstances, sometimes even in countries with relatively advanced planning traditions. Thus the replanning of the historic French city of Reims, devastated in World War One, became an exercise in scientific American city planning. United States' wartime relief had brought a leading American planner, George B. Ford, on the scene who quickly assumed technical dominance in the reconstruction debate (Bédarida, 1990; Wright, 1991). Ironically, very similar circumstances at almost exactly the same time allowed the French *urbaniste*, Ernest Hébrard, to assume an even more dominant role in the replanning of Thessaloniki in Greece, following its destruction by fire in 1917 (Yerolympos, 1996). The resultant plan was a grand exercise in French urban design. Such uncritical absorption of external models was not usual in either country, however, especially France.

An example of a more habitual uncritical borrowing that did not depend on imperial ties or emergency situations, was early twentieth century Japan. Here Western, especially German, British and American, planning practices were borrowed and applied with a surprising lack of adaptation to Japanese conditions (Hein and Ishida, 1998; Watanabe, 1988, 1992). The context was the rapid modernization of Japan from the later nineteenth century, which encouraged a fairly systematic trawling of the advanced Western countries for progressive practices which could be adopted.

What was particularly striking, however, was the rather imperfect conceptual grasp of the models that were being received. Having only a weakly developed reformist movement and professional skills in planning, the possibilities of conscious selection or synthesis were quite limited. Initially, at least, Western planning was copied quite slavishly, the only adaptations arising unconsciously from misunderstanding. In some aspects, what was borrowed was an even purer version of Western ideas than was actually adopted in the West. The land readjustment proposals incorporated in the first Japanese planning legislation of 1919, for example, were a more radical version of the widely admired German *Lex Adickes* than the German parliament had been prepared to adopt (Ishida, 1988).

Over time Japanese planning began to assume a more distinctive character, not least because Japanese planners, unlike those in Australia, had always looked to the West as a whole. Even so, as late as the 1950s, Tokyo's planners were still directly mimicking a planning model drawn directly from another country

(TMG, 1994). This was the archetypal British metropolitan planning solution, with encircling green belt and planned decentralization. Ironically, the Tokyo plan was very similar to proposals for Sydney adopted a decade earlier and on the point of being substantially abandoned (Winston, 1957). The Japanese plan proved even more short-lived. In both cases the failures reveal the weakness of over reliance on imported models. They failed entirely to grasp political and growth realities that were quite different to those of British cities.

DIFFUSION BY IMPOSITION

The dangers of inappropriate transfers were (and are) much higher when the balance of power is such that the exporting country can exert more control than the importing country over the diffusion process. Thus, instead of authorities in the importing countries deciding themselves what they wish to borrow from foreign planning repertoires, relevant agencies in the exporting countries make the key decisions. This inherently limits the opportunities for local participation in the planning process within the countries receiving planning. Even more than in cases of undiluted borrowing, the importing countries will typically have very underdeveloped planning and reform movements. They will also be heavily, often almost totally, reliant on imported planning expertise. In most cases, of course, imposition is symptomatic of colonial or neo-colonial relationships. Beyond these underlying characteristics, there have been several distinct varieties of diffusion through imposition.

Negotiated Imposition

The post-colonial period for many former colonies in some cases encouraged greater scepticism about external planning ideas. The more affluent and determined former colonies were quickly able to cross that critical divide between imposition and borrowing. In Singapore, for example, the post-independence State constructed a distinctive form of planning that borrowed freely from Western planning practice (Perry, Kong and Yeoh, 1997). The original base was British, yet ideas and practices were seemingly drawn from a variety of sources. They were applied, however, with a relentless discipline rarely matched in the West, inspired by distinctively Asian social and political ideologies.

Yet Singapore was exceptional, not least because of an extraordinary material progress that no other recent ex-colonies have matched. Many other ex-colonies in Asia and, above all, in Africa experienced a mode of diffusion that can be called negotiated imposition. It was characterized by continued dependence on external technical expertise and, often more importantly, material aid. Nominally the independent governments actually took the decisions about the acceptance of external aid and assistance. Yet such was the extent of dependence that the process went beyond borrowing. If these countries wanted the aid, the technical assistance went along with it. The offer could scarcely be refused.

It was a diffusional relationship that perpetuated some aspects of colonialism. At the outset, at least, there were slightly more options. In cases where the transition to independence had been peaceful and without active resentment, planning and aid flowed smoothly from former colonial masters. Thus from the late 1940s many British and French consultants began to find work in former colonies in this way, funded substantially by their own governments or United Nations agencies (e.g. MLCERG, 1997). Independence could also mean seeking aid and expertise from sources other than (or at least, in addition to) former colonial masters. Indonesia, for example, turned sharply away from the Netherlands for fifteen years after independence (van der Heiden, 1988). While maintaining fairly strong connections with Britain, India began to develop other linkages (Evenson, 1989). Its great size and economic and strategic potential gave it a negotiating power in external dealings that few other underdeveloped ex-colonies possessed. Yet smaller countries such as Tanzania also avoided some of the worst features of neo-colonialism by seeking aid and expertise from several countries (Armstrong, 1987).

These developments had implications for the international planning influence of Western countries which had not been formal colonial powers. Here was one of the principal ways in which the international planning influence of the United States grew (Cody, 1998). Yet less dominant affluent countries such as Canada, Australia, the Scandinavian countries and Western Germany, not lately major colonial powers, similarly exported their own conceptions of planning from the 1960s. Countries from the communist bloc sometimes used comparable methods to increase their influence in parts of Africa and Asia.

More recently there has been a growing tendency for international bodies such as the United Nations Centre for Human Settlements (Habitat) and the World Bank to play important roles (Okpala, 1990). These have purveyed generalized Western-determined planning and development solutions, premised increasingly on economic liberalization. The disappearance in the 1980s of a communist alternative and the wider effects of global economic change have strengthened this latest variant of imposition. Although at the technical level, the export of foreign planning aid is now being undertaken more sensitively than ever before, with growing emphasis on indigenous expertise, the wider sense of imposition remains very strong. The scope for negotiation, apparently so great in the bright confident morning of independence, has narrowed.

Contested Imposition

Countries where planning has been externally imposed often, though not invariably, have underdeveloped civil societies. This, with the absence of both elective democratic government and indigenous reformist movements, has been a key part of the colonial experience. Yet such formidable obstacles, although they seriously weakened the possibilities of any formal negotiation of what was proposed by colonial powers, did not condemn the recipient population to absolute passivity. A more typical situation within long established colonial empires was indigenous obstruction of externally imposed planning projects. Measures that were particularly repugnant to local interests and sensitivities might well provoke rioting or other forms of protest. More typically, indigenous populations might simply abuse or superimpose their own meanings or customary uses on external planning forms.

The case of colonial Singapore, which has been meticulously documented (Yeoh, 1996), provides a particularly good example of this. A British colony, its highly urbanized character made it an early target for British sanitary and planning ideas. From 1913, when municipal elections were abolished, authority was vested in a nominated body which proved more amenable to the 'progressive' principles of British-style municipal management (undertaken by British professionals). Yet this apparent authoritarianism was actually rather more subtle in operation. Nominations included a growing number of the indigenous population. This approach gave influence to those Europeanized Singaporeans who could be expected broadly to support the British town planning model. At the same time they also had to mediate between the indigenous population as a whole and colonial authority. In effect, this usually meant trying to contain opposition.

Reality, however, was more complex. Wholesale zoning powers sought for the Singapore Improvement Trust founded in 1927 were compromised by property interests of all races. However, there was strong and persistent Asian opposition by both owners and occupiers to the more modest proposals to open up what to British eyes were congested districts. Similarly attempts effectively to anglicize public space by limiting street trade proved practically impossible to enforce. Conflicts were particularly strong where planning proposals affected indigenous sacred spaces.

The pattern in French colonies, at least showpieces such as Morocco or Indo-China, was supposed to be different (Wright, 1991). By the time of World War One, the official imperial ideology had become an enlightened approach called 'associationism'. In planning terms, this involved modern colonial built forms, yet planned in styles that were supposed to defer to indigenous culture. Moreover they were developed as new settlements alongside indigenous traditional communities, without any direct intention to replace them. The rigid racial segregation that was typical of British imperial planning was also rejected. The intention was that the indigenous population would gradually realize the superiority of the new settlements, planned by leading French *urbanistes*, and increasingly use and occupy these spaces themselves. It promised, at least, a different social geography to that of the British colonial city, which rested on the separatist concept of the 'dual mandate' (Home, 1997).

Yet the French strategy also acknowledged the potential tensions with the colonized peoples and sought to avoid them, though without actually involving indigenous viewpoints. Despite physical results that were often impressive, the overall outcomes scarcely lived up to the ideals. Invariably the financial benefits of these planned urban developments went overwhelmingly to French interests. Only small numbers of the indigenous elite were able to embrace fully the spirit of 'associationism'. And, though it took different forms, colonial planning was challenged, actively and passively, by the majority of indigenous people. Thus in Indo-China, enlightened *urbanisme* could not tame the rising indigenous challenge to the colonizers in major cities such as Saigon and Hanoi. Around Moroccan cities such as Casablanca, burgeoning indigenous *bidonvilles* (shack communities) were soon challenging the colonial planning process.

As the case of Indo-China shows, the contesting of externally imposed planning was an integral part of a much wider process of challenging imperial dependence. This link with struggles for national self-determination confers a wider resonance on the planning history of these countries. It contains, too, some faint glimmerings of that conscious exercise of critical selectivity that was the hallmark of diffusion in the Western heartlands of planning innovation and their immediate neighbours.

Authoritarian Imposition

The most acute type of imposition has occurred in situations of extreme repression. Characteristic of this type of diffusion would be externally imposed planning proposals and methods of enforcement which grant few if any concessions to established indigenous interests. This can sometimes lead to a curious phenomenon whereby 'purer' expressions of one country's planning approach may appear elsewhere, in lands appropriated as colonies. Invariably, conquering powers are far less compromised in overriding indigenous democratic or property rights than in the imperial homeland.

Perhaps the most extreme example of this phenomenon was the short-lived Nazi replanning of the provinces of Poland incorporated into the German Reich in the early 1940s (Fehl, 1992). Seeking to impose a new regional order, the planners of the SS (amongst them the geographer Walter Christaller) adopted a version of Ebenezer Howard's social city that had been entirely stripped of its original social reformist meanings. In this form it became the basis for a new Germanic ethnic template, to be implemented by seizure of indigenous property, forced depopulation, slave labour, and extermination. Jews alone suffered such wholesale dispossession within Germany proper, where most existing property rights were treated more respectfully.

Another example was the imposition of Japanese notions of planning (in effect, the rather imperfectly assimilated pastiche of Western planning ideas) on its colonies, Taiwan, Korea and Manchuria (Hein, 1998). Again, such planning efforts were frequently far more interventionist than anything undertaken in the imperial homeland. Thus Taiwan had building controls from 1895 and housing laws from 1900, well in advance of comparable developments in Japan. By the 1930s, land readjustment in the Japanese colonies was also being undertaken in a far more draconian fashion, without compensation, than would have been tolerated in Japan itself.

There were comparable episodes of planning in other colonial settings. Thus the early planning of Algeria, though skilfully undertaken by French military engineers, was a heavy handed exercise of imperial power, without concession to indigenous society or culture (Malverti and Picard, 1991). Similar charges have been laid against the British, especially in the most grandiose exercises in imperial planning such as New Delhi (Irving, 1981). Yet extreme authoritarianism, without reference to indigenous sentiments, has actually been quite rare in twentieth century colonial planning. As we have seen, indigenous populations were not usually meekly accepting of these alien forms of planning, imposed through imperial power.

CONCLUSIONS AND SPECULATIONS

Amongst the impositional forms of diffusion, context – the power relationship between the 'exporter' and the 'importer' of planning – has clearly assumed the central role in shaping the diffusion experience. The relationship was absolute in authoritarian imposition but increasingly modified by other factors as we ascend the categories in table 3.1. Effectively, the influence of context would tend to increase as the power relationship between the originating and receiving countries became more uneven. Where the power relationship was more even or favoured the receiving country, however, indigenous reformist milieux could be expected to play a more important and autonomous role, often adapting what was imported. Further down table 3.1, however, the likelihood of adaptation was much lower. There was instead a stronger likelihood that purer versions of the exported model might appear than had been found in its country of origin.

The role of individuals as independent variables in the diffusion process remains less amenable to generalization, however. Key individuals could become significant if not primary determinants in all diffusion types. Clearly, though, the way they could exert influence would vary according to context. The more authoritarian nature of colonial power probably gave the individuals in whom that power was vested the greatest scope to shape outcomes. Where indigenous reformism was more established, the extent of the individual's role would depend on the persuasive appeal of his/her proposals, particularly within reformist and professional milieux.

A more human way of expressing this explanatory problem is to go back to the story in the introduction. If Cerdà *had* sold his work to Haussmann, or at least made greater efforts to promote his pioneering *Teoría* outside Spain, perhaps the course of international planning history might have been different. Barcelona might have become a mecca of late nineteenth century planning, alongside Paris, Frankfurt, Birmingham and Chicago. On the other hand, it is doubtful whether one individual can ever surmount the limits imposed by context. Spain's terminal decline as a world power in the late nineteenth century would almost certainly have prejudiced reformers in other countries against adopting any Spanish ideas, whatever their intrinsic merits. It certainly ruled out colonial demonstrations of the kind that played such a key part in the international spread of British and French planning during the twentieth century.

Today, paradoxically, Barcelona's planning and other lessons are being widely studied, borrowed and, to varying degrees, adapted in both the post-industrial and Hispanic developing worlds (Borja, 1996). The context has now shifted to remarkable global economic success, underpinned by progressive reformism and dazzling professional expertise. Cerdà's modern equivalent has been the city's charismatic and visionary recent Mayor, Pasqual Maragall. Unlike Cerdà, however, Maragall has played on the world stage, importing and adapting external planning models (for example, from Baltimore) and, even more, promoting the international spread of the Barcelona model.

Now, of course, we accept globalization as a reality of everyday life. The jet airliner, satellite telecommunications, and the internet have accelerated an internationalization of information that was originally made possible by the steamship, the railway, the mechanized printing press, the postal service, and the telegraph. Throughout the world the lessons of Barcelona (or indeed any other admired planning model) can today be examined more conveniently than ever before. Even past, unrecognized lessons can finally be diffused – Cerdà would doubtless be astonished to learn that it is now possible to buy video cassettes with commentaries in several languages, explaining his great plan for the *Eixample*.

Despite all this, the likelihood is that, in most respects, the diffusion of planning will continue as before. Modern communications certainly allow borrowing to be more rapid, more comprehensive, and less spatially bounded. Yet the process is being driven (as it always was to some extent) by economic imperatives. Throughout the world, cities are now exposed to the full rigours of global market forces. Their leaders seek planning models from successful cities everywhere in the continual battle to win or retain highly mobile international capital. In Beirut, for example, now rebuilding itself after a destructive civil war to regain its role as the Middle East's main international centre, planners have quite deliberately drawn on the widest possible range of Western planning expertise and models (Gavin and Maluf, 1996). Despite occasional echoes of French colonial links, this borrowing process is, however, largely orchestrated by Lebanese interests.

More generally, the effective ending of all but the final vestiges of formal colonialism in the last years of the century may perhaps reduce the extent of impositional diffusion. We may, at least, see a shift to more negotiative forms than have been typical for much of the twentieth century. Yet colonialism continues to cast a long shadow of dependence over the poorest parts of the world, especially in Africa. For many countries, saddled with huge external debt burdens and lacking sufficient indigenous reformist and professional resources, diffusion will necessarily continue to be an externally-determined process. Nor is it certain even that the more authoritarian forms of imposition have entirely gone. Thus the chronic instability of post-imperial or post-communist nationalism in disputed regions of the Balkans and Middle East is bringing a reordering of settlement, enforced by military power.

Yet, while such trends are regrettable, there is also an emergent type of impositional diffusion by consent that can be welcomed. The European Union is the clearest example, where powerful affluent countries have lately begun to defer (rhetorically, at least) to mutually agreed models for environmental management and planning that transcend the nation state. Such initiatives recall the spirit of internationalism that characterized many European planning movements at the beginning of the twentieth century. There are much weaker signs that the same spirit has also found global expression in the recent Earth Summits. If (and this is a very big 'if') tangible achievement follows, then one of the promises of twentieth century planning may finally be fulfilled. Whether or not this occurs, however, there is much about how planning diffusion occurs that will not change. Fundamentally, it will continue to be shaped by the endlessly fascinating mix of context, reformist and professional milieux, and individual action that largely determined its course during the twentieth century.

ACKNOWLEDGEMENT

My thanks to my colleague Roger Zetter for advice on literature on aid-related planning.

REFERENCES

Albers, G. (1997) *Zur Enwicklung der Stadtplanung in Europa: Begegnungen, Einflüsse, Verflechtungen*. Wiesbaden: Vieweg.

Armstrong, A. (1987) Tanzania's expert-led planning: An assessment. *Public Administration and Development*, 7, pp. 261-271.

Barker, J.M. (ed.) (1997) *Old Institutions – New Images*. Proceedings of the International Conference, John Curtin International Institute, Perth: Curtin University.

Bédarida, M. (1990) La 'renaissance des cités' et la mission de Geo B. Ford, in Gaudin, J. P. (ed.) *Villes Réfléchies: Histoire et Actualité des Cultures Professionelles dans l'Urbanisme*. Dossiers des Seminaires, Techniques, Territoires et Societes, no. 11/12, Paris: Délégation à la Recherche et à l'Innovation, Ministère de l'Equipement, du Logement, des Transports et de la Mer, pp. 33-42.

Borja, J. (ed.) (1996) *Barcelona: An Urban Transformation Model 1980-1995*. Urban Management Series Volume 8. Quito: Urban Management Programme.

Bosma, K. and Hellinga, H. (eds.) (1997) *Mastering the City 1: North-European City Planning 1900-2000*. Rotterdam: NAI Publishers/EFL Publications.

Breen, A and Rigby, D. (1996) *The New Waterfront: A Worldwide Success Story*. London: Thames and Hudson.

Buder, S. (1991) *Visionaries and Planners: The Garden City Movement and the Modern Community*. New York: Oxford University Press.

Claude, V. (1989) Sanitary engineering as a path to town planning: The singular role of the *Association générale des hygiénistes et techniciens municipaux* in France and the French-speaking countries, 1900-1920. *Planning Perspectives*, **4**, pp. 153-66.

Cody, J. (1998) Private Hands and Public Gloves: Options for globalizing US planners, 1945-1975, in Freestone (ed.), pp. 95-100.

Collins, G.R. and Collins, C.C. (1965) *Camillo Sitte and the Birth of Modern City Planning*. London: Phaidon.

Estapé, F. (1996) Ildefons Cerdà i Sunyer, in Palà, M. and Subirós, O. (eds.) *1856-1999 Contemporary Barcelona Contemporánea*. Centre de Cultura Contemporània de Barcelona: Barcelona, pp. 53-55.

Evenson, N. (1989) *The Indian Metropolis: A View Toward the West*. New Haven: Yale University Press.

Fehl, G. (1992) The Nazi Garden City, in Ward, S.V. (ed.) *The Garden City: Past, Present and Future*. London: E & FN Spon, pp. 88-106.

Freestone, R. (1989) *Model Communities: The Garden City Movement in Australia*. Melbourne: Nelson.

Freestone, R. (1997) The British connection: Convergence, divergence and cultural identity in Australian urban planning history, in Barker (ed.), pp. 61-70.

Freestone, R. (ed.) (1998) *The Twentieth Century Urban Planning Experience: Proceedings of the 8th International Planning History Society Conference and 4th Australian Planning/Urban History Conference*. Sydney: University of New South Wales.

Gavin, A. and Maluf, R. (1996) *Beirut Reborn: The Restoration and Development of the Central District*. London: Academy Editions.

Gold, J.R. (1997) *The Experience of Modernism: Modern Architects and the Future City 1928-1953*. London: E & FN Spon.

Hall, P. (1996) *Cities of Tomorrow: An Intellectual History of Urban Planning and Design in the Twentieth Century*. Updated edition. Oxford: Blackwell.

Hall, T. (ed.) (1991) *Planning and Urban Growth in the Nordic Countries*. London: E & FN Spon.

Hall, T. (1997) *Planning Europe's Capital Cities: Aspects of Europe's Nineteenth Century Urban Development*. London: E & FN Spon.

Hardy, D. (1991) *From Garden Cities to New Towns: Campaigning for Town and Country Planning , 1899-1946*. London: E & FN Spon.

Hass-Klau, C. (1992) *The Pedestrian and City Traffic*. London: Belhaven.

Hein, C. (1998) Japan and the transformation of planning ideas: Some examples of colonial plans, in Freestone (ed.), pp. 352-357.

Hein, C. and Ishida, Y. (1998) Japanische Stadtplanung und ihre deutsche Wurzeln. *Die Alte Stadt*. **3**, pp. 189-211.

Home, R. (1997) *Of Planting and Planning: The Making of British Colonial Cities*. London: E & FN Spon.

Irving, R.G. (1981) *Indian Summer: Lutyens, Baker and Imperial Delhi*. New Haven: Yale University Press.

Ishida, Y. (1988) Some Failures in the Transference of Western Planning Systems to Japan, in TIPHC, pp. 543-555.

King, A.D. (1976) *Colonial Urban Development: Culture, Social Power and Environment*. London: Routledge.
King, A.D. (1990) *Urbanism, Colonialism and the World-Economy: Cultural and Spatial Foundations of the World Economic System*. London: Routledge.
Lock, M. (1947) *Reconstruction in the Netherlands: An Account of a Visit to Post-War Holland by Members of the Town Planning Institute*. London: Jason.
Lorange, E. and Myhre, E. (1991) Urban Planning in Norway, in Hall, T. (ed.), pp. 116-166.
Malverti, X. and Picard, A. (1991) Algeria: Military genius and civic design. *Planning Perspectives*, **6**, pp. 207-236.
MLCERG (Max Lock Centre Exhibition Research Group) (1996) *Max Lock 1909-1988: People and Planning. An Exhibition of His Life and Work*. London: University of Westminster.
Okpala, D.C.I. (1990) The roles and influences of external assistance in the planning, development and management of African human settlements systems. *Third World Planning Review*, **12**, pp. 205-229.
Parsons, K.C. (1992) American influence on Stockholm's post World War II suburban expansion. *Planning History*, **14**(1), pp. 3-14.
Perry, M., Kong. L. and Yeoh, B. (1997) *Singapore: A Developmental City State*. Chichester: John Wiley.
Sanyal, B. (1990) Knowledge transfer from poor to rich cities: A new turn of events. *Cities*, **7**, pp. 31-36.
Schubert, D. (1995) Origins of the neighbourhood unit idea in Great Britain and Germany: Examples from London and Hamburg. *Planning History*, **17**(3), pp. 32-40.
Schubert, D. (1997) *Stadterneuerung in London und Hamburg: Eine Stadtebaugeschichte zwischen Modernisieriung und Disziplinierung*. Wiesbaden: Vieweg.
Sidenbladh, G. (1964) Planning Problems in Stockholm, in Planning Commission of the City of Stockholm, *Stockholm Regional and City Planning*. Stockholm: Planning Commission of the City of Stockholm, pp. 55-64.
Simpson, M.A. (1983) *Thomas Adams and the Modern Planning Movement: Britain, Canada and the United States, 1920-1940*. London: Mansell.
Smets, M. (1977) *L'Avènement de la Cité-Jardin en Belgique: Histoire de l'Habitat Social en Belgique de 1830 à 1930*. Bruxelles: Pierre Mardaga.
Sutcliffe, A. (1981) *Towards the Planned City: Germany, Britain, the United States and France, 1780-1914*. Oxford: Blackwell.
Tetlow, J.D. (1959) Sources of the neighbourhood idea. *Journal of the Town Planning Institute*, **45,** pp. 113-5.
TIPHC (Third International Planning History Conference) (1988) *The History of International Exchange of Planning Systems*. Tokyo: City Planning Institute of Japan/ Planning History Group.
TMG (Tokyo Metropolitan Government) (1994) *A Hundred Years of Tokyo City Planning*. TMG Municipal Library no 28. Tokyo: TMG.
van der Heiden, C.N. (1988) Foreign influence on Dutch planning doctrine, in TIPHC, pp. 89-112.
Ward, S.V. (1997) A paradoxical persistence? British influences on Canadian and Australian urban planning, in Barker (ed.), pp. 51-60.
Ward, S.V. (1999) The international diffusion of planning: A review and a Canadian case study. *International Planning Studies*, **4**, pp. 53-77.
Watanabe, S.J. (1980) Garden City Japanese style: The case of Den-en Toshi Company Ltd, 1918-1928, in Cherry, G.E. (ed.), *Shaping an Urban World*. London: Mansell, pp. 129-143.

Watanabe, S.J. (1988) Japanese vs Western urban images: Western influences on the Japanese architectural profession, 1910-1920s, in TIPHC, pp. 568-84.
Watanabe S.J. (1992) The Japanese Garden City, in Ward (ed.), pp. 69-87.
Williams, R.H. (1996) *European Union Spatial Policy and Planning.* London: Paul Chapman.
Winston, D. (1957) *Sydney's Great Experiment.* Sydney: Angus and Robertson.
Wright, G. (1991) *The Politics of Design in French Colonial Urbanism.* Chicago: Chicago University Press.
Wynn, M. (ed.) (1984) *Planning and Urban Growth in Southern Europe.* London: Mansell.
Yeoh, B.S.A. (1996) *Contesting Space: Power Relations and the Urban Built Environment in Colonial Singapore.* Kuala Lumpur: Oxford University Press.
Yerolympos, A. (1996) *Urban Transformations in the Balkans (1920-1920): Aspects of Balkan Town Planning and the Making of Thessaloniki.* Thessaloniki: University Studio Press.

CHAPTER 4

Quasi Utopias: Perfect Cities in an Imperfect World

Dennis Hardy

Town planning, historically, has experienced an uneasy relationship with utopian ideas, seduced by the prospect of a shimmering city in the sun yet rebuffed by the sheer impossibility of achieving it. This contrasting pattern of hope and disappointment proved no less evident in the twentieth century than in earlier periods.

During this past century, the relationship between town planning and utopianism has, at different times and in different contexts, varied quite markedly. To illustrate these variations, three models can be identified. At one extreme, there has been little obvious connection between the practice of planning and the idea of extensive if not total change, signalling a tacit acceptance of imperfect cities in an imperfect world. At the other end of the spectrum, attempts have been made not simply to remake the city but also to remake society at large, a vision of perfect cities in a perfect world. And somewhere in between were plans for perfect cities without comparable plans for the rest of society, balancing idealism with pragmatism. For this last pattern, the term '*quasi* utopia' has been coined.

To demonstrate the contrasting nature of these three models, examples are drawn primarily from a specific cultural *milieu*, that of England, but comparative material is also used and generic lessons will, in any case, be of international interest. Two conclusions are drawn. One is that in the twentieth century, *quasi* utopias (even though the extent of their influence has been limited) have led to more positive outcomes for planning than have other models of utopianism. The second conclusion is that *quasi* utopias provide a useful basis for exploring a continuing role for utopianism in the century to come.

IMPERFECT CITIES IN AN IMPERFECT WORLD

After a century of unprecedented urbanization, most, if not all, of the world's urban population is a long way from inhabiting the mythical 'cities in the sun', the eternal quest of utopians. Arguably, urban conditions are better now than they were at the beginning of the century, but it would be difficult to claim that they are anywhere near as good as they could be. Compared with parallel developments in, say, telecommunications or aircraft design, in medical treatment or motorized transport, progress in town planning has been minimal; cities have failed to deliver more than a small fraction of what is socially and technologically possible.

Planning historians will be able to point to some exceptions: some gems in the world urban landscape, and parts of cities or even whole settlements which are directly the result of a visionary approach combined with political will. But these are mere fragments in a wider picture in which it might seem that cities (in so-called developed as well as developing nations) have emerged wholly without the benefit of concerted action. Neither capitalist nor communist régimes have succeeded in creating cities that meet standards which are theoretically possible, while abject conditions in third world cities have become a telling icon of the age.

In England (where, at the start of the century, more people were already living in towns than in the country), there have been marked improvements in some aspects of city life. Smoke pollution, which was endemic in industrial centres, is now very largely a thing of the past; so too are the insanitary housing conditions and water-borne epidemics that were typical of all the major cities. But if some problems have been erased, others have replaced them. Social observers throughout the century have found no shortage of urban ills to describe; the record is one of consistent critique, in which cities reflect the problems of society at large.

At the start of the century, the American journalist and novelist, Jack London, donned workmen's clothes to observe life in the capital city's notorious East End. There he witnessed the 'nightmare [of] the miserable and despised and forgotten', described in his book, *The People of the Abyss* (1903, pp. 114-115). In the following decade, recruits were enlisted from city slums, such as those visited by London, to fight in World War One, but instead of returning to promised 'homes for heroes' most of those who survived the conflict found themselves back in the familiar slums of industrial Britain (Reiss, 1918). To traditional problems were added those of mass and persistent unemployment. Observers like J.B. Priestley and George Orwell recorded scenes of poverty, social misery and environmental conditions not unlike those witnessed in the previous century (figure 4.1). When Priestley visited the depressed shipbuilding town of Jarrow, he proclaimed it 'dead', and what he saw 'far outran any grim expectations of mine' (Priestley, 1934, p. 313). Orwell drew similar conclusions from another industrial town, Wigan, with its 'filthy slums round belching foundries and stinking canals and slag-heaps that deluge them with sulphurous smoke', and life all but extinguished by 'the deadening, debilitating effects of unemployment' (Orwell, 1937, pp. 46-47, 75).

Figure 4.1. The type of urban environment described by George Orwell and J.B. Priestley was portrayed in propagandist posters by Arthur Wragg for the garden cities and town planning movement. (*Source*: Town and Country Planning Association, London)

Attempts to deal comprehensively with this legacy of problems after 1945 resulted too often in not a new urban utopia, but a depressing townscape with problems that were different but no less in their effects. A new generation of observers discovered, in what was from the 1960s termed 'the inner city', a residue if not a resurgence of severe poverty, a new spectre of unemployment as traditional occupations declined and footloose industries moved away, sub-standard housing in new as well as old developments, racial disharmony, and a young, disaffected generation turning increasingly to widespread drug use and crime (e.g. Scarman, 1981; Townsend, 1973). Terms like 'the underclass' were introduced in the 1980s to describe a submerged strata of society, without work or even hope. Meanwhile, even for the more affluent, living in cities meant increasingly the daily experience of gridlocked traffic, pollution from exhaust emissions, collapsed public services, and a context of social tension along dividing lines of race and class.

Two centuries of technological progress and an unimaginable growth of aggregate wealth had led inexorably to this. In spite of the enormous potential to create better cities, and in spite of the benign aims of town planning, the prospect of cities in the sun was still obscured by a persisting mist of political obstruction and perversity. Boundless opportunities to find a way through this mist had been spurned. At the end of the twentieth century, as at the start, it was hard to avoid the conclusion that the dominant picture remained one of imperfect cities in an imperfect world.

Perfect Cities in a Perfect World

This continuing imperfection of cities is not an isolated failing in the record of the twentieth century, and in some ways it pales into relative insignificance in the broader scheme of things. Two world wars, numerous regional conflicts, political upheavals, global recessions and widespread unemployment, poverty and economic exploitation, resource depletion and planetary pollution, are just some of the century's related, and similarly self-imposed, sources of suffering.

Problems of this magnitude might, in some ways, be seen as fertile ground for utopian thought, with its simplistic offer of redemption in a time of crisis: illustrated by the much-cited biblical choice between the lake of fire and brimstone or the eternity of the New Jerusalem. In some respects, this has, in fact, been the case, with the emergence of political blueprints to save the world from itself. Free market capitalism, Marxist socialism, and national socialism have each provided mutually exclusive panaceas, including the specification of ways in which cities should be planned and functions they should perform. But, by the end of the century, of these only capitalism remained dominant, with China the latest recruit to the new world order. Economic turbulence in world markets might yet upset this balance, but there is no immediate sign of that. As a key feature of the twentieth century, planning historians might find scope for some interesting work in tracing more extensively than has been done to date, connections between these overarching ideologies and urban form and development (e.g. Fehl, 1992).

It is these political ideologies which are the century's main utopian legacy, each offering a framework for total change. In contrast, utopian ideas which start with a vision of the city and extrapolate from that to a plan for society as a whole, have been relatively few. An important exception, however, is to be found at the start of the century, when utopians, like H.G. Wells in England and the Futurists in Continental Europe, were excited by the potential of technological invention and by the prospect of worldwide control and order. Though of different origins, and based on different values, they shared a belief that the form of cities could not reasonably be considered in isolation from wider social change. Theirs was a classic vision of perfect cities in a perfect world.

On the eve of the new century, H.G. Wells, the English novelist and socialist campaigner, presented an early version of his vision of the future (some two hundred years hence), in the form of a novel, *The Sleeper Awakes*. In this, far from disaggregating London into smaller settlements (as others were advocating at that time), most country towns and villages had disappeared and people lived in ever larger urban centres. There, people could be better controlled, and any lingering individuality could more easily be suppressed. London itself had grown to a population of 33 million, a city of 'overwhelming architecture', its skyline dominated by huge, curving 'Titanic buildings' and great iron wheels to generate wind power (Wells, 1899, p. 52). At ground level in deep chasms, streets and squares were covered, lit in a white light by powerful electric globes. Most urban dwellers were transported by moving platforms, three hundred feet across with parallel platforms at different speeds. People were perpetually in a crowd, inhabiting subterranean worlds, ruled by a world leader called Ostrog; the modern world had arrived. 'We are making the future . . . and here it is!' (Wells, 1899, p. 70).

Wells wrote as many as one hundred books, social commentary as well as fiction, but consistently he returned to the theme of new technologies enabling the emergence of a world order as well as transforming town and country. In *A Modern Utopia* (Wells, 1905), he points again to large urban centres with efficient transport systems. But he spells out in more detail the global context, and the nature of a political world order. A huge database in Paris contains personal details of the entire population, and social behaviour is strictly controlled; marriage, for instance, can occur only if the partners have appropriate qualities. The social structure is stratified on the basis of qualities of imagination and selection, and a select band of especially able and dedicated people, the *samurai*, provide unquestioned leadership. Planning and order are consistent themes in all of Wells' works. A fitting epitaph for Wells is to be found in the words of one his own fictional heroes: 'I meant to leave England and the empire better ordered than I found it, to organise and discipline, to build up a constructive and controlling State out of my world's confusions' (in *The New Machiavelli*, in Morton, 1952, p. 239).

The imagery evoked by Wells was compelling, but his ideas were not translated in this period into plans for the ideal city. In fact, as John Gold has shown in his comparative study of the development of modernism, at a time when modernity brought forward 'an astonishing array of ideas for future cities', the English preference continued to be for garden cities and their like (Gold, 1997, p. 21). Not until the late 1920s, does Gold detect the 'earliest stirrings' of interest in modernism in urban projects in Britain (Gold, 1997, p. 12).

In contrast, a modernist view of the city in Continental Europe, more Wellsian than that of Wells, led not only to wide interest but also, indirectly, to plans to translate the ideas into reality. Centred around the Italian Futurists (though with related movements in France, Germany and Russia), the first decade of the century saw an explosion of ideas designed, sometimes in the most violent ways, to reject the old and usher in the new. The Futurists expressed themselves in a wide range of arts, including (most notably through Antonio Sant'Elia) architecture and city planning (e.g. Gold, 1997; Tod and Wheeler, 1978). Their vision of future urban form was shaped by soaring skyscrapers and systems of rapid transportation, with new construction materials and techniques to produce

Figures 4.2. The futurist vision of Antonio Sant'Elia (*above*) stands in contrast to the poster advertising the garden city idyll (*right*). (*Source illustration above*: John Gold)

a cityscape of steel, concrete and glass. The past was to be totally obliterated (figures 4.2 opposite and below). While some of the force of the movement was absorbed in, first, a world war and, later, the rise of totalitarianism (both of which were embraced with enthusiasm by these advocates of total change), the essential message was not lost on a new generation of more serious architects, such as Le Corbusier. Through mediators such as this, ideas which at first seemed outrageous were introduced in a more palatable form and gradually applied in most of the world's cities during the rest of the century (Gold, 1997).

PERFECT CITIES IN AN IMPERFECT WORLD

Somewhere between the two extremes of an absence and totality of utopian ideas are examples of ideas for perfect cities that did not at the same time seek, as a pre-condition, to change the rest of the world. It was anticipated that improvements in society would follow, but this would be an organic process without the degree of control and coercion that uncompromising utopian schemes implied. These are the *quasi* utopias, in the sense that their vision of the future city was utopian but their vagueness on what would happen to the rest of society rendered them incomplete; they envisioned perfect cities in an imperfect world. Examples are centred on the experience of England in the first quarter of the century: the archetypal garden city of Ebenezer Howard, and a lesser known variant on the garden city theme developed by a group of Quakers.

The garden city movement, which originated in England and dates from the very end of the last century, is extensively documented (e.g. Beevers, 1988; Hardy, 1991; Ward, 1992). In the context of a discussion of utopianism, it is the story of a search – part dream, part reality – for the idyll of a green and pleasant land. As such it is intimately connected to a long cultural tradition that recalls lost rural values and the virtues of a world that never in fact existed. Indeed, perhaps it could only have attracted so much interest in a country that had for centuries mourned the absence of a peasant proprietorship, and which perpetually dreamt of ways of 'returning' the people to the land. Although an urban solution, garden cities offered the prospect of community ownership of land for housing, in a countryside setting.

Ebenezer Howard, who first proposed garden cities, was wise enough not to present his ideas as utopian. He went to great trouble even before the publication in 1898 of his seminal book, *To-Morrow: A Peaceful Path to Real Reform*, to ensure that it would appeal to middle opinion. Not only did the title itself undergo various drafts to remove any suggestion of extremism, but he was always to stress the practical application of his ideas. He was ready to acknowledge a debt to earlier utopians, but he set his own ideas in a different mould (Howard, 1898, 1902). In spite of his implicit denial of utopian tendencies, however, the actual text demonstrates far-reaching goals. Howard was not one to await an overturn of the political system prior to reconstructing the cities; instead, he believed that his own ideas could trigger changes in society at large. Each of his garden cities would be a building block in the reconstruction of society as a whole, and from small beginnings 'will spring a new hope, a new life, a new civilisation' (Howard, 1946 edition, p. 48). Metropolitan areas like London would largely disappear, in the wake of their populations choosing to move to more attractive locations in the countryside; the urban map would be redrawn.

The scheme's inherent utopianism was acknowledged at the time in critical reviews of the book, the term in this context used dismissively to describe the scheme as dangerous in its implications as well as being merely fanciful. Thus, one critic wrote that 'Mr Howard is not content with half measures; like Sir Thomas More he builds a Utopia . . . The only difficulty is to create it, but that is a small matter to Utopians'. Others referred to the 'unpalatable dough of his Utopian scheming' (in Hardy, 1991, p. 26). Later writers have recognized its utopianism in a more benign way. Lewis Mumford, for instance, saw the scheme as 'transformative . . . Howard's ideas have laid the foundation for a new cycle in urban civilisation' (in Howard, 1946 edition, p.40). In a similar vein, Peter Hall has acknowledged that Howard's vision took him well beyond the normal limits of physical planning: 'his garden cities were merely the vehicles for a progressive reconstruction of capitalist society into an infinity of cooperative commonwealths' (Hall, 1988, p. 87).

Figure 4.3. The opening of Welwyn Garden City railway station in 1926 by the Rt. Hon. Neville Chamberlain MP (left) illustrates official recognition of the first garden cities. (*Source*: Welwyn Garden City Library)

In fact, from the outset, in spite of Howard's denial, the first garden city, Letchworth, attracted a utopian following, so much so that the settlement as a whole acquired a reputation for 'crankiness'. Spiritual and political explorers were among those who made their way to the pioneer settlement, in search of kindred souls and a chance to rebuild society. There they donned sandals and smocks, and contemplated an alternative future. The rest of the world looked on with amusement (Miller, 1989).

But garden cities in practice were, of course, far from being utopian in the true sense of the word. Howard was above all a pragmatist, and in order to get the first two experiments, at Letchworth and Welwyn, established, he was forced to make one compromise after another. One recalls in this context the words of a contemporary of Howard, Nellie Shaw, who in attempting to live by the exacting doctrine of Leo Tolstoy, confessed that 'it was impossible to live in a capitalist world and not become mixed up in it' (Shaw, 1935, p. 228). The simple fact was that in order to raise sufficient capital, and to convince industrialists to locate in the new settlements, conventional procedures had to be followed. In the event, Letchworth and Welwyn (figure 4.3) evolved as pleasant enough settlements, and were certainly a marked improvement on what most of their population had experienced before, but utopian they were not.

Inspired by Howard's ideas, this period saw another practical experiment with utopian pretensions, and one that at least in theory was more radical in intent. Against a backdrop, not only of prevalent urban problems but also of the horrors of World War One, a group of Quakers envisioned plans for 'a new beginning'. One of their group, W.R. Hughes, took on the task of articulating these plans in a book, published in 1919 as *New Town: A proposal in agricultural, industrial, educational, civic and social reconstruction* (figure 4.4). The plan was (if not a contradiction of terms) a 'practical utopia'. It was rooted throughout in idealism, seeking as it did to outline how society could be reconstructed, yet at its heart were detailed specifications for building a new town. Much of the latter lent heavily on the garden city idea, but it was more explicit than the former in its hope that a new type of settlement might, in turn, lead to a new social structure. Unlike the Wellsian approach to utopias, Hughes was deliberately less prescriptive about the form this might take, other than asserting that education was the key to lasting change. The rhetoric spoke of a new beginning made with high hope in the days that came after the war. Hughes acknowledged that they 'were aiming very high . . . our ultimate object is to provide the right conditions of life for the full development of human personality' (Hughes, 1919, p. 10).

Utopianism owes a significant debt to the Quaker movement, which from the time of its origins in the seventeenth century, as well as instilling a belief in the individual, soon became associated with community ventures. In 1669, for instance, John Bellers proposed a national network of colleges of industry, organized on principles of primitive Christianity (e.g. Darley, 1975). *New Town* was part of this tradition. What was proposed was a 'Country-Town in England in such a spirit as shall stir the hearts of all who are seeking after freedom and fellowship' (Hughes, 1919, p. 9). The formation of a new settlement was preferred as it was thought that this would allow all aspects of life to be remodelled. As with garden cities, it was proposed that land should be held for community benefit; at the same time the scheme avoided extremism, seeking instead 'the reconcilement of the complementary ideals that we label individualistic and socialistic' (Hughes, 1919, p. 16).

New Town was planned to embody many novel features, with radical changes to domestic life offered through a People's Kitchen and Restaurant and what was

Figure 4.4. Cover page for the book *New Town* by W.R. Hughes, 1919.

termed the Household Auxiliary Corps. The People's Kitchen was not only to provide restaurant meals and a 'take away' service, but was also to hold a stock of basic cooking utensils for use within the community. In turn, the Household Auxiliary Corps was to consist of 'experts in household arts, the majority being all-round people, others being specialists as cleaners or needlewomen, cooks or waitresses' (Hughes, 1919, p. 127). Both innovations were designed to free women from the home, but it has to be said that both appear to be based on an assumption that women were to continue to perform the same kind of work as they had previously done (albeit in a different setting). Thus, 'the girl who would refuse to go back into the home [after the experience gained in World War One] as drudge for her parents, or even for her boy, would readily join the staff of a People's Kitchen working in shifts with regular hours, or enlist in a uniformed corps of Household Auxiliaries' (Hughes, 1919, p. 126).

But, like garden cities, the high ideals of its founders were compromised by the realities of raising sufficient capital and by the practical obstacles to building a perfect city in an imperfect world. The proponents were determined, but failed to raise more than a quarter of the capital required. Instead of buying their own estate, they resorted instead to leasing a section of Welwyn, the second garden city. There they made progress on some fronts, especially with experiments in cooperative housing, agriculture and industry, but these were, compared with the grand plan, piecemeal achievements. Probably the greatest contribution of those who carried the New Town banner was in injecting additional idealism into the life of Welwyn as a whole. The great hopes of social reconstruction were never entirely lost, and in critical areas like education, health and housing, the New Town contingent became actively involved as local residents. Subsequently, in 1934, the experiment was formally merged into the general management of the garden city.

A further step in distancing experiments from the utopianism of the original garden city concept came at the end of World War Two, with the introduction in Britain of an official programme of new towns (see Burton and Hartley, 1997). What might have been the apotheosis of Howard's ideals proved, instead, to be an even more compromised version of the original scheme. New towns were not without their merits in providing a decent environment in the postwar years, but what they demonstrated was that the very idea of 'official utopias' is a contradiction of terms. As models of perfect cities they were even further from the original concept than were Letchworth and Welwyn Garden City.

UTOPIANISM AND TOWN PLANNING

The above examples are few, and by no means comprehensive, but sufficient to illustrate different ways in which utopianism continued in the twentieth century to exercise an influence on town planning (see also e.g. Meyerson, 1961; Mumford, 1922). Reflecting on this experience, one might ask, first, whether it has been a productive relationship, and, based on that evaluation, whether, at the start of a new century, there might still be some value in drawing on utopian ideas. Has utopianism made a useful contribution in the past century, and does it have a continuing role in the future?

Looking Backward

Three types of relationship have been illustrated: one where there is no obvious contribution of utopianism; one where there is a total relationship, embracing visions of the perfect society as well as the perfect city; and one where there is a partial relationship, the case of *quasi* utopias.

Of the first of these relationships, where there is no obvious link with utopianism, it might be asked whether town planning is more, or less, effective without this link. Those who argue that market forces provide the only realistic basis on which to manage a nation's affairs would be in favour of an absence of utopian involvement. So, too, would those who ridicule utopian ideas for being, by their very nature, unrealistic. In some ways, however, these are both specious arguments, as they obscure the fact that even *laissez faire* is ideologically driven. Market forces are by no means value free. Instead, as Adam Smith's classic argument for a free market economy, *Wealth of Nations* (1776), showed more than two hundred years ago, the idea of limited interference is as much a utopia as any other set of ideas with a prescription for society. Thus, in a market-dominated situation, town planning assumes a role of enabler for capitalism, driven in directions required by the market.

The evidence of imperfect cities in an imperfect world is, therefore, less a result of society not embracing utopianism and more one of an over-dependence on capitalism. Market forces are not expected to remove inequalities in society, and depressed conditions in cities are merely a physical manifestation of the unevenness of wealth. Similarly, social democracy, though well-intentioned in seeking to ameliorate the worst effects of the market, has been, by its very nature, partial in its application (constantly trimmed by compromise) and flawed in its vision. Some of the worst examples of urban development have resulted, not through leaving cities wholly to market forces, but through crass interference. In her indictment of misguided intervention, Alice Coleman (1985) has concluded that 'Utopia is not synonymous with progress' (p. 6), and that, far from liberating people from slum housing, it has 'come to represent an even worse form of bondage' (p. 180).

If there is a simple lesson to draw from the evidence of imperfect cities in an imperfect world – the dominant outcome of urbanization in the twentieth century – it is that ideas designed to benefit everyone should be explicit and directed towards measurable improvement. Utopian ideas *per se* do not automatically lead to social improvement, but utopian ideas, democratically based and openly understood, are more likely to.

Of the second type of relationship, perfect cities in a perfect world, critics have always been wary of attempts to impose too much control and order. The experience in many countries of totalitarianism and, in Britain, the more limited threat of a Wellsian vision, justifies this caution. There has not in Britain been anything like the imposition of a blueprint for a perfect society as well as cities, and, because of the degree of control required, any attempt to do so would undoubtedly have been resisted. The works of Aldous Huxley (1932) and George Orwell (1948) were simply warnings of latent political concerns.

In mitigation, it might be argued that the twentieth century has revealed other models of utopia, less dependent on the creation of massive cities and total control. In the second half of the century, for instance, anti-urbanism as an ideal has been articulated through a strong 'alternative' movement, linked increasingly with environmental action groups (e.g. Pepper, 1991). More than one hundred years after William Morris's *News from Nowhere*, anti-urbanists still evoke images of a land of small towns and villages, of people living cooperatively, and of a rejection of world trade (Morris, 1890). The vision is as attractive now as it was when it was first presented, but those still lured by it ignore one essential contradiction: in order to create this idyllic world, society (as Morris recognized in the nineteenth century) must first engage in what would inevitably be a revolutionary process, the consequences of which no-one could predict. The creation of perfect cities in a perfect world could not occur without a strong degree of control and coercion at some stage, if not all the time.

Finally, of the third type of relationship, perfect cities in an imperfect world, one way to demonstrate its value is to compare it with the previous approach. One might conclude that if a total utopia were shown to be totally beneficial, it would follow that a partial utopia would be less beneficial. But total utopias are not totally useful, and, paradoxically, it might be that partial utopias are of greater value to society. Of two contrasting utopias, modernism and garden cities, it is concluded that the latter resulted in greater social benefit. This was not because the idea of the garden city was necessarily any better in itself, but rather because it matched more closely a dominant English cultural tradition which espoused rural imagery and favoured political evolution rather than revolution. The fact that garden cities fell short of an ideal model of utopianism is of no more than intellectual interest. What matters in practical terms is that garden cities succeeded in combining elements of idealism and realism, and benefits for society flowed from that combination.

Two further lessons follow from this. The first is that society might, as a rule, be better advised to look to proposals that are idealistic but which also offer a practical chance of measured improvement, rather than await a panacea that promises everything but delivers nothing. The second is that, no matter how good ideas are in the abstract, society has to be in harmony with their basic assumptions; in other words, there has to be cultural compatibility. Garden cities matched both of these criteria. They were, in intellectual terms, merely *quasi* utopias, but in practical terms probably more beneficial than fully-fledged utopianism. They show that the goal of creating perfect cities in an imperfect world is by no means as misguided as would at first appear. Indeed, the very partiality of some schemes might at the same time be seen as a source of strength; they can offer society a realistic chance of progress. A *quasi* utopia is better than no utopia at all; but a *quasi* utopia is also better than utopia *in toto*.

Looking Forward

In spite of its value at times in the past, it might be asked whether, at the start of a new century, a relationship of any kind between utopianism and town planning is any longer relevant or even desirable. The connection is longstanding, but that is not in itself a reason for one to continue; instead, there are reasons why this tie might now be broken.

One reason is that circumstances now are wholly different from those when some of the classic utopias were first presented: when, say, in the sixteenth century Thomas More (1516) created a vision of utopia as a way of shaming his own society, or when, in the following century, Tomasso Campanella (1623) wrote of the legendary cities in the sun. Indeed, circumstances are even a world apart from those of Ebenezer Howard, at the end of the nineteenth century. One obvious difference now is the globalization of activity, which calls into question attempts to create isolated pockets of perfection.

A second reason is that the track record of the utopian contribution to town planning is variable. From the evidence of Britain in the twentieth century, some utopian ideas were actually or potentially damaging. Given the link between the ideas of Wells and the Futurists, and the later application of bastardized versions of 'cities in the sky', an earlier break would almost certainly have been socially beneficial. The most charitable observation on the experience of the twentieth century is that the relationship between utopianism and planning worked best when it was only partial.

More generally, utopianism is unable to free itself from two inherent weaknesses. One is that schemes for the total reconstruction of society, although purporting to perfection, are most likely to be flawed in practice. The other is that blueprints are, in any case, almost by definition impossible to implement in full. Evidence of these weaknesses in the twentieth century has been particularly damaging, and the reputation of utopianism has been tarnished, perhaps beyond redemption, by its associations in various countries with failed ideologies. Not only did the Marxist dream, the ultimate utopia, become a nightmare, but utopianism has also been linked with other forms of totalitarianism. A blueprint approach to social change can lead to the legitimization of control by one group over another, as critics like Huxley and Orwell warned in England in the 1930s and 1940s.

Although compelling in themselves, these reasons might equally be interpreted as a challenge: lessons from history to inform the definition of a new and more productive role. There are empirical and theoretical arguments to support this. Empirically, the fact remains that utopianism still attracts considerable interest amongst leading planners. Peter Hall, in *Cities of Tomorrow* (1988), has acknowledged the importance of vision in the history of modern planning, and he has since, with Colin Ward, marked the centenary of Ebenezer Howard's *To-Morrow*, with a modern reassertion of the original work

(Hall and Ward, 1998). 'Howard's century-old prescription remains extraordinarily useful', they conclude, in proposing an updated version of his concept of the social city (Hall and Ward, 1998, p. 209).

Theoretically, the incentive to renew a link with utopianism is twofold: to encourage planners to think beyond the immediacy of everyday problems, and to provide an explicit set of aspirations. But, first, the very concept of utopia must be redefined, in postmodernist rather than modernist terms. As a modernist concept, utopianism – based characteristically on certainties and finite boundaries, with a presumption that the future can be shaped and a further presumption that it will be shaped by those with most power – is contradictory in every way to contemporary tendencies. It has become an outmoded, if not an anti-social, concept. In contrast, postmodernism represents a rejection of scientific prediction and regular patterns of events and outcomes. It embraces diversity rather than uniformity, and flexibility is a keyword. If a traditional utopia is typified by a blueprint, a postmodernist version will be defined, instead, by pluralities, perhaps as shifting and overlapping images, multi-dimensional, and global in scope. One must think not of *the* future but more in terms of *multiple* futures.

The conceptual leap from modernist to postmodernist is formidable in its range, and a debt is acknowledged to Leonie Sandercock (1998) for most clearly showing the way. With a critical eye on past applications, she reasserts her faith in utopianism, with the view that 'the Utopian impulse is, and will hopefully remain, an irresistible part of the human spirit' (Sandercock, 1998, p. 1). Breaking with the past, she looks to a future, postmodernist model of utopia, *cosmopolis*. This will evolve following the demolition of the 'pillars of modernist planning wisdom', and their replacement by 'new concepts of social justice, citizenship, community, and multiple publics' (Sandercock, 1998, p. 5). It will be an ever evolving project, that by definition can never be completed, 'an always unfinished and contested construction site, one characterized above all by its space for difference' (Sandercock, 1998, p. 198).

CONCLUSION

Cosmopolis is in one sense more than a *quasi* utopia, but it shares with the latter a powerful rejection of a blueprint approach. Both embrace a willingness to see things evolve, yet both also share an underlying faith in such enduring values as hope, compassion and aspiration. In this evolving, new world openness, tolerance, sustainability, access, freedom, uncertainties, humanity are all concepts that should have a place. It is a novel approach to utopia, yet also one that (if only through learning not to repeat mistakes) has its roots in the past. This juxtaposition of historical evidence and future ideas is complex and intriguing, yet it is here, if at all, that a continuing link between utopianism and town planning might most effectively be redefined.

REFERENCES

Beevers, R. (1988) *The Garden City Utopia: A Critical Biography of Ebenezer Howard.* London: Macmillan.

Burton, A. and Hartley, J. (eds.) (1997) *The New Towns Record, 1946-1996: 50 years of UK New Town development.* Glasgow: The Planning Exchange (CD-ROM).

Campanella, T. (1623) *The City of the Sun.* 1981 edition. Berkeley: University of California Press.

Coleman, A. (1985) *Utopia on Trial: Vision and Reality in Planned Housing.* London: Hilary Shipman.

Darley, G. (1975) *Villages of Vision.* London: Architectural Press.

Fehl, G. (1992) The Nazi Garden City, in Ward (ed.), pp. 88-106.

Gold, J. (1997) *The Experience of Modernism: Modern Architects and the Future City, 1928-1953.* London: E & FN Spon.

Hall, P. (1988) *Cities of Tomorrow: An Intellectual History of Urban Planning and Design in the Twentieth Century.* Oxford: Blackwell.

Hall, P. and Ward, C. (1998) *Sociable Cities: The Legacy of Ebenezer Howard.* Chichester: John Wiley.

Hardy, D. (1991) *From Garden Cities to New Towns: Campaigning for Town and Country Planning, 1899-1946.* London: E & FN Spon.

Howard, E. (1898) *To-Morrow: A Peaceful Path to Real Reform.* London: Swann Sonnenschein. Revised and republished in 1902 as *Garden Cities of Tomorrow.* 1946 edition, with introductions by Lewis Mumford and F.J. Osborn, London: Faber and Faber.

Hughes, W.R. (ed.) (1919) *New Town: A Proposal in Agricultural, Industrial, Educational, Civic and Social Reconstruction.* London: Dent.

Huxley, A. (1932) *Brave New World.* London: Chatto and Windus.

London, J. (1903) *The People of the Abyss.* 1977 edition. London: Journeyman Press.

Meyerson, M. (1961) Utopian traditions and the planning of cities. *Daedalus*, 90, pp. 180-193.

Miller, M. (1989) *Letchworth: The First Garden City.* Chichester: Phillimore.

More, T. (1516) *Utopia.* 1965 edition. Harmondsworth: Penguin.

Morris, W. (1890) *News from Nowhere* (1970 edition edited by James Redman). London: Routledge.

Morton, A.L. (1952) *The English Utopia.* London: Lawrence and Wishart.

Mumford, L. (1922) *The Story of Utopias: Ideal Commonwealths and Social Myths.* New York: Boni and Liveright.

Orwell, G. (1937) *The Road to Wigan Pier.* 1986 edition. Harmondsworth: Penguin.

Orwell, G. (1948) *Nineteen Eighty-Four.* 1984 edition. Oxford: Clarendon Press.

Pepper, D. (1991) *Communes and the Green Vision: Counterculture, Lifestyle and the New Age.* London: Green Print.

Priestley, J.B. (1934) *English Journey.* London: Gollancz.

Reiss, R.L. (1918) *The Home I Want.* London: Hodder and Stoughton.

Sandercock, L. (1998) *Towards Cosmopolis: Planning for Multicultural Cities.* Chichester: John Wiley.

Scarman, Lord (1981) *The Scarman Report.* 1982 edition, Harmondsworth: Penguin.

Shaw, N. (1935) *Whiteway: A Colony in the Cotswolds.* London: Daniel.

Smith, A. (1776) *Wealth of Nations.* 1976 edition. Chicago: Chicago University Press.

Tod, I. and Wheeler, M. (1978) *Utopia.* London: Orbis.

Townsend, P. (1973) *The Social Minority.* London: Allen Lane.

Ward, S.V. (1992) *The Garden City: Past, Present and Future.* London: E & FN Spon.

Wells, H.G. (1899) *The Sleeper Awakes.* 1954 edition. London: Collins.

Wells, H.G. (1905) *A Modern Utopia.* Undated edition. London: Collins.

CHAPTER 5

Learning from Twentieth Century Urban Design Paradigms: Lessons for the Early Twenty-first Century

Jon Lang

Urban design is an activity of long-standing, although the label itself dates back only to the 1960s. It is a term that has come to have a multiplicity of meanings as different design disciplines have laid claim to it. To many architects, it simply refers to designing buildings that are responsive to their urban context; to landscape architects, to the detailing of the public realm; and to planners, urban beautification. In this chapter urban design refers to the design of the three dimensional qualities of cities and their precincts and how they are perceived and change over time. It thus embraces the 'city design' of Kevin Lynch (Lynch, 1984) and the design of the public realm (Lang, 1994).

An urban design paradigm is a generic solution to a set of problems that is set up as an exemplar of good practice. There are two types of paradigms: substantive and procedural. The substantive specify good patterns, or layouts, of built form and the purposes they serve in terms of human activities and aesthetic values. The procedural deal with how urban design should be done: the processes of analysis, synthesis evaluation, implementation and post-implementation assessment (Shirvani, 1985). The concern in this chapter is with the substantive paradigms that have had a major impact on urban design thought during the twentieth century.

This chapter examines the nature of the set of substantive paradigms that hold sway in the professional practice of firms who are concerned about the quality of their urban design work in terms of what it affords people. This body of work stands in strong contrast to that of architects whose concern is with the designed environment as a work of self-expression – as art *per se*. The models of interest here were self-consciously designed primarily by leading architect-planners in response to the conditions of the city they saw around them and/or predictions of how technological and social changes would, in turn, change the character of the world. The discussion of the recurrent patterns of urban form that have evolved as the result of the simultaneous working of market forces, taxation policies, political exigencies, and the entrepreneurial skills of property developers (i.e. developers' paradigms), fundamentally important though they are, falls outside the domain of concern here.

The discussion in the chapter is organized around (1) the two philosophical traditions – Empiricism and Rationalism – that have shaped the models propounded by architect-planners, and (2) the spatial level of their concern – city and precinct forms. Based on these distinctions, the utility and limitations of some significant urban design paradigms are explored and what needs to be done to make them more user-friendly is discussed. They need to be user-friendly both in the sense of enhancing the professional's work and in the sense of providing more livable environments for their inhabitants.

TWENTIETH CENTURY PARADIGMS: EMPIRICISM AND RATIONALISM

Empiricist paradigms can be regarded as regressive utopian and Rationalist as progressive utopian. Empiricists tend to look at the present and/or the past for inspiration. The Rationalists invent futures based on assumptions and/or proposals for new technologies and new social orders, i.e. new ways of life. They focus on changing the world to futures for which there are no precedents. These new orders are advocated rationally, based on idealistic assumptions rather than on life experiences – on how the world *should* function and how people *should* behave, rather than how it actually does and they do. In developing their paradigms, the Rationalists focused on eliminating problems they saw in the world around them. Often the baby was thrown out with the bath water: the good with the bad.

Although a number of proposed urban design paradigms, particularly Rationalist ones, have fallen by the wayside, few should be seen as the whimsical ego-trips of self-indulgent architects and city planners dreaming up social and physical futures. They have been, in contrast, carefully reasoned responses to the problems their proponents saw in the world around them. These problems included polluting industries, long journeys to work, inadequate housing, poor educational and recreational facilities (especially for the poor), the growth of cities into the countryside consuming agricultural land, and the general distancing of the artificial from the natural (or, rather, rural) world. Above all, the early twentieth century proposals were concerned with eliminating the insalubrious environments of the contemporary industrial city. As the century progressed, the question of how to deal with the human love affair with automobiles became a major worry.

Most urban design paradigms fit neatly into the bi-categorization of Rationalism and Empiricism but a number straddle the line (e.g. the deconstructionist thinking of the 1980s). Most urban design theorists can also be classified as Rationalists or Empiricists, although a few have shifted from one camp to the other as their careers have developed. Leon Krier's early work, for instance, was highly Rationalist, but his later work has relied heavily on precedents, particularly in the design of Poundbury at Dorchester in England (Broadbent, 1990). To confuse matters more, embedded in the design paradigms of each camp are lines of thinking drawn from the other.

Within the framework of Rationalism and Empiricism, the twentieth century has seen a number of major movements in urban design. Each represents an attitude towards the city – a set of problems which are perceived to be in need of addressing and a set of design patterns to solve them. The city beautiful and the garden city ushered in the century. They were followed during the first four decades of the century by the international modern movement in all its variant forms. After the intellectual hiatus of World War Two and the spate of applications of the ideas of modernist ideologies came the post-modernism reaction in all its variants.

A number of what might be regarded as post-modern paradigms have been developed and are competing for attention in the market place of urban design ideas (Broadbent, 1990). The post-modernists include such diverse groups as the Archigram movement of the 1960s in Britain, a 'discrete' architecture and urbanist movement based in Spain and Italy, and a classical revival movement in the hands of European architects such as Ricardo Bofill (figure 5.1). The post-modern movements can be grouped under the rubrics of either Neo-Rationalism or Neo-Empiricism. A set of movements that might be regarded as Empiricist falls outside the concerns of the discussion here; they might be called cosmological revival movements, such as those being advocated by small groups in India (Lang *et al.*, 1997) and more pervasively in China.

THE EMPIRICIST PARADIGMS

The Empiricist paradigms of the early part of the century consciously and explicitly drew on precedents that were perceived to be working well, and those which were believed or known to have worked well in the past. They were either known or believed to afford a good way of life and a supportive physical environment for that way of life. Recently, the growth of environment-behaviour studies as an intellectual discipline has provided the basis for the formulation of a new set of design principles. Few practitioners heed the research unless the design principles that result from it are obvious, as in the work of Kevin Lynch on urban imagery (Lynch, 1960) or Oscar Newman on defensible space (Newman, 1974, 1996).

Three sets of precedents have been particularly powerful for early twentieth century empiricists: the medieval city, the baroque city, and the small country town. These precedents have been drawn on repeatedly. The second shaped the

Figure 5.1. Les Arcades du Lac, St. Quentin en Yvellines, France. This proposal by Ricardo Bofill is as much large-scale architecture as urban design. It illustrates the classical revivalist approach of a sub-set of neo-Rationalist architects in both its overall composition and architectural style. (*Photo*: Richard Fitzhardinge)

city beautiful movement (Wilson, 1989), the third the garden city (Howard, 1902), and the first and third the thinking of the townscape movement (Cullen, 1961). More recently, during the last quarter of the twentieth century, the urban and suburban landscapes of the first half of the century have influenced new urbanist thought (Katz, 1994; Kelbaugh, 1989). These observations do not deny the inventiveness involved in creating the major Empiricist paradigms of the twentieth century for none were simply replications of the past. All recognized that social and technological changes had created a new world. Nevertheless, their proponents sought and saw ways of dealing with the future in existing schemes or those that existed in the past.

Philosophically, the city beautiful falls within the Empiricist school because of its heavy reliance on precedents but it also involves much rationalist thought. It is often regarded as the climax of baroque planning. It is better, however, to think of it as the baroque's latest evolution. The city beautiful was arguably the major paradigm for civic design – the design of the municipal, or governmental, sector of cities – during the first three decades of the century. It was the basis of almost all the competition entries for the design of Canberra (1912) and for the almost contemporaneous design for New Delhi (1913). Although it is much disparaged, its use is still revived from time to time, as in the proposal for the new town of Bandar Baru Nusajaya in Malaysia by Jean-Paul Viguier in the 1990s (Prolink, 1997). The generic model has the basic elements of the baroque

– axial avenues terminating at focal points, grand plazas, wide streets and monumental, ideally classical, buildings enclosing open spaces. Its direct antecedents were, however, the World's Columbian Exposition in Chicago in 1893, Haussmann's work in Paris, and L'Enfant's before that in Washington (Wilson, 1989). Washington too has its own family tree (Peets, 1927).

In terms of urban renewal, the city beautiful paradigm has been applied in bits and pieces around the world. The Benjamin Franklin Parkway in Philadelphia in the 1920s is a prime example of only one piece of a grand plan being executed (Benevolo, 1980). The plans for Hitler's Berlin by Albert Speer were never implemented (Krier, 1985). All these plans were developed before mid-century. In the 1970s, the urban designs for central Bucharest (figure 5.2) and, less obviously, Pyongyang in North Korea are clearly within the city beautiful ideology, although Paris was the referent in mind for both.

Figure 5.2. The Avenue of the Victory of Socialism, Bucharest. City beautiful concepts are still used to achieve an urban design of grandeur but not on as grand a scale as Ceausescu's Bucharest. The buildings designed to frame the public realm lie largely vacant. (*Photo:* Ruth Durack)

While the original garden city paradigm of Ebenezer Howard (1902) was highly utopian and displayed considerable rationalist thought, the garden city model as applied in towns was more down to earth. This characteristic is clear in places such as Stevenage and Harlow in Britain during the late 1940s and 1950s, and in the privately developed new towns in the United States, ranging chronologically from Columbia, Maryland in the 1960s to Las Colinas, Texas in the 1980s. The paradigm consists of a city centre surrounded by neighbourhoods

– based largely on Clarence Perry's neighbourhood unit model – with an industrial area on the periphery, and the whole scheme being well-treed and surrounded by a green belt. Rural precedents informed its architecture. This model is the basis for even the most recent developments of the garden city ideal (e.g. Simonds, 1994).

Frank Lloyd Wright's Broadacre City is its ultimate expression in landscape architecture terms, although many critics feel uncomfortable in discussing the British and American ideal under the same rubric (Fishman, 1982). While it has not been self-consciously adopted as a design paradigm in practice there are segments of American suburbia that approximate it. The reason is simple: it drew on values deeply embedded in Anglo-American culture including the desire for individualism and a deeply held suspicion of cities.

Neo-Empiricism

A number of Neo-Empiricist lines of thought arose in response to the perceived limitations of the garden city ideal. They are all reflective of the past but less hindered by romantic images. They include the pattern language of Christopher Alexander and his colleagues (1977), the works of Charles Moore as exemplified by Kresge College at Santa Cruz, Herman Hertzberger's housing schemes in the Netherlands, and Lucien Kroll's reworking of mass housing projects (Kroll, 1994). These works have served as examples for other urban designers to follow but they have yet to be developed into clear models which can be mimicked. The new urbanism, in contrast, has attracted considerable attention.

The new urbanism, or neo-traditionalism as it was first called, is a regressive utopian movement, which looks back at towns of the early twentieth century and even earlier for its inspiration (Katz, 1994). The new urbanist paradigm is based on the observation that the traditional forms of cities have much to be admired and can be replicated with success. In this sense the movement is a direct descendent of the garden city movement. Perhaps the greatest contribution of the new urbanism, as we now know it, is that it recognizes that streets are the seams for life if properly designed and not something to which, as the Modernists thought, one's back should be turned. Lively streets make lively cities. While seen as primarily an American movement of the 1980s and 1990s, with Seaside in Florida as its exemplar, there are earlier and European demonstrations of the line of thinking. Port Grimaud (1963) in southern France is one and the new university town of Louvain-la-Neuve in Belgium (1970s), is another (figure 5.3). They were both, however, behaviourally and aesthetically, more influenced by the medieval town than any early twentieth century model.

The Deconstructionist view is that there are layers of life to a city. Solving the requirements of each layer, layer by layer independently and then synthesizing a series of layers into a whole, will result, they perceive, in an environment which has the loved qualities of traditional cities while meeting the demands of modern

Figure 5.3. Louvain-la-Neuve, Belgium. The precedent for the design of Louvain-la-Neuve is clearly the medieval town. The pedestrian spine threading through the mixed-use town links various elements of the university to each other and to the railway station. (*Photo*: Jon Lang)

life. As such, the paradigm is both a procedural and a substantive one. The full implications of this line of thinking remain to be explored but they echo a process (morphological analysis) advocated for product design over fifty years ago (Zwicky, 1969).

A simpler way of looking at the city as the basis for urban design work is also becoming apparent. Given the precedent of existing, and much enjoyed, Victorian and earlier centres of European cities, a number of urban designers are viewing cities simply as an agglomeration of streets and squares (Krier, 1979; Moughtin, 1992) or more generally, links and places defined by building facades and uses (Lang, 1994). They share the new urbanist concern with the nature of streets as the defining element of cities. What has been generated is a series of models for street and open space types. The view is neo-empiricist and highly pragmatic – the concern is with what works.

THE RATIONALIST PARADIGMS

Empirical attitudes to urban design were perceived by the international modernists to be a highly romantic view of city and urban design and irrelevant to the modern world. The label 'international modernism' is too gross a rubric to do justice to the variety of paradigms that are subsumed under it. Between the World Wars it included the Italian Futurists and Rationalists, the De Stijl group in the Netherlands, the cubists of France and the Bauhaus in Germany (Broadbent, 1990). In the post-World War Two era, it is exemplified by the megastructure – the city-as-a-building advocates, the Archigram group in Britain, the Metabolists in Japan and the work of Americans such as Paul Rudolph. The label will, nevertheless, have to suffice for this exposition. Le Corbusier's Radiant City (1934) and the ideas of the Congrès Internationaux d'Architecture Moderne (CIAM) in the 1930s and 1940s, as developed by Team10 in the 1950s and 1960s (Smithson, 1969), will serve as exemplars of the thought processes of the various groups. What unifies the set is, indeed, their way of thinking.

Rationalist generic urban design proposals combine a progressive, if, in retrospect, often poorly reasoned, social organization for the city with Platonic geometric shapes for its physical form. They display no fear of large cities, modern technology, tall buildings, cars, the creation of new land ownership patterns, or of developing a new aesthetic. Rationalists embraced change. They have provided two major analogies for the city and its development: (1) the city as a machine, and (2) the city as an organic living form. The former is explicit in the work of Le Corbusier and implicit in the Bauhaus models; the latter is exemplified by the thinking of the Metabolists in Japan. The Metabolists claim a higher moral integrity for their model on the grounds that their analogy is more humane because it refers to biological variables. Their attitudes towards life and people's lives, however, deviate little from other Rationalist thought especially that of the megastructure proponents. Indeed their views of human life might well be regarded as a caricature of Rationalist thinking and their designs a caricature of human cellular structures.

A number of Rationalist city form models were developed during the century (Broadbent 1990; Lang, 1994, amongst many other commentaries). They include the City for Three Million and the Radiant City of Le Corbusier (Le Corbusier, 1960). Rush City, proposed by Richard Neutra, and Ludwig Hilbersheimer's theoretical city layouts, which can be considered to be examples of Bauhaus thinking. Le Corbusier's schemas were most thoroughly applied in Brasilia and, to a lesser extent, in the design of Chandigarh with which he himself was closely associated. The Metabolist line of thought is, perhaps, most closely associated with Kenzo Tange's proposal for a new city for Tokyo Bay (*c*1955). With its orthogonal geography, buildings and highways on pylons and cellular-formed architecture, it presented a radical image of what a city might be. However, the ideas have seen their application at a building rather than an urban scale.

The greatest application of Rationalist urban design paradigms has been in mass housing. There are few countries without a building complex showing the power of the paradigm in capturing the imagination of architects. Throughout the world, public housing, or social housing, as it is known in many countries, for better but, unfortunately, often for worse, shows a major debt to international modernist thought (figure 5.4). The strength of the basic ideas depends on the validity of the premises on which they were built. Too often, however, rationalist thought has demonstrated a greater concern for the aesthetics of its internal logic than its external validity, its applicability in the inhabited rather than the symbolic world.

Despite the variety of Rationalist design concepts, the predominant image of their ideal world is one of slab or tower buildings one after the other, surrounded by large open spaces with the buildings linked to but removed from highways. This paradigm emerged from the Bauhaus's concern for sunshine reaching all the habitable rooms of a housing unit. This simplistic image, born of the repeated application of the paradigm in housing schemes around the world, is unfortunate for it does a disservice to the basic ideas of the Rationalists. The image remains, however, a pervasive one.

Figure 5.4. Southside Chicago. Many of the principles of Rationalist thinking of what cities should be were applied in a highly diluted if extensive manner. The bleak housing environments which have resulted hardly test the paradigm. (*Photo*: Jon Lang)

Neo-Rationalism

The Neo-Rationalists of the past four decades have, like their predecessors, had a major interest in formal geometries and in what they regard as progressive politics. These concerns have led to the group being called neo-modernist although a number of Neo-Empiricists who believe that 'the modernists almost got it right' would also claim that label. To generalize, Neo-Rationalists have explored the use of the grid both in plan and in three dimensions. Although they label it urban design, much of the work of people who might be classified as

Neo-Rationalists has been confined to urban architecture. It does, however, include such building complexes as the Rauchstrasse housing scheme directed by Rob Krier in Berlin, the Cemetery at Moderno by Aldo Rossi, and the housing projects of Carlo Aymonimo (Broadbent, 1990). Although, these schemes have attracted much attention, no clear substantive paradigms have emerged from them. What they do offer is a more complex, visually pleasing geometry than that sought by international modernism in urban design.

With its concern for the layering of geometric grids in design, Deconstruction can also be seen as a Neo-Rationalist enterprise. While it has had more of an impact on architecture than urban design, there have been a number of noteworthy explorations in park and suburban planning (Rowe, 1991). The best-known deconstructionist scheme is Bernard Tschumi's Parc de la Villette in Paris with its three layers of geometric patterns being brought together on the ground plane. His proposal for a new residential suburb at Flushing Meadows, New York, has remained on the drawing board but does represent the application of the paradigm to a specific site.

PARADIGMS FOR CITIES AND THEIR PRECINCTS

Almost all the paradigms of city form divide the city in a hierarchical manner. Both the Rationalists and the Empiricists have considered the city at essentially the same spatial levels. The majority of substantive paradigms can thus be classified into three spatial groups: (1) models for general city organization, (2) models for the central city (or central business district) and/or small town centres, and (3) models for residential area designs. A fourth set of paradigms looks at the components of the city in a different way. Its members consider single issues such as designing for climate (e.g. Givoni, 1998), for transportation, for energy efficiency, for reducing the reliance on cars, environments for the elderly, and so on. This focus on single issues does not necessarily eliminate their utility for they bring designers' attentions to frequently neglected variables of importance in achieving high quality design. It does mean, however, that the issues they address are frequently neglected as designers deal with variables which they are accustomed to considering and/or the more pressing fiscal concerns of the development community and the quantifiable variables of traffic engineers. It is, nevertheless, the more global models that are of interest here.

Rationalist and Empiricist urban form paradigms share many characteristics. All have what Gertrude Stein noted that Oakland, California lacked, a 'there there' in the form of a city centre, and subsidiary residential and industrial areas. There is a tendency for Rationalists to favour the linear city as a type while Empiricists, as the garden city model suggests, favour an hierarchical set of nested sub-areas within an approximately circular boundary. The superblock is common to both. The models of the two lines of thought differ in their details: in the way they deal with city centres, with residential areas, and with their buildings. The first two of these differences are of most interest here.

The City Centre

Empiricist and Rationalist concepts for city centres have much in common. The basic paradigm is that of the centre as a superblock surrounded by a ring road with structured carparks feeding off it. Major residential and industrial precincts would lie beyond the ring road. The 'streets' in the central area are reserved for pedestrians. The difference between the two lines of thought has to do with the scale of the built environment. The Empiricists proposed relatively low buildings horizontally spread out and a traditional (even revivalist) approach to architecture. In contrast, the Rationalists had no fear of – indeed they embraced – high-rise buildings harnessing the latest in building technology and a multiplicity of transportation modes organized in horizontal layers.

In Europe, the Lijnbaan in Rotterdam and the Coventry city centre are mid-century examples of the low scale Empiricist application of the urban superblock idea when those cities were rebuilt immediately after the devastations of World War Two. The high-density central city mixed-use superblock of Rationalist origins with its layered transportation systems has also been used in a number of situations where it is possible to build *de novo*. Examples include the Barbican (1966-83) in London and Paris's La Defense (1960s+) which has become a precedent for later work such as the building of the new central city for Naples (1980s+). The Nelson D. Rockefeller Empire State Plaza (1968-1975) in Albany, New York, is a close relative. What might be regarded as the most ambitious plan – Victor Gruen's scheme for Fort Worth (figure 5.5) – has remained on the drawing board (Gruen, 1964). On a smaller scale, Rockfeller Center in New York of the 1930s might be seen as a precedent for Gruen's scheme. It certainly was in the 1960s for Philadelphia's Penn Center and Baltimore's Charles Center. The latter itself became a model for centre city urban renewal projects in the United States (Lang, 1994)

On a smaller scale, a widely used urban design paradigm is that of the main street of a small town and suburban shopping district as a pedestrian mall. The two parallel streets on each side of the pedestrianized main street become one-way streets with the parking located off them. Passages then link parking lots to the mall. Examples can be found in many countries; a large number were built in the United States alone with mixed success (Houston, 1990).

Residential Areas

At the residential area scale the ideas of the Empiricists and the Rationalists have diverged widely. What they have in common is a link to the central city. The neighbourhood unit concept, as exemplified in the design of Radburn, New Jersey, is the classical Empiricist paradigm (revised into the pedestrian pocket concept in the 1980s; see Kelbaugh, 1989) while the form of Le Corbusier's high-rise slab block Unité d'Habitation (Le Corbusier, 1960) exemplifies the Rationalists solution. The design principles of both have been widely adapted in diverse environments, both socially and climatically.

Figure 5.5. Fort Worth Plan. Victor Gruen's plan for Forth Worth may well be the exemplar of the Central Business District as a superblock. Never implemented, its basic principles have been applied in bits and pieces around the world. (*Photo*: Jon Lang)

The neighbourhood unit's applications include those in the British new towns (particularly the first generation), many post-independence company towns in India (Lang, Desai and Desai, 1997), privately developed new towns in the United States (Brooks, 1974), and suburban development in Australia (Freestone, 1989). Adaptations of Le Corbusier's Unité include many social housing schemes in France, the British Isles (Marmot, 1981) and the United States, and even more widely in Eastern Europe, Asia and Latin America. Perhaps the design is best exemplified by the 1950s housing schemes in Brasilia and by the contemporary 23 de enero development in Caracas. Designed by Carlos Raul Villanueva, the latter project consists of over forty residential blocks set in ample open space. It is now notorious for its hostile social environment. The design does show how a design paradigm can capture the imagination of a highly creative architect. In the 1940s Villanueva had designed the very sensitive highly Empiricist El Silencio housing development in Caracas. Today it might well be regarded as being based on a new urbanist paradigm.

The new urbanists advocate looking at traditional much-loved neighbourhoods and adapting their patterns for present day life. This line of thought is exemplified in the design of Seaside in Florida. The houses have little in the way of set-backs from the street, have porches facing it, picket fences, and garages in the back (Katz, 1994). New urbanist thought is also evident in the design of high-rise housing in Battery Park City in New York. There the building types were borrowed from prestigious areas of Manhattan such as Gramercy Park and Morningside Heights. In India, a number of designers have turned to precedents such as the 'pol' (a traditional gated caste and clan neighbourhood) as the basis for residential area design. The designers of the capital of Gujarat, Gandhinagar (1960s) claim they did likewise, although the city's plan seems to be more a mixture of Rationalist and Empiricist thought with Chandigarh as its precedent (Lang, Desai and Desai, 1997).

In residential area design, the problem of how to deal with the car remains paramount in the eyes of paradigm designers. The Radburn plan with its culs-de-sac and carefully segregated movements systems was one way. However, the cul-de-sac intended for vehicles only, has proven also to be a good playground for children. Consciously or not, the Dutch-inspired '*woonerf*' system, as a paradigm for residential street design with its lack of differentiated space for cars and children, recognizes that the two can share a space if it is properly designed (Francis, 1987).

Working within the Rationalist tradition of devising an ideal way of life and then designing the physical environment to afford it easily, have been a number of recent explorations in the provision of communal living arrangements. Based on a feminist and Marxist view of life, Dolores Hayden has identified a new residential type based on a number of precedents (Hayden, 1984). It involves much communal sharing of the chores of everyday life. The co-housing movement (McCamant and Durett, 1988) has suggested similar arrangements. There are a number of small-scale applications of the latter around the world, but they are really architectural rather than urban design ideas. Attractive to a number of people, they have yet to demonstrate the ability to capture the public imagination at large.

THE UTILITY AND LIMITATIONS OF PARADIGMS

It can be argued that the significant twentieth century paradigms have not been truly tested because they have been whittled away during the design process when applied. After all, architects have to show they are creative. Unfortunately, it is too often a superficial creativity that is displayed; the appearance only is novel. The various adaptations of Le Corbusier's Unité, for instance, are spatially tighter and have fewer communal facilities than those in the Unité in Marseilles (Marmot, 1981). The same observation can be made about the application of the Radburn plan. Its descendants are meaner.

Each of the urban design paradigms has had and still has strong proponents who are generally disparaging about their ideological opposition. What we need to do now at the dawn of the twenty-first century is to understand what the patterns that each advocates do and do not afford people in terms of their ways of life and aesthetic values and what they afford for the biogenic health of the planet. We should also look at the design patterns used repetitively by developers – at their paradigms. All have some utility; all have limitations.

It is important to have this understanding because property developers, urban planners and architects rely heavily on a relatively restricted set of design paradigms as the basis for urban design. These paradigms are regarded as exemplars of good solutions or, at least, good enough solutions. While the urban design related research in areas such as environment-behaviour studies has an increasingly strong theoretical base (Lang, 1987; Rapoport, 1977), all indications are that most practitioners will continue to eschew coming to an understanding of it. They will, rather, continue to formulate, consciously or subconsciously, design decisions based on the paradigms they know (Francescato, 1989). The result is that the design situations that practitioners face, although unique, are perceived in terms of their generic, not unique, qualities. They then adapt the paradigms to the market, and the topographical and political conditions of specific locations. The climatic conditions of specific places and the ways of life within specific cultural contexts tend to be neglected. So too are the needs of the less vocal in society, the children and the poor amongst them.

The application of many urban design paradigms has indeed led to environments that have met their users' needs very well and have been economically viable in some situations, while in other cases they have failed on one or other dimension. They have been viable when the problems they address match those of the context in which they were applied. They have been considerably less successful when applied in cultural and climatic contexts for which they were not developed. This observation is particularly, but by no means uniquely, true of those paradigms proposed in the predominantly temperate climate, industrialized world and then imported into developing countries. More generally, what have seldom been understood are the economic, social and political side effects that the application of specific paradigms has had on the world around them.

One of the continuing limitations of generic solutions is that their proponents are not nearly clear enough about the nature of the cultures within which they can operate. Design philosophers in a very few countries have created the paradigms mentioned in this chapter. The applications of these paradigms have overridden potential local approaches in other countries in the hand of both international and local architects and planners. While much criticized by indigenous observers, design practitioners in such countries have little other than imported ideas on which to rely. Feng-Shui in the Chinese world and variants of the Shilpa-Shastras in India, for example, do provide urban design principles and paradigms on which practitioners can base designs (Lang, Desai and Desai,

1997). These models of how to approach urban design, however, have yet to demonstrate how to deal with the world of cars, high-density living and the ways of life to which many people in those countries nowadays aspire. Despite repeated statements on the need for new paradigms over the last twenty years, city and urban design practitioners in countries such as China and India – the two most populous countries in the world – still await their academic (or professional) colleagues to develop them.

The application, consciously or subconsciously, of existing paradigms has resulted in built environments which have been regarded by those who inhabit them as wonderful places in some cases and as miserable in others. Radburn itself, remains much cherished by those who live there but the neighbourhood unit as developed in Chandigarh is largely irrelevant in terms of activity patterns because of the social structure of Indian society (Brolin, 1976). Even so, it still provides an easy basis for forming a cognitive image of the city. The Australian experience with the Radburn ideal is especially salutary. Neighbourhoods based on Radburn principles worked well for their original inhabitants but now, inhabited by newcomers, the once hospitable environments have become inhospitable without their layouts being changed. They have now been de-Radburnized into traditional dwelling-and-garden estates fronting access streets. The paradigm that has been used so unsuccessfully in the 23 de enero scheme in Caracas works well in Singapore. Much depends on the nature of the social organization of life of the people who live in each of the two places. The broader lesson that changing the layout of environment will not change social behaviour patterns unless there is a strong predisposition to change, has yet to be learnt.

Another example of disparate success and failure (in terms of their designer's objectives) has been with the pedestrianized suburban or small town shopping street. Some have been so successful that they have been extended (e.g. Kalamazoo, Michigan). Others where failure was predicted have proven to be highly successful (see Gehl, 1987 on Copenhagen, Denmark). A considerable number of these malls, however, have now been returned to vehicular traffic because of their failure to engender economic life in the face of competition from suburban strip shopping areas and shopping malls – the economically driven paradigms. They continue to operate in much the same way but the main street now has both pedestrian sidewalks and vehicular traffic lanes (Houston, 1990). This pattern had already been adopted as a logical way of organizing a central business district in places such as Glendale, California.

This mixture of success and failure enables some generalizations to be made about all the paradigms developed during the twentieth century. None deals with the full richness of life for the variety of people that might inhabit a place. The model of the human being that has been adopted is, universally, much too limited. None is explicit about the circumstances to which it does and does not apply. In addition, they have a strongly deterministic quality; they were supposed to shape the social behaviour of people. Their authors (often highly esteemed) have frequently been strong advocates for the models they developed on social engineering grounds. We know better now. What we have learnt is that the affordances of the environment make specific behaviours possible and others not. Whether the possibilities are perceived and acted on depend on people's motivations, predispositions and competences (Gibson, 1977; Lang, 1987).

LOOKING AHEAD: PARADIGMS FOR THE EARLY TWENTY-FIRST CENTURY

Many planning theorists, including myself, have argued strongly for urban designers to base their work on the accumulating abstract theoretical knowledge that the systematic research of the last four decades, in particular, has yielded. The political upheavals of the 1960s led to a significant shift in the urban design thinking which coincided with the coining of the term itself. Urban designs, if they were to serve human purposes well, could no longer afford to be based on the individual belief systems and experiences of an architect or planner; they had to be based on knowledge. This knowledge base has indeed been significantly advanced at the interstices of the behavioural and physical sciences and aspects of built form. What it has created are increasingly complex models of the world. Implicit in this new knowledge base are many urban design principles, but they require ferreting out by designers dealing with specific situations. Most design practitioners are reluctant to follow a process of deductive logic in creating designs and, often quite rightly, argue that they have few fiscal or temporal resources to do so. They have to produce designs quickly and economically so they rely on a process of mimesis.

As long as practising urban designers are reluctant to draw their design principles from empirical theory (Rapoport, 1977) – what Lynch (1984) called 'functional theory' – and a problem-oriented approach to the process of designing, design paradigms will form an important part of their repertoire. Experiences with the application of paradigms during the twentieth century enable one to make two contradictory observations about the early part of the next. Firstly, changes in ways of life, for the middle classes in the developed world, have had fewer consequences for the design of the environment than we are often wont to think, and secondly, that predictions about future changes are fraught with difficulties. In other words, any conclusion drawn from the first observation should be made with caution!

Ever since the development of ENIAC, the first electronic computer, in the 1940s, predictions have been made about how new information technology will transform life styles. The changes, as Isaac Asimov (widely recognized as the foremost science fiction writer and futurist of the second half of the twentieth century) acknowledged shortly before his death, have come much more slowly than he and others predicted in the 1960s. It would, however, be foolhardy to assume that the prolonged *status quo* will continue forever. Any statement about the nature of the design paradigms needed for future practice must recognize the need for them to be robust – tolerant of change. Even if this is done, any claim for their utility unchanged beyond a generation ahead must constantly be challenged.

It is important for urban designers to understand how the economically driven paradigms have evolved because urban development, unless subsidized by governments or philanthropy, has to be profitable for developers. What has emerged is an increasing public-private partnership in development (DeSeve, 1986). This change has heralded in a series of new procedural paradigms but has yet to be reflected in substantive ones.

Rethinking the Nature of the Next Generation of Paradigms

At the start of the twenty-first century, the perception of the problems that should be addressed by urban designers has changed. Their goals in the early 1900s were to bring a more salubrious life to the inhabitants of the cities of the 'dark satanic mills'. Those at mid-century were to improve transportation systems. Now they are often to create a salubrious biogenic environment (e.g. Barnett, 1998). No existing paradigm deals with the full complexity of life nor can it, but we can do better than we have. While it is important to understand what each existing paradigm offers (i.e. to understand its affordances) in specific contexts, it is also important to heed the lessons of how to use paradigms in the designing process.

The substantive models need, in particular, to deal with cultural and climatic diversity – life as it is led in specific terrestrial environments. Procedural models need to be specific about how designers should recognize problems and the nature of urban development processes and the diversity of actors engaged in them. The prescriptive procedural models which advocate empirical theory-based, problem-solving modes of thought need to be reshaped to recognize that few design professionals have the desire or the confidence to work that way nor the ability to keep apace of current research. The substantive models at their disposal must, therefore, keep apace. Whose responsibility is it to develop them? Surely it is the academy's. Yet academics are loath to take stands on what *should* be done except in very general terms such as in this chapter. This attitude has to change.

There are two characteristics that all urban design paradigms for the future must possess. They must be multi-dimensional and it must be made clear what the problems and issues are that they do and do not address in what socio-cultural and terrestrial contexts. In this sense, the purposes they serve must be transparent in much the same way that Christopher Alexander and his colleagues have made transparent the statements in their pattern language – the problems they address and the logic of the pattern is fully expounded (Alexander, Ishakawa and Silverstein, 1977). What is required, however, is that they prescribe responses to human needs and be organized in that way and not in terms of presumptions about the scale of solutions: towns, precincts or buildings.

What is needed is not only a greater range of paradigms on which urban designers can draw but richer paradigms. They must be multi-layered in much the way Bernard Tschumi used in the design of Parc de La Villette in Paris (Broadbent, 1990) although the neo-Rationalist basis for his design needs rethinking. What is required are more layers, each dealing with a specific issue/ problem. Tschumi used three. Three may be sufficient for a park, but not for a city. Each layer should represent the interests of a particular segment of the population of concern and would be designed with a particular climatic and cultural environment in mind. One layer, for instance, could deal with the ideal transportation network, a second with the distribution of parks for recreation, a third with open space for the biological health of the city, a fourth for the educational environment of children, and so on. Synthesizing them into wholes is the creative task. With how many such layers can the human mind cope? Inevitably, putting the layers together into a single paradigm will lead to trade-offs between solving one problem well and another problem well. The trade-offs need to be made clear to those who use the paradigms. If we can achieve that end we shall have achieved much in improving the quality of urban designs and urban designing.

Conclusion

There is a tendency, when design paradigms are perceived not to work well when applied, to throw them out *in toto*. This action is foolhardy because the twentieth century paradigms, for all their warts, offer designers much. We need to recognize the problems they address and the nature and reasons for their likely successes or failures (and in whose terms) when they are translated into practice. It is their performance in the real world – their external validity – rather than the beauty of their internal logic that is important. Where they have been unsuccessful, they need to be altered or refined based on the empirical evidence of their performance. Most professionals are on the firing line and have little time in which to redevelop the paradigms with which they work. It really is for the academic community to rework old paradigms and to generate and evaluate new ones in the symbolic world prior to their testing in the 'real' world. What paradigms do and do not afford politicians, developers and users alike, needs to be clearly based on knowledge, not hopes nor dreams. Yet if we do not dream where will we get?

REFERENCES

Alexander, C., Ishakawa S. and Silverstein M. (1997) *A Pattern Language: Towns, Building, Construction.* New York: Oxford University Press.

Barnett, W. (1998) *Re-defining the Australian Dream: Towards a Sustainable Future.* Sydney: The Author.

Benevolo, L. (1980) *The History of the City.* Translated from the Italian by G. Culverwell. Cambridge, Mass: MIT Press.

Broadbent, G. (1990) *Emerging Concepts in Urban Space Design.* New York: Van Nostrand Reinhold.

Brolin, B. (1976) *The Failure of Modern Architecture.* New York: Van Nostrand Reinhold.

Brooks, R.O. (1974) *New Towns and Communal Values: A Case Study of Columbia, Maryland.* New York: Praeger.

Cullen, G. (1961) *Townscape.* London: Architectural Press

DeSeve, G. E. (1986) Financing urban development: The joint efforts of government and the private sector. *The Annals: The American Academy of Political Science*, **53**, pp. 58-76.

Francescato, G. (1989) Paradigm lost: exploring possibilities in environmental design research, in Hardie, G., Moore, R. and Sanoff, H. (eds.) *Changing Paradigms: Proceedings of EDRA20.* Oklahoma City: Environmental Design Research Association, pp. 63-67.

Francis, M. (1987) The making of democratic streets, in Vernez Moudon, A. (ed.) *Public Streets for Public Use.* New York: Van Nostrand Reinhold, pp. 23-39.

Freestone, R. (1989) *Model Communities: The Garden City Movement in Australia.* Melbourne: Nelson.

Fishman, R. (1982) *Urban Utopias of the Twentieth Century: Ebenezer Howard, Frank Lloyd Wright and Le Corbusier.* Cambridge, Mass: MIT Press.

Gehl, J. (1987) *Life Between Buildings.* New York: Van Nostrand Reinhold.

Gibson, J. J. (1977) *The Ecological Approach to Visual Perception.* Boston: Houghton Mifflin.

Givoni, B. (1998) *Climate Considerations in Architecture and Urban Design.* New York: Van Nostrand Reinhold.

Gruen, V. (1964) *The Heart of Our Cities.* New York: Simon and Schuster.

Hayden, D. (1984) *Redesigning the American Dream.* Cambridge, Mass: MIT Press.

Houston, L.O. (1990) From street to mall and back again. *Planning*, **56**(4), pp. 4-10.

Howard, E. (1902) *Garden Cities of Tomorrow.* London: Swan Sonnenschein.

Katz, D. (1994) *The New Urbanism: Toward an Architecture of Community.* New York: McGraw Hill.

Kelbaugh, D. (1989) *The Pedestrian Pocket Handbook: A New Suburban Design Strategy.* New York: Princeton Architectural Press.

Krier, L. (1985) *Albert Speer: Architecture 1932-1942.* Bruxelles: Archives d'Architecture.

Krier, R. (1979) *Urban Space.* Translated from the German by C. Czechowski and G. Black. New York: Rizzoli.

Kroll, L. (1994) Vers une écologie urbaine, in Neary, S. J., Symes, M. S. and Brown, F.E. (eds.) *The Urban Experience: A People-Environment Perspective.* London: E. & FN Spon, pp. 19-29.

Lang, J. (1987) *Creating Architectural Theory: The Role of the Behavioral Sciences in Environmental Design.* New York: Van Nostrand Reinhold.

Lang, J. (1994) *Urban Design: The American Experience.* New York: Van Nostrand Reinhold.

Lang, J. Desai, M. and Desai, M. (1997) *Architecture and Independence: The Search for Identity. India 1880 to 1980.* New Delhi: Oxford University Press.

Le Corbusier (1960) *My Work.* London: Architectural Press.

Lynch, K. (1960) *The Image of the City.* Cambridge, Mass: MIT Press.

Lynch, K. (1984) *Good City Form.* Cambridge, Mass: MIT Press.

Marmot, A. (1981) The legacy of Le Corbusier and high rise housing. *Built Environment*, **7**(2), pp. 82-95.

McCamant, K.M. and Durett C.R. (1988) *Cohousing: A Contemporary Approach to Housing Ourselves.* Berkeley: Ten Speed Press.

Moughtin, C. (1992) *Urban Design: Street and Square.* Oxford: Butterworth Architecture.

Newman, O. (1974) *Defensible Space: Crime Prevention through Urban Design.* New York: Macmillan.

Newman, O (1996) *Creating Defensible Space.* Washington, DC: U.S. Department of Housing and Urban Development.

Peets, E. (1927) The genealogy of L'Enfant's Washington. *The Journal of the American Institute of Architects*, **15**(5,6,7), pp. 115-118, 151-154, 187-191.

Prolink Development Sdn. Brd. and Jean-Paul Viguier SA d'Architecture (1997) *Concept Development Plan for Bandar Nusajaya.* Kuala Lumpur: The Authors.

Rapoport, A. (1977) *Human Aspects of Urban Form: Towards a Man-Environment Approach to Urban Form and Design.* New York: Pergamon.

Rowe, P. (1991) *Making a Middle Landscape.* Cambridge, Mass: MIT Press.

Shirvani, H. (1985) *The Urban Design Process.* New York: Van Nostrand Reinhold.

Simonds, J. O. (1994) *Garden Cities 21: Creating a Livable Urban Environment.* New York: McGraw Hill.

Smithson, A. (1969) *Team10 Primer.* Cambridge, Mass: MIT Press.

Wilson, W. (1989) *The City Beautiful Movement.* Baltimore: Johns Hopkins University Press.

Zwicky, F. (1969) *Discovery, Invention, Research through the Morphological Approach.* New York: Macmillan.

CHAPTER 6

Rethinking the Significance of the City Beautiful Idea

Gilbert A. Stelter

'Plan making is in the air' explained William Langton to his fellow architects in Toronto in 1905: 'we have caught it from our generation' (Langton, 1905). Langton was referring to the popularity of one of the great planning movements of the early twentieth century, that of planning in the 'grand manner', based on traditional European urbanism and usually referred to in North America as the 'city beautiful' movement. Langton would also have known about a rival vision of planning, the 'garden city' movement, which sprang from the housing reform and land reform movements. These two approaches to the crisis of the large city at the turn of the century differed considerably in goals and in how to achieve them, but they did share some basic assumptions. Both saw the necessity of state intervention into the city building process; both argued for the desirability of segregating the various functions of the city into discrete zones; and both assumed that urban design could be an active, positive force in social improvement.

This chapter deals with the worldwide strategy for re-establishing beauty and a sense of order into *fin-de-siècle* cities, with particular attention to the city beautiful movement in Canada. Many planning historians either ignore the grand manner and its city beautiful component or dismiss it as an approach that worked only when some dictatorial authority forced it on to a city (Kostof, 1991). Others see it as an approach which narrowly concentrated on aesthetics and collapsed under the weight of its expensive and unrealistic schemes (Scott, 1969; Van Nus, 1984). Certainly, anyone who examines the North American portion of the movement will quickly see its shortcomings. Many plans were hastily drawn up with little chance for success; some were ruinously expensive; some put forward by outside consultants ignored local traditions and even geography; most did not appear to address larger social issues such as poverty and the lack of housing.

In the light of these critiques and glaring weaknesses, it might seem surprising that there was so much international interest in this approach during the latter part of the nineteenth and the beginning of the twentieth centuries. European cities were swept by replanning in the grand manner, primarily influenced by the example of Baron Haussmann's transformation of Paris. Cities outside Europe followed the same fashionable trend. In North America this movement took on a rather populist tone. The major impetus was in the United States, where significant projects were planned and completed in dozens of cities and smaller interventions took place in hundreds of other communities. Almost every major city in Canada also drew up elaborate plans in the prevailing style.

Was all this activity merely misguided reformism or local boosterism? To what extent was and is the negative image of the movement the product of those contemporaries who represented competing visions of a good city? In retrospect, the critiques of the garden city and the 'city practical' advocates seem only partially justified or even misleading. When looking at the contemporary sources of the day, I was struck by the quality of many of the planning documents, by the amount of serious newspaper coverage and discussion, and by the extent to which the concerned public was involved in those discussions. As a result, I will argue that the city beautiful movement embodied some very positive planning ideas, some of which were unfortunately lost when the movement declined after World War One. Three of those positive characteristics will be looked at in turn: first, the idea that a city could be a work of art, based on traditional European urbanism as a model; second, a broadening of the leadership involved in the planning process; third, a theory of the city as an organism that required holistic rather than piecemeal planning. In a final section I will assess the degree to which the movement was successful, not only in terms of projects actually built but also of its influence on subsequent city planning.

THE CITY AS A WORK OF ART

At the turn of the last century much of the world looked to Europe for examples of how to remake cities. What seems to have attracted those outside Europe was the long tradition of planning in the grand manner, which emphasized geometric order; long, straight streets culminating in vistas; formal squares and places; elaborate parks; and the ability to link all of these elements into one coherent whole. Observers were aware of classical precedents or were influenced by the master plan of Rome in the sixteenth century, and the French baroque tradition of the seventeenth century exemplified by Versailles and its descendants, St. Petersburg in Russia and Washington in the United States. The greatest interest, however, was provided by the Paris that was transformed between 1853 and 1870 under the leadership of the Emperor, Napoleon III, and his *de facto* mayor of Paris, Baron Haussmann. Parts of Paris had been reworked, especially by Louis XIV and Napoleon I, but Haussmann's effort was the first to subject the whole city to modernization. Haussmann's Paris became the ideal to which other cities aspired, more significant in this respect than two other nineteenth century urban projects which did not deal as comprehensively with entire cities – the Vienna Ringstrasse and the reorganization of Rome in the 1880s based on the plans of Allesandro Viviani.

What was there about the Haussmannization of Paris that made it the urban ideal, the modern city *par excellence*? A key aspect seems to have been the boldness of the conception, for the transformation of this royal capital was done on the grandest scale (Jordan, 1995). Haussmann gutted what had still been a medieval city at mid-century and made it into a modern city in less than two decades. Another factor was Haussmann's successful blending of the practical with the beautiful and symbolic. The extensive system of new boulevards was effectively connected to an older ring of boulevards by means of *places-carrefours*, plazas where several streets met. A new sewage system alleviated some of the conditions blamed for a series of cholera epidemics. The demolitions wiped out much of the decaying medieval core and replaced it with expensive apartments and shops. And Haussmann continually emphasized that the cost of urban renewal was justified by the enhanced property values along the new streets. The result was a city with a bourgeois flavour, a city fit for the new capitalism of railroads, commerce and banking (Harvey, 1985).

Closely corresponding to the practical was Haussmann's concern for beautification (Olsen, 1986). Based on his mandatory guidelines, boulevards were lined with uniform buildings designed in the French classicism of the previous century. Vistas were created to give prominence to monumental structures such as Garnier's new Opéra, which became the symbolic centre of the new city, replacing the older icons of palace and church (figure 6.1). Several new parks were created, such as the Bois de Boulogne.

Figure 6.1. Charles Garnier's Paris Opera House. The Opéra became the symbol of Haussmann's transformed Paris, partly for the elegance of its architectural style but also because of its dominant setting as the focal point of several major radiating streets. (*Source*: Anthony Sutcliffe)

The spectacular setting apparently met with the approval of well-to-do Parisians. A vibrant cultural life was lived in public, in full view on the boulevards and in the cafés, exemplified by painters like Manet and writers like Baudelaire (Reff, 1982). The busy street life of the new city attracted visitors from around the world and gave Paris its special charm. However, as anyone who has wandered along some of Haussmann's boulevards knows, beyond this public activity the streets and facades by themselves can appear sterile and boring. The international fascination with the central core unfortunately masked the brutal destruction of entire neighbourhoods, the displacement of inhabitants without compensation, and the zoning of what were regarded as the dangerous classes into isolated neighbourhoods in the eastern part of the city and the suburbs. What Haussmann had created was a sanitized zone, the model city, with the unsavoury elements hidden from view. Only a few individuals like Emile Zola attacked this 'immense hypocrisy, a colossal Jesuitical lie' (quoted by Reff, 1982, p. 140). This underside of Parisian society occasionally asserted itself, as in the short-lived Commune of 1870, but the official cynical disregard for a portion of the population probably also proved attractive to those foreigners who felt they might be able to take similar steps with some of their own people.

By the early 1900s Haussmann's transformation of Paris had become the most emulated method of urban renewal in the world. The following examples represent only a portion of the cities that looked to Paris as an example of what a city could be. The leaders of several major South American cities, especially Rio de Janeiro and Buenos Aires were Francophiles. In Rio, the chief engineer, Pereira Passos, had studied in Paris and consciously adopted Haussmann's methods in major works between 1902 and 1906. These included the demolition of a working class section of the port area and the construction of new wide avenues such as the Avenida Central, lined with buildings with facades in the *Beaux Arts* manner. In keeping with the Parisian spirit, Passos' son designed the Teatro Municipal as a smaller version of Garnier's opera house. A shock of recognition greets today's visitor to the Avenida, even though the buildings seem strangely out of scale. This kind of Europeanization of a New World city was clearly designed to demonstrate a 'civilized' Rio, distinct from the Afro-Brazilian culture which Rio's elite considered to be backward and barbaric (Needell, 1987). In Argentina as in France, the president appointed an *intendant* with the power to remodel Buenos Aires. The redesign of the Plaza de Mayo, and the cutting of an avenue west from the plaza, the Avenida de Mayo, symbolized the rejection of the Hispanic tradition and the imitation of French civilization (Scobie, 1974). The upper class quickly followed with a wholesale acceptance of the Parisian *Beaux Arts* style in their mansions. The result was Parisian-looking streets although Haussmann's height limitations were not applied.

The Parisian influence was also powerful in Istanbul where the Ottoman elite partially modernized the city by adopting new regulations and design principles for street networks, and planned monumental public squares, attempting to insist on a uniform urban fabric based on *Beaux Arts* designs. In 1901 the Ottoman government had a master plan drawn up by the Director of the Department of Architecture in Paris, Joseph-Antoine Bouvard. Without actually going to Istanbul, Bouvard put forward an extravagant series of proposals, with beautiful paintings depicting a Parisian-like cityscape. Local bureaucrats applauded the plan but the state was too impoverished to put it into practice (Çelik, 1986). In Tokyo, the Meiji era's programme of Westernization in the late nineteenth century resulted in vast changes by 1914 to a city that seemed chaotic by Western standards. New developments included the widening of some major streets, the establishment of a water and sewerage system, the building of a new commercial district in brick (the Ginza), the construction of a business district with buildings like the Bank of Japan (1896) in Second Empire style, the construction of a major park (Hibiya), and the centralization of some of the government buildings. While some Westerners still find Tokyo's complexity intimidating, the result of the modernization movement was a degree of centralization unknown in Japan previously, with a core made up of the Imperial Palace, government buildings and business districts adjacent to each other even though authorities had no single grand design in mind (Fujimori, 1987; Jinnai, 1995).

In the United States, what has been termed 'the discovery of Europe' in the late nineteenth century saw a succession of Americans who visited Europe come away impressed by clean and beautiful cities which contrasted with their own large cities like Chicago, which had become symbols of disorder and danger (Smith, 1994). In the vanguard were American architectural students at the famed Parisian Ecole des Beaux-Arts. One of the results was a resurgence of the neo-classical style on the East Coast which reached national proportions with the World's Columbian Exposition in Chicago in 1893. Most of the fair's architects had studied in Paris and they convinced Daniel Burnham, the Chicago architect in charge of the fair's construction, to accept neoclassicism as the style for the major buildings (figure 6.2). Particularly spectacular was the ensemble of buildings, especially the central Court of Honour, which provided an example of a monumental core or civic centre, an arrangement of buildings that was designed to inspire by the combination of beauty, grandeur, and a sense of harmony (Hines, 1979; Wilson, 1989).

While the Chicago Exposition may not have directly created the city beautiful movement, it played an important role in demonstrating the possibilities of replanning the centre of cities on a grand scale. Burnham had made his architectural reputation by that most American of building types, the skyscraper, but he now became the leading proponent of the city beautiful approach. The first explicit project in this vein was his heading of a commission to redesign the monumental core of Washington, the Mall, in 1901 and 1902. Burnham and other commission members made a pilgrimage to the great cities of Europe in

Figure 6.2. The Court of Honour at the World's Columbian Exposition, Chicago, 1893. The unified ensemble, with its neoclassical styles, was designed to promote a sense of dignity and harmony. It became an influential model for North American civic centres. (*Source*: Chicago Historical Society)

order to get ideas, reflecting, as Thomas Hines describes it, 'the reverence of the City Beautiful mentality for the culture of the Old World' (Hines, 1979, p. 146). The commission was particularly impressed with Paris, seeing it as a 'well articulated city – a civic work of art'. The commission's design of the Mall was in many ways a revival of Pierre L'Enfant's original baroque plan. The central area had become a site for cow pastures, a lumber yard, and the city's railroad centre. This was cleaned up, and L'Enfant's axes of the Capitol and the White House re-emphasized with a crossing at the proposed Washington monument (figure 6.3).

The Washington plan was the first actual demonstration of the city beautiful movement's approach, although the term was in common use by 1899, promoted by Charles Mulford Robinson in books like *The Improvement of Towns and Cities* (1901). Robinson had made the obligatory trip to Europe before publishing his book. Some of the major American projects were those in Kansas City, Seattle and Denver, but Daniel Burnham became the superstar of the movement with civic centre projects in Cleveland (1903) and San Francisco (1905). His influence peaked with the publication of his magnificent volume, the *Plan of Chicago* (1909), which summarized the ideas and aims of the movement. The first chapter of this influential work outlined some of the basic principles of European cities as works of art. The illustrations for the Chicago plan by Jules

Guérin showed a giant civic centre, focused on an impressive dome, with great diagonal avenues radiating from the centre, and Parisian-looking commercial buildings of uniform, limited height (Burnham and Bennett, 1909).

In turn-of-the-century Canada, those searching for models occasionally looked to continental Europe. Percy Nobbs, an Englishman who was Professor of Architecture at McGill University in Montreal, gave numerous public addresses, complete with the usual slides of Greek, Roman, Medieval and Early Modern European planning examples. He was not an admirer of the grand manner in itself for he felt that Paris 'was ruined from the picturesque point of view when Haussmann drove its great boulevards for strategic reasons through the town', but he later admitted that 'it has come to be accepted as a commendable example of city layout' (Nobbs, 1906, p. 39; 1911, p. 140).

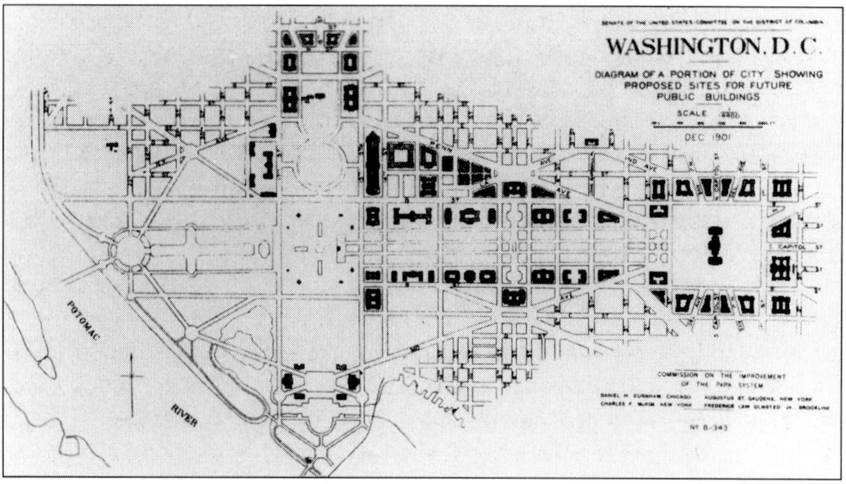

Figure 6.3. Plan of the Washington Mall, 1902. This plan represented the first American attempt to redesign the centre of a city on City Beautiful principles. (*Source*: Federal Planning Commission, National Archives, Washington DC)

Canadians tended to look primarily to the United States for examples of reintroducing order and beauty into their cities. This amounted to seeing the European tradition through American eyes. The Chicago World's Fair was a central event in this regard. Canadian architects travelled to Chicago and returned to regale their colleagues about the spectacular architecture they had seen. By 1905, with the American city beautiful movement in full bloom, Toronto architect William Langton concluded:

> The point of departure was, without doubt, the Chicago Exhibition, which gave the public a new idea of the possibilities in the building of cities, and a greater opinion of the importance not only of architecture but of architects. It is since then that the wave of civic improvement has swept over the country. (Langton, 1905)

Burnham himself was often referred to in relation to the fair and to the civic improvement projects he supervised. He was thought to be more than just an architect of individual buildings; he was the architect of cities. Langton commented that 'the newspapers mention Mr. Burnham's name with the same simplicity as they would a general's, in full confidence that everyone will know who he is' (Langton, 1906).

The first city beautiful proposal in Canada seems to be the Toronto plan of 1905, put forward by a group of local architects. William Langton, a leader in the group, admitted that: 'Toronto in taking up this plan and carrying it out will be merely following a movement and following it a good way behind' (Langton, 1906). Certainly the diagonal plans, a ceremonial waterfront entrance, and connected parks owed a great deal to current American thinking, and especially to Burnham's recent Cleveland civic centre project. But when questions were raised about the high costs of cutting streets through the existing city, Langton referred to 'the classic example of Paris in which, through roads were run regardless of cost, they were found to have cost less than expected because of the opening up of new frontages'. In this vein, Langton reported on a local stockbroker's favourable references to Rio de Janeiro's experience in building the Avenida Central: 'The street is made and they have money in hand'. By 1909, a more detailed version of the 1905 plan was put forward, and in 1911 a third plan was drawn up by John Lyle, trained at the Ecole des Beaux-Arts in Paris, with prestigious New York experience. This plan featured a massive civic centre complex, diagonal avenues, and a major new avenue cutting through the business district from the civic centre to a proposed union railroad station (figure 6.4). In Montreal, the local architects' association drew up a 'tentative scheme' to invite discussion about introducing city beautiful ideas. A more highly developed proposal was published by G.-A. Nantel, a journalist and provincial politician, as *La Métropole de demain: avenir de Montréal* (1910). Nantel admired Haussmann's achievements and hoped that Montreal might be remade into the Paris of North America. Other proposals in Canadian cities also referred favourably to American and European precedents, as in the plans for Calgary and Ottawa.

Not all concerned Canadians were enthralled with the examples of European or American modernization. Percy Nobbs was critical of those who designed excessively long streets, especially 'the Civic Improvers of American cities, who are sometimes influenced too much by the cult of the Grand Manner and ignore what the Middle Ages have taught' (Nobbs, 1911). The traditional grand manner continued to provide the inspiration, however, especially for a capital city. In the late 1930s, Ottawa's earlier plan was partially revived, this time by Prime Minister Mackenzie King, who personally hired a French planner, Jacques Gréber, whose views coincided with King's on introducing a sense of grandeur into the national capital. After getting general agreement on some changes involving the location of the national war memorial, King recorded in his diary: 'I at once saw that I had my Champs Elysées, Arc de Triomphe and Place de la Concorde all at a single stroke' (quoted by Tomovcik, 1977, p. 40).

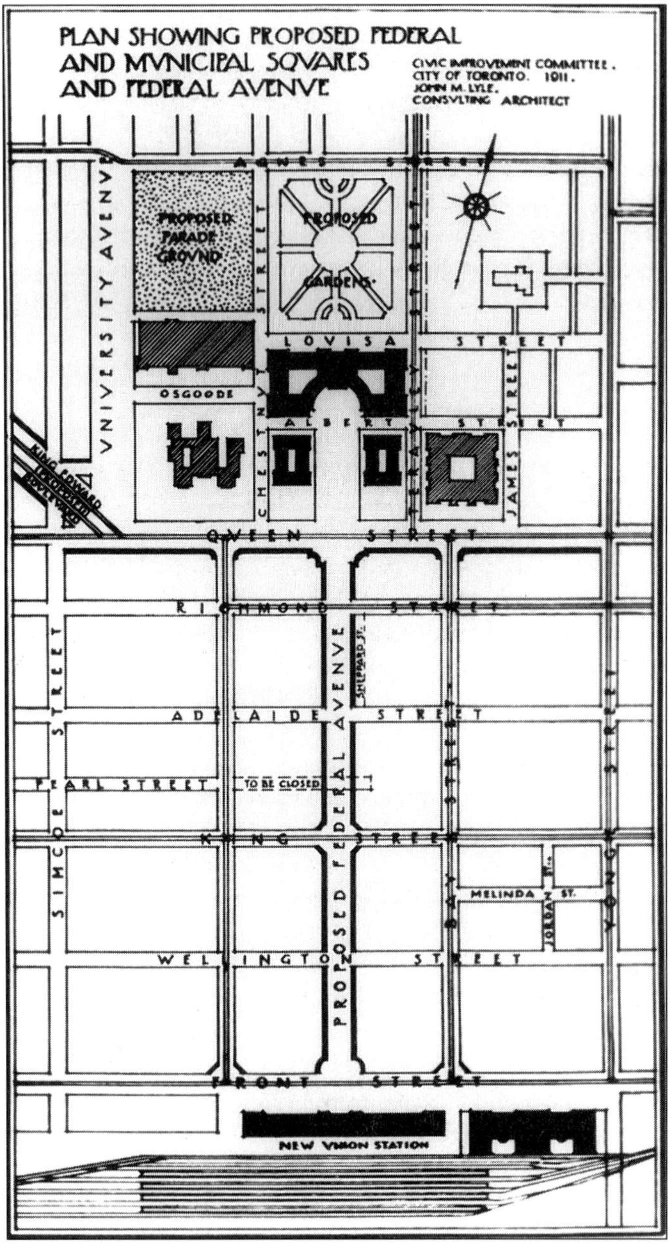

Figure 6.4. Plan for central Toronto, 1911. Shown are John Lyle's plans for a grand civic centre and a new street, Federal Avenue, to connect the civic centre to the city's entry point, Union Station. (*Source*: Report of the Civic Improvement Committee for the City of Toronto, 1911)

In Australia, as in Canada, planning discourse early in the century combined city beautiful ideology and traditional European urbanism. The major example was Walter Burley Griffin's plan for the new federal capital, Canberra, in 1911-1912. Griffin's winning design incorporated the geometry of Washington, with grand axes and vistas, and a strong central focal point. Griffin was a Chicago landscape architect who had been impressed with the Chicago Fair as a young man and who greatly admired Burnham's Chicago plan (Peisch, 1964). However, Australia's leading proponent of the city beautiful paradigm was John Sulman (Freestone, 1996), who respected the American experience and also the Parisian precedents of uniform building heights and grand boulevards, as outlined in his influential book, *An Introduction to the Study of Town Planning in Australia* (1921).

BROADENING THE DECISION-MAKING POWERS OF THE PUBLIC

The physical design goals of planning in the grand manner paradigm were similar in many parts of the world, but the decision-making process differed considerably from country to country, and even from city to city within the same country. The public interest could be served by several types of leadership. While this approach is often associated with 'an authoritarian impulse', the reality is much more complex. The key question is – who initiated the planning, set the goals, and saw the projects through to completion? It is possible to identify at least three types of leadership, although in most cities at least two of the three types were combined in some manner.

The first type, the baroque cult of power, in Lewis Mumford's terms, was the prevalent type in traditional European urbanism. Political power was exercised from the top down, as in the case of Napoleon III and Haussmann in Paris or in local leadership in Rio de Janeiro and Istanbul. Capital city planning in North American had elements of this type. The second type was professional and bureaucratic, as in Australia and some North American cities. In most cases this involved architecturally inspired planning, but eventually architects were replaced by those with training in land use and engineering. The third type was a more populist initiative, generated by civic and commercial elites who formed civic improvement associations, sometimes with fairly broad public support. The American city beautiful movement is the best example. It was, according to its major historian, William Wilson, a political movement that forged a politics of accommodation between planning professionals, city officials, entrepreneurs, and various reformers (Wilson, 1989). The lay dimension of the movement was exemplified by Robinson, who was a journalist, a publicist, and a poet. The foreword to his first book included a 'partial' list of over 75 local, state and national civic improvement societies (Robinson, 1901). At the national level a major force was the American Civic Association, which united civic activists and professionals. Laymen provided the necessary public support, but it was always understood that specific projects were to be planned by professionals.

Washington was a model of one kind of American city beautiful planning. Even in a national capital, where a president appointed a national commission, planning involved a complex political process which included compromises between powerful corporate interests (a railroad company), congressional leadership, the Washington Board of Trade, the local newspapers, civic leaders, the advocates of civic art and parks, the national association of architects, and the aesthetic authority of the commission headed by Burnham (Peterson, 1985). Kansas City represented another type of leadership structure involved in planning. A coalition of park advocates, the City Council, a newspaper editor, a landscape architect, and some business leaders successfully fended off the opposition of a taxpayer's league to begin a comprehensive park and boulevard system by 1900. This early initiative was greatly extended during the height of the city beautiful era (1900-1915) in the city, apparently with considerable public support (Wilson, 1989).

Those who supported the city beautiful idea in Canada never organized at the national level as was the case in the United States. The earliest impetus came from architects who had been to the Chicago Exposition, like Andrew Taylor, who obviously wished for the kind of power that Haussmann and others had in Europe:

> It is not often the happy lot of any members of our profession to be called upon to design an ideal city. The average modern city is not planned – like Topsy, it just grows, and we are only allowed to touch with the finger of beauty a spot here and there. One longs for the days of Pericles or Caesar, or even those of the First Empire, when cities were laid out with beauty and effect and were exquisite settings for noble gems of architecture. (Taylor, 1893)

As in the United States, a variety of leadership types were represented. The Toronto plan of 1905 was a partnership between the city's architects who had initiated the formation of the Ontario Association of Architects (1889) and the city's elite who had formed the Toronto Guild of Civic Art (1897). The Guild was made up mostly of businessmen who were leaders of a new corporate elite and who also were patrons of art and learning. They funded the project; the architects drew it up. Guild members included Byron Walker, the head of the city's chief bank, the Bank of Commerce, who argued that 'it is not a question of the city beautiful, it is just a question of common sense . . . We do not always want to remain a wooden backwoods place with narrow provincial ideas. We aim to be cosmopolitan, to have a larger outlook' (Walker, 1905-6, p. 139). This plan and a subsequent proposal now seem politically naive, for no effort was made to get municipal government implementation. Guild members finally pressed the City Council to create a Civic Improvement Committee, made up of members of the Council and the public (mainly the Guild executive) to pursue this kind of planning. This committee chose John Lyle to draw up a new plan in 1911. Like previous Toronto plans, this was widely discussed in the newspapers and interested members of the public gathered in Clubs to hear and discuss the latest proposals.

In several western Canadian cities, variations on the Toronto theme were introduced in efforts to invoke 'the noble idea of city planning.' Winnipeg's City Council took the lead in producing a comprehensive report in 1911, through the agency of a City Planning Commission, a body similar to those in several other western cities such as Calgary, Regina, Edmonton and Saskatoon. Vancouver's initiatives were more populist, similar to many in the United States. The Local Council of Women inspired a movement, which included prominent members of the Board of Trade, to form the Vancouver Beautiful Association. Their immediate goals were quite modest: planting trees and window boxes, securing City Council support for a downtown civic park, and providing a forum for Thomas Mawson's proposals for beautifying Stanley Park (Roy, 1980).

A plan for Ottawa, the nation's capital, had origins similar to the plan for Washington (figure 6.5). In 1913, Prime Minister Robert Borden appointed a Federal Plan Commission chaired by Herbert Holt, a railway executive and Conservative Party organizer. The selection of Edward Bennett, Burnham's partner, as planner evoked considerable criticism, partly because he was considered to be an American, although British by birth and Ecole des Beaux-Arts trained. Bennett was certainly the best known name in city beautiful circles after Burnham, and displayed his organizational as well as aesthetic skills by the

Figure 6.5. The proposed municipal plaza of Ottawa, 1915. This drawing by Jules Guérin pictured possible changes to Ottawa as described in E.H. Bennett's plan. The existing Parliament Buildings are at the upper left. (*Source*: Federal Plan Commission, 1915)

preparation of a very impressive volume, complete with drawings by Jules Guérin (Federal Plan Commission, 1915).

The coming of the war was a major blow to most Canadian urban planning schemes, including those for Ottawa. Equally damaging was the high-placed criticism of several authorities with garden city credentials. In Ottawa, a local engineer, Noulan Cauchon, had hoped to get the commission and attacked the Bennett plan in a series of newspaper articles (Gordon, 1998). At the national level, the opposition of Thomas Adams was crucial. Adams was one of the leading garden city advocates, and he had been appointed as Town Planning Advisor to the Federal Commission of Conservation in 1914. He argued that the city beautiful movement was 'merely an aesthetic fad,' and suggested that his own approach would 'produce beauty without seeking beauty as an end in itself' (Adams, 1920, p. 35). Ironically, the garden city approach generally led only to suburban planning in Canada, and the problems of the central city were ignored. The professional planners who organized as a national group in the Town Planning Institute of Canada in 1919 were sympathetic to the garden city principles. Professional planning thus came to be equated with suburban planning. By this time, planning literature had become highly technical and inaccessible to the public. Judging by the coverage of planning questions in the local press and in periodicals such as the popular *Canadian Magazine*, the public apparently had lost interest, and control as well.

A THEORY OF THE CITY AS AN ORGANISM

For both celebrants and detractors of the city at the turn of the century, the analogy often used to describe the city was that of 'organism'. The organic or biological conceptions of the city had been developed in Britain into the equivalent of a scientific theory and had become a favourite tool of those promoting conservative and free-market ideas (Davison, 1983). City beautiful enthusiasts, however, used the concept of the city as a living organism to explain the possibility of curing or fixing it. To Canadian social reformer, J.S. Woodsworth, a city was 'an immense and highly developed organism in which each minutest part has a distinct function' (Woodsworth, 1911, p. 22). This kind of general theory may be simplistic, but it may also give meaning to human experience. The point here is that this kind of organicism provided a view of the city as more than just an agglomeration of individuals and buildings, but as an entity that must be manipulated and designed as a whole (Auster, 1990).

The analogy of the city as an organism, perhaps a diseased organism that could be treated, had at least two major planning implications. The first was that the organism had a heart which should be strengthened and made more easily identifiable. At the hearts of the new Paris and Rio de Janeiro were symbols of bourgeois luxury and culture, the opera house. In more democratic North America, the new heart of the city was the civic centre, the emblem of local autonomy and the public interest in local decision-making. Most proposals involved the construction of a formal ensemble of buildings to house the operations of local politicians and bureaucrats. Stateliness and dignity would best be achieved by the use of neoclassical architectural styles. The horizontal orientation of these public buildings would distinguish them from the more vertically oriented temples of commerce. These visible manifestations of local, public identity would promote civic pride and a sense of community. Some of the most highly publicized American civic centres were those in Cleveland, San Francisco, and Denver, while that of Chicago was influential in planning circles even though it never left the proposal stage. As many as seventy-two civic centre plans were drawn up but the majority of these were not built (Wilson, 1989, p. 294).

At another political scale, that of the capital city, the same principles applied, as in the Mall in Washington, the Parliament Buildings in Ottawa, and the government core of Canberra. In Washington, in particular, the scale was national, not municipal, with the lengthy, two mile Mall tying together the monumental federal institutions and the country's main memorials. The plan is best appreciated from the air or on maps, for the immensity of the place defies comprehension on the ground. This was not designed for the individual citizen, but rather for the citizen in the aggregate, for the hundreds of thousands, for major national occasions.

In Canada, elaborate civic centre plans were drawn up by local architects for Toronto and Maisonneuve, and by Thomas Mawson for Calgary and Regina, complete with stunning paintings of neoclassical ensembles. Less complete proposals were put forward for Montreal, Winnipeg, Edmonton, Hamilton, London, Windsor, and possibly several more, indicating the ubiquitous nature of this idea. Two examples will give an indication of the nature of these proposals. In Toronto, Lyle argued that his 1911 plan would be 'not only aesthetic in character but practical to a high degree', and it would give 'a head, a heart, and a pulse to Toronto' (Lyle, 1911, p. 56). An equally elaborate civic centre scheme was proposed and constructed in Maisonneuve, a new suburban city created just to the east of Montreal, between 1910 and 1915. Maisonneuve was conceived by local landowners and widely promoted as a city beautiful project; unlike most other schemes, it was not a reshaping of an existing city but the construction of a new one on largely open space. Four public buildings formed the core complex – a town hall, a market, and public baths, all in the neoclassical style, and a fire station modelled on Frank Lloyd Wright's Unity Temple in Oak Park, Illinois (Linteau, 1985).

A second implication of the city as an organism was the perceived need for comprehensive rather than piecemeal planning. At one scale this involved a synthesizing of beauty and utility which was generally spelled out in the introductions to many plans. At another, it meant dealing with many aspects of a city's problems – beautifying the ugly, creating a splendid civic centre, improving the transportation system, and establishing a system of parks and playgrounds. The comprehensive planning of the era usually did not include detailed recommendations for housing, the major concern of the garden city advocates. An example of how city beautiful planners approached the concept of comprehensive planning was the work of Thomas Mawson, a promoter of the notion of town planning as civic art in many projects in Britain and Canada. In what was acclaimed as 'the first great work of 20th Century European city planning' (quoted by Cherry, Jordan, Kafkoula, 1993, p. 324), Mawson produced a plan for Thessaloniki in Greece in 1918. This plan was based on a detailed economic and social survey and included proposals for transportation links, water and sewage networks, green space for parks, gardens and playgrounds as well as the expected monumental structures and diagonal streets.

In the United States, some of the best known city beautiful plans were comprehensive in scope. Jon Peterson argues that the plan for Washington was the first and most influential comprehensive city plan in the United States, fusing 'all forms of physical design into a single, coordinated plan applicable to the whole city' (Peterson, 1985, p. 134). The plan included a union railroad terminal, a bridge across the Potomac River, major sites for public buildings, the eradication of a slum area, a public recreation complex, a vast system of parks and parkways, and a scenic preservation programme. Burnham's Chicago plan is generally regarded as the mature statement of the comprehensive plan idea. The most publicized aspect of the plan was the grand civic centre, which would stand as the symbol of civic order and unity, but the plan went far beyond this to incorporate a concept of the metropolitan region (up to 60 miles from the city's core) linked to the city by improved passenger and freight rail services, great diagonal avenues, and a huge lakefront park scheme.

The Chicago plan directly influenced subsequent Canadian plans. An example was the plan produced in 1911 by Winnipeg's City Planning Commission which included the usual components – a civic square, widened streets, parks – but was unusual in that it included a detailed analysis of the shortage of workers' housing and proposals for the building of model suburbs for workers (City of Winnipeg, 1911). Edward Bennett's plan for Ottawa was equally wide-ranging, although it did not deal with the housing problem directly. Essential to its aesthetic considerations was a central core made up of the existing federal government buildings and a municipal presence through a city hall. The 'practical' aspects were a rationalization of the railway system and a central station; improved traffic circulation through street widening and the opening of some short diagonal streets; the continuation of an improvement programme for parks and a parkway system; while district development control, an early term for zoning, would concentrate industry in segregated areas, with residential areas protected from the intrusions of industry or commerce.

As late as 1927, Montreal sociologist C.H. Dawson was using the 'city as organism' analogy to champion the idea of a comprehensive plan (Dawson,

1927). It is difficult to know if the analogy was taken literally, that is, that the city could be regarded as *being* a natural organism, rather than being *like* one. The use of the analogy, nevertheless, appears to have aided in the development of a more systematic and holistic approach to the city, both during the height of the city beautiful's influence and later. Harland Bartholomew's later comments provide a fitting climax: 'Unfortunately,' he wrote in 1943, 'most of these first comprehensive city plans were considered too visionary or impractical or were misconstrued to be schemes chiefly for beautification of the city. Their more fundamental objectives, to correct mistakes and bring about a more orderly growth, were seldom appreciated' (quoted by Wilson, 1989, p. 299).

Success and Relevance

Of the two great alternative approaches to urban planning in the twentieth century, that of the garden city has obviously been more influential than that of the grand manner and its city beautiful segment. Planned and unplanned decentralization has been the hallmark of the century. The ultimate goal for most people, evidently, has been a life centred on a home set in green, low-density surroundings. Less appealing has been the idea of the city in a historic sense as a concentration of people and institutions who benefit by constant interaction. In this concluding section, I will assess the success of planning in the grand manner and discuss the relevance of this approach for the way we now plan cities.

First, how successful was the grand manner in rebuilding existing cities or in building new ones? There is no doubt that many of the grand plans remained unrealized. The reasons for these failures have been discussed at length by many critics and include the enormous costs involved, the greater pressure of other priorities, the lack of political will, and especially the strength of a privatistic ideology. And yet, there was an incredible amount of urban reconstruction and construction in this vein. As described in earlier sections, many European cities followed the lead of Paris, hoping to provide a unified appearance and a sense of social order and national pride. This impulse could be magnified and distorted to gross levels as demonstrated by Mussolini in Rome and in Hitler's plans for Berlin. Outside of Europe, this approach was often part of a larger desire to westernize, as in the Muslim world, Asia, and South America. The two great approaches were not entirely exclusive, of course, for the grand manner and the garden city were often combined in some interesting hybrids. Notable examples were Edwin Luytens' New Delhi, Griffin's Canberra, and even Brasilia, with its consciously modernistic architecture.

William Wilson has brilliantly demonstrated that the American city beautiful movement was a great deal more successful than its critics have charged. He cites the completion of projects in Denver, Kansas City, Harrisburg, Dallas, and other cities, and partial victories in San Francisco and even Chicago, where many elements of Burnham's plan were eventually realized. Wilson also points to less spectacular beautification goals that were achieved in hundreds of places, from putting utility wires underground, to improving street furniture, restricting signage, and the extraordinary expansion of park, parkway and boulevard systems (Wilson, 1989).

The Canadian balance sheet seems less impressive, for the only large-scale scheme completed during this period was the construction of the new town of Maisonneuve. The vast public expenditures involved bankrupted the town, and a reluctant Montreal was forced by the province of Quebec to annex it. The picture looks somewhat brighter, on the other hand, if we include postwar developments such as partial completions, hybrids which were combinations with garden city principles, or even later comprehensive plans that were influenced by city beautiful ideas. Toronto's earlier plans had been 'lost in management', in James Lemon's phrase (Lemon, 1989), but by 1929 a new proposal for Parisian avenues and places was brought forward. Because of the Depression, only parts of this plan, such as the embellishment of University Avenue, were eventually completed, leaving unresolved some major design issues (Greenberg, 1981). In Regina, Thomas Mawson's ambitious plan for an elaborate civic centre and an enhanced cityscape were partially implemented because succeeding planners shared some of Mawson's vision (Brennan, 1994).

Hybrids of the Canberra type were quite common, such as the 'Model City' of Mount Royal in Montreal, a corporate suburb built by the Canadian Northern Railway. The parentage of the plan seems quite clear: city beautiful ideas were promoted in Montreal in 1910 by Frederick Law Olmsted, Jr., and garden city ideas by Henry Vivian, a British housing activist then touring Canada. As planner, the railway company chose Frederick Todd, a former assistant of Olmsted's, who was enamoured by both approaches (McCann, 1996). Ironically, one of the chief producers of hybrid plans was the redoubtable Thomas Adams, as in his plan for Kipawa (now Témiscaming) in northern Quebec in 1917 (Saarinen, 1986), and in other designs that often bore a strong hint of the grand manner (Bloomfield, 1982).

Some of the later comprehensive plans for Canadian cities illustrated the continuing attraction of the city beautiful approach, even though the term no longer continued to be used. Harland Bartholomew's Vancouver plan of 1928 projected a civic centre, street widenings and openings, parks and a strong zoning system (Bartholomew, 1928). Much of this was eventually implemented (Perks, 1985). In Ottawa, some of the main components of the Bennett plan re-emerged when the Gréber plan of 1950 was officially adopted (Taylor, 1989).

Even if the movement was partially successful, how relevant for current planning are the three principles explored here? Certainly they would appear to run counter to much of the prevailing ideology and practice of twentieth century urbanization. The first basic principle was a celebration of the idea of the city itself. Cities and civilization were equated; humans were at their most creative and productive in a city setting. The implications for planning were that the solutions to the urban crisis were to be focused on the reform of the city itself, rather than a withdrawal to new towns and the abandonment of the city. An important aspect of this first principle was the emphasis on the city as a work of art, a place of beauty, reflecting in physical form the highest aspirations of a society. The reality of cities in the industrial age contrasted sharply with this vision, as do the banal and ugly cities of our day. Closely associated with the idea of the city as a work of art was a strong sense of history. Even the relatively young cities of the New World could symbolically connect with their civilization's past by reviving classical traditions of planning and building. Most of the twentieth century, however, has been dominated by a modernist rejection of history, perhaps most extremely exemplified by Le Corbusier's *Plan Voisin* for Paris in 1925, which in theory at least proposed the razing of the existing city and its replacement by a series of uniform towers. Our present age appears to reflect a greater appreciation of the past and a recognition that heritage is a significant dimension of our urban identity.

A second principle was that of broadening the base of decision-making in urban planning. This applied primarily to the movement in North America, and especially to the United States where it was associated with urban political reform and civic activism. The experts, often architects, shared the stage with commercial-civic elites. As professionalism in planning increased, the role of the layman was slowly diminished. In Canada, the professional expert and property interests dominated the planning of cities after the decline of the city beautiful movement. After World War Two, planning became bureaucratized through city halls, seemingly impervious to lay activists. With the advent of populism in civic politics since the 1970s, planning has again become a political as well as a technical act, the product of negotiated outcomes between competing interests (Smith and Moore, 1993).

A third principle was that of looking at the city as a whole, as an organism that could not be cured on a piecemeal basis. These ideas were formulated in the age of the industrial city and were based on the assumption that the central city should have precedence over the periphery. The basic trend everywhere during the twentieth century, however, has been that of decentralization, the development of multi-nucleated metropolitan regions not dominated by a central city. Historians have attributed this phenomenon to public demand, but also to planning policies stemming partly from the garden city tradition, involving transportation, the location of industry, the provision of open space, and the reduction of required residential densities (Sies and Silver, 1996). Some observers now argue that the problems of sprawl and deconcentration cannot be solved without a strong central city as the heart of region.

In retrospect, the legacy of the city beautiful movement seems mixed at best, yet its accomplishments were considerable and deserve more recognition from planning historians. And its ideals merit serious consideration by planners seeking to re-establish a sense of order and beauty into our urban world.

REFERENCES

Adams, T. (1920) The future of Ottawa. *Town Planning and the Conservation of Life*, **6**(2), pp. 33-37.

Auster, M.L. (1990) 'Not without art, but yet to nature true'. A re-evaluation of organic thought in urban planning and geography. *Australian Geographical Studies*, **28**, pp. 178-86.

Bartholomew, H. (1928) *A Plan for the City of Vancouver, British Columbia*. Vancouver.

Bloomfield, E. (1982) Reshaping the urban landscape? Town planning efforts in Kitchener-Waterloo, 1912-1925, in Stelter, G.A and Artibise, A. (eds.) *Shaping the Urban Landscape: Aspects of the Canadian City-Building Process*, Ottawa: Carleton University Press, pp. 256-303.

Brennan, W. (1994) Visions of a city beautiful: The origin and impact of the Mawson plans for Regina. *Saskatchewan History*, **46**, Fall, pp. 19-33.

Burnham, D. and Bennett, E.H. (1909) *Plan of Chicago*. New York: Da Capo Press reprint (1970).

Çelik, Z. (1986) *The Remaking of Istanbul: Portrait of an Ottoman City in the Nineteenth Century*. Seattle: University of Washington Press.

Cherry, G., Jordan, H. and Kafkoula, K. (1993) Gardens, civic art and town planning: The work of Thomas H. Mawson (1861-1933). *Planning Perspectives*, **8**, pp. 307-332.

City of Winnipeg (1911) City Planning Commission, in Artibise, A. (ed.) (1979), *Gateway City: Documents on the City of Winnipeg, 1873-1913*. Winnipeg: Manitoba Record Society Publications, pp. 225-264.

Davison, G. (1983) The city as a natural system, in Fraser, D. and Sutcliffe, A. (eds.) *The Pursuit of Urban History*. London: Edward Arnold, pp. 349-70.

Dawson, C.A. (1927) The city as an organism, reprinted in Rutherford, P. (1974) (ed.) *Saving the Canadian City, 1880-1920*, Toronto: University of Toronto Press, pp. 275-282.

Federal Plan Commission (1915) *Report of the Federal Plan Commission on a General Plan for the Cities of Ottawa and Hull*. Ottawa.

Freestone, R. (1996) Sulman of Sydney: Modern planning in theory and practice, 1890-1930. *Town Planning Review*, **67**, pp. 45-63.

Fujimori, T. (1987) Urban planning in the Meiji Era. *Japan Echo*, **14** (Special Issue), pp. 45-49.

Gordon, D. (1998) A City Beautiful Plan for Canada's capital: Edward Bennett and the 1915 Plan for Ottawa and Hull. *Planning Perspectives*, **13**, pp. 275-300.

Greenberg, K. (1981) Toronto: The unknown grand tradition. *Trace: A Canadian Magazine About Architecture*, **1**(2), pp. 37-47.

Harvey, D. (1985) *Consciousness and the Urban Experience*. Oxford: Blackwell.

Hines, T. (1979) *Burnham of Chicago: Architect and Planner*. Chicago: University of Chicago Press.

Jinnai, H. (1995) *Tokyo: A Spatial Anthropology*. Berkeley: University of California Press.

Jordan, D. (1995) *Transforming Paris: The Life and Labors of Baron Haussmann*. Chicago: University of Chicago Press.

Kostof, S. (1991) *The City Shaped: Urban Patterns and Meanings Through History*. Boston: Little Brown.

Langton, W. (1905) Editorial. *Canadian Architect and Builder*, **18**, p. 18.

Langton, W. (1906) The plan of improvements to Toronto. *Canadian Architect and Builder*, **19**, p. 26.

Lemon, J. (1989) Plans for early 20th-century Toronto: Lost in management. *Urban History Review*, **18**, June, pp. 11-31.

Linteau, P.-A. (1985) *The Promoter's City: Building the Industrial Town of Maisonneuve, 1883-1918*. Toronto: Lorimer.

Lyle, J. (1911) Proposed federal and municipal scheme for Toronto. *Construction*, **4**, July, pp. 51-53.

McCann, L.D. (1996) Planning and building the corporate suburb of Mount Royal. *Planning Perspectives*, **11**, pp. 127-153.

Needell, J. (1987) *A Tropical Belle Epoque: Elite Culture and Society in Turn-of-the-Century Rio de Janeiro.* Cambridge: Cambridge University Press.

Nobbs, P. (1906) The official architecture of European capitals. *Canadian Architect and Builder*, **19**, March, pp. 38-40.

Nobbs, P. (1911) City planning. *Canadian Municipal Journal*, **7**, April, pp. 140-141.

Olsen, D. (1986) *The City as a Work of Art: London, Paris, Vienna.* New Haven: Yale University Press.

Peisch, M. (1964) *The Chicago School of Architecture: Early Followers of Sullivan and Wright.* New York: Random House.

Perks, W. (1985) Idealism, orchestration and science in early Canadian planning: Calgary and Vancouver re-visited, 1914-1928, *Environments*, **17**(2), pp. 1-28.

Peterson, J. (1985) The nation's first comprehensive city plan; A political analysis of the McMillan plan for Washington, D.C., 1900-1902. *Journal of the American Planning Association*, **51**, pp. 134-150.

Reff, T. (1982) Manet and the Paris of Haussmann and Baudelaire, in Sharpe, W. and Wallock, L. (eds.) *Visions of the Modern City*. New York: Heyman Center for the Humanities, pp. 131-63.

Robinson, C. (1901) *The Improvement of Towns and Cities, or, The Practical Basis of Civic Aesthetics*. New York: Knickerbocker Press.

Roy, P. (1980) *Vancouver: An Illustrated History.* Toronto: James Lorimer.

Saarinen, O. (1986) Single-sector communities in Northern Ontario: The creation and planning of dependent towns, in Stelter, G.A. and Artibise, A. (eds.) *Power and Place: Canadian Urban Development in the North American Context*, Vancouver: University of British Columbia Press, pp. 219-264.

Scobie, J. (1974) *Buenos Aires: Plaza to Suburb, 1870-1910.* New York: Oxford University Press.

Scott, M. (1969) *American City Planning since 1890.* Berkeley: University of California Press.

Sies, M. and Silver, C. (1996) Planning history and the new American metropolis, in Sies and Silver (eds.) *Planning the Twentieth-Century American City*. Baltimore: Johns Hopkins University Press, pp. 449-473.

Smith, C. (1994) *Urban Disorder and the Shape of Belief: The Great Chicago Fire, the Haymarket Bomb, and the Model Town of Pullman.* Chicago: University of Chicago Press.

Smith, P.J. and Moore, P. (1993) Cities as a social responsibility: Planning and urban form, in Bourne, L.S. and Ley, D. (eds.) *The Changing Social Geography of Canadian Cities*. Montreal/Kingston: McGill/Queens University Press, pp. 343-366.

Taylor, A. (1893) Brief notes on the architecture of the World's Fair. *Canadian Architect and Builder*, **6**, October, p. 104.

Taylor, J.H. (1989) City form and capital culture: Remaking Ottawa. *Planning Perspectives*, **4**, pp. 79-105.

Tomovcik, V. (1977) The Greber Plan for Ottawa. M.A. thesis, Planning, University of Waterloo.

Van Nus, W. (1984) The fate of City Beautiful thought in Canada, 1893-1930, in Stelter, G.A. and Artibise, A. (eds.) *The Canadian City: Essays in Urban and Social History.* Ottawa: Carleton University Press, pp. 167-186.

Walker, B. (1905-6) A comprehensive plan for Toronto. *Canadian Club Proceedings, Toronto*, **3**, pp. 134-9.

Wilson, W. (1989) *The City Beautiful Movement.* Baltimore: Johns Hopkins University Press.

Woodsworth, J.S. (1911) *My Neighbor: A Study of City Conditions, A Plea for Social Service.* Toronto: University of Toronto Reprint (1972).

Chapter 7

The Neighbourhood Paradigm: From Garden Cities to Gated Communities

Dirk Schubert

Hopes have long been raised by the planning idea of 'the neighbourhood' – the idea that manipulation of the local built environment and social interaction can positively influence urban space and foster community life. The concepts of neighbourhood have no clear definition and are vague socially and spatially. The terms 'neighbourhood unit' and 'neighbourhood concept' tend to denote planning and built aspects, whereas 'neighbourhood relations' is concerned with social interaction. This 60 year old vision is undergoing a renaissance; stripped of its true meaning and often used as an empty phrase, neighbourhood terminology is again found in urban development and planning literature, as well as real estate brochures.

In this chapter, some of the meanings of neighbourhood in planning terms will be examined in their historical context. The origins of the idea and its development and reception in England, the United States and Germany will be considered. The relevance of neighbourhood theory as a planning concept for the new millennium in times of globalization and increasing social disintegration is also examined.

Decentralization, Urban Reform, and Planned Communities

The phenomenon of large cities in Europe and North America has proven permanent and irreversible since the mid-nineteenth century. Problems of hygiene in cities were resolved and national and international organizations and institutions concerned with housing, urban hygiene and urban development grew rapidly (Sutcliffe, 1983). Reformers motivated by social pedagogy, religion and philanthropy tried to support the poor financially and practically. In the slums of London's East End, social reformers such as Reverend Samuel Barnett, created social 'settlements' with neighbourhood centres like Toynbee Hall. Barnett, his wife and others wanted to mix the poor people of the slums with those who were better off in the hope that they would learn from each other (Schubert, 1998). This movement had many followers and before World War One, England had over 50 such settlements.

From around the turn of the century, social scientists, especially in Germany, identified the phenomenon of 'losing one's roots' in large cities. The German sociologist Ferdinand Toennies made an important distinction between community (*Gemeinschaft*) and society (*Gesellschaft*), the latter dominant in the post-industrialization period. Toennies defined community as reliant on blood ties, neighbourhood and friendship, whilst social interaction was based on evaluation of advantages, disadvantages, and expectations of reward. Family, clan, village and friendship are forms of communities, whereas city and state are categories of society. 'In large cities, that is in capitals and in the metropolis, the family is in decay . . . Large cities typify society as such . . . Therefore the city and the condition of society is the decline and death of the people', wrote Toennies (1887, pp. 244, 242, 246), laying the foundations for a hostile perception of large cities.

The American sociologist Charles Horton Cooley developed and based his differentiation of 'communal' (primary) and 'social' (secondary) groupings on Toennies' work. But the model was most influential in the Chicago School of human ecologists in the 1920s. The Chicago School was concerned with organic relationships between human communities and their physical environment (Blowers, 1973). Sociologists, like Robert Park and E.W. Burgess, did extensive survey research on the city, social organization, 'natural areas', 'community units', and how to establish neighbourhoods as an important part of urban life. Urbanization in this sense was seen as tearing people from the land, and cities were represented as the dangerous massing of disconnected individuals. Stanton Coit (1891) published work based on his experience in London and New York that encouraged the theory of neighbourhood guilds. The ideas of neighbourhood guilds, the settlement movement and similar concepts were developed to organize 'the social life of all people in one small district [bringing] neighbors together, families together, different interests together' (Bliss, 1908, p. 821).

Social problems did accumulate in large city environments, but many solutions were also developed as they became laboratories for the latest economic, social and cultural trends. Fear, inexperience and insecurity would be challenged, if not replaced, by experience and knowledge. The damage done by cities came to be regarded as 'curable'. By the end of the nineteenth century, several reforms had fused into one. The housing reform movement dealt with bad housing conditions and tried to improve these mainly by controls and legislative measures. Model flats and model estates were to show realistic alternatives. The discipline of urban planners largely sought to control the processes of urbanization spatially. Lower densities and structuring of the urban configuration became a standard approach. The profession was in the forefront of the search for ordered, structured, zoned environments to replace the chaos of the unplanned city. The main opportunity arose when problems of ownership and compulsory purchase in built-up areas became insurmountable and the planning of new settlements on the urban periphery commenced.

Both the housing and planning movements were unified by the aim of decentralization. Suburbanization was one means; the more radical step was the planning of new towns with their own places of work and service institutions outside the sphere of influence of large cities. The garden city movement tried to solve rural and urban problems of employment and housing simultaneously by clearing away inner city slums. It also combined various ideals, such as land reform and the co-operative movement. Although the concepts of reform had differing priorities, they were all meant to be synthesized in practice. This would illustrate the possibilities of a new spatial order, better housing, healthier cities and living conditions, which would in turn lead to a better civilization. The subtitle of Ebenezer Howard's book *To-Morrow: A Peaceful Path to Real Reform* (1898) is an accurate summary. Howard's vision was of 'social cities', a comprehensive socio-spatial reform achievable without revolution. With so many theoretical ideas of decentralization, model communities, lower urban densities, slum clearance and housing reform in the air, actual physical experimentation was inevitable.

Before World War One, employers in Germany had built factory housing, which served as model settlements for their employees. Often these paternal employers were open to ideas of housing reform. Industrialists, such as Alfred Krupp in the Ruhr district, were celebrated as 'fathers of the new settlements'. Suggestions for housing and land reform came from diverse and sometimes unpalatable sources. Two years before Ebenezer Howard, the anti-Semite Theodor Fritsch, had said in his book *Die Stadt der Zukunft* how the severe damage of large cities can be rectified through garden cities with different zones and neighbourhoods (Schubert, 1982). Whilst Howard was concerned with social reform, Fritsch criticized (Jewish) land speculation and tried to use his new town concept to advance the 'renewal of the German race'.

Factory estates like Paterson in New Jersey, Humphreysville in Connecticut, and Pullman in Chicago were built in the United States (Crawford, 1995). The integration of estate and landscape with curving residential streets had been mastered in the planning of Riverside (1869) in Chicago by Frederick Law Olmsted and Calvert Vaux. In 1913, the City Club of Chicago held a competition for neighbourhood centres. Drawing on theories of settlement, decentralization and urban expansion, the architect William E. Drummond had developed a concept of neighbourhood units by 1912. He defined neighbourhood units as physical and social fabric strictly in the context of Daniel Burnham's Chicago Plan and the local urban improvement scene (Johnson, 1998).

In England, new garden settlements were occasionally built. They were often philanthropically motivated, like the industrial housing estates of 'gentlemen reformers' like William Lever (Port Sunlight) and George Cadbury (Bournville). The founding of Letchworth (1903) and later Welwyn Garden City (1919) proves that the idea of the garden city could be turned into reality. But the magnetism of large cities prevailed, and the garden suburb emerged as a more realistic vehicle for the spatial reform of community. One example serving as a widely used model was Hampstead Garden Suburb (1906) in London, planned by Raymond Unwin and Barry Parker.

AMERICAN EXPERIMENTS IN THE 1920S AND 1930S

Unlike Britain, city planning in the United States was primarily based on the city beautiful, 'city scientific' concepts, and private initiatives. The issue of planned urban expansion was explored primarily in a suburban context with garden city overtones. The Russell Sage Foundation formed the Sage Foundation Homes Company to implement a model development project, Forest Hills Gardens in Queens, New York, between 1908 and 1917. Frederick Law Olmsted Jr. replaced the original rectangular grid with curving streets, following the Letchworth precedent. The relatively high cost of land led to the construction of expensive housing and the idea of a socially mixed settlement was relinquished for economic reasons. In contrast to prewar England, postwar United States saw two important changes which influenced planning. The increase in income made it possible for a larger segment of the population to finance a home and growing mobility afforded by the automobile made it possible for many citizens to fulfil the dream of the 'American way of life' on the urban periphery.

Much of the experience of planners in the United States had been obtained in the context of war-induced housing programmes (Scott, 1969). They then aimed to continue this form of unitary planned development motivated by housing and social ideals. In 1923, the Regional Planning Association of America (RPAA) was formed in New York through the driving force of Henry Wright (Lubove, 1963; Schaffer, 1992; Sussman, 1976). An informal interdisciplinary 'think tank' of no more than a dozen individuals, including Catherine Bauer Wurster, Lewis Mumford, Benton MacKaye, Frederick Ackerman, Stuart Chase, Edith Elmer Wood and Clarence Stein, discussed the idea of state-wide regional planning, low-income housing and, especially, the concepts which could lead to the construction of such a settlement representing their goals. Anticipating increasing social disintegration in the metropolis, problems of unplanned sub-urbanization, and urban sprawl, the RPAA met these with a vision of planned decentralization by regional planning and the establishment of neighbourhoods.

In 1923, a RPAA delegation including Clarence Stein, Henry Wright, and real estate developer Alexander M. Bing, visited Ebenezer Howard and Raymond Unwin to see English garden city projects. They returned converted and decided to create an American version. In 1925, the RPAA hosted the first International Conference of the International Federation for Town and Country Planning and Garden Cities outside Europe. Howard, as well as Unwin, Barry Parker and Patrick Geddes joined the conference and endorsed RPAA ideals. In 1924, the RPAA founded the City Housing Corporation, which was, with limited assets, to master the synthesis of theory and practice to produce an American model settlement. In 1924, the corporation bought a piece of land in Queens and Stein and Wright began to plan and build Sunnyside Gardens, constrained by the pre-existing grid street system (Stein, 1966). In 1928, the project was completed, including about 1200 housing units and garden areas inside the blocks. A community centre was bought and the Sunnyside Association organized communal activities. Residents of this planned garden community included Lewis Mumford, popular singer Perry Como, jazz musician Bix Beiderbecke, and other intellectuals and artists.

Also in the early 1920s, a private organization, the Advisory Commission on City Planning, commenced work on a Regional Plan of New York and Its Environs to be sponsored by the Russell Sage Foundation. Thomas Adams, with experience in the British garden city movement, assumed the responsibility of 'General Director of Plans and Surveys' (Johnson, 1996). While working on the Regional Plan of New York, Clarence Perry (a member of the RPAA) formulated the first definitive expression of the neighbourhood unit. Perry predicted the need for new urban planning solutions because of the increase in private car ownership: 'The cellular city is the inevitable product of the automobile age' (Perry, 1929, p. 31). He identified six key principles revolving around size, boundaries, open spaces, institution sites, local shops, and the internal street system. He set a ceiling population of 5000 for an area surrounding an elementary school, placing services required on a daily basis within easy walking distance along streets on the edge of the estate, re-routing through-traffic, and segregating modes of transportation (figure 7.1). Perry had tried to transpose the positive experience of the settlement movement – with its strong social links and networks seen as strengthening a sense of community – to a flexible planning concept for built-up areas and new developments (Silver, 1985). Some technical planning principles which had been an integral element of urban design textbooks since the nineteenth century resurfaced in a new package. Perry lived in Forest Hills, and also drew from this positive personal experience in developing his ideas. He said that the social mix of the population in the neighbourhood units should be 'a wide range of income classes', but meant a socially homogeneous population (Silver, 1985).

In 1928, after completion of Sunnyside Gardens, the City Housing Corporation bought a site in Fair Lawn, New Jersey to translate Perry's theoretical framework into built reality as Radburn. Stein and Wright were again

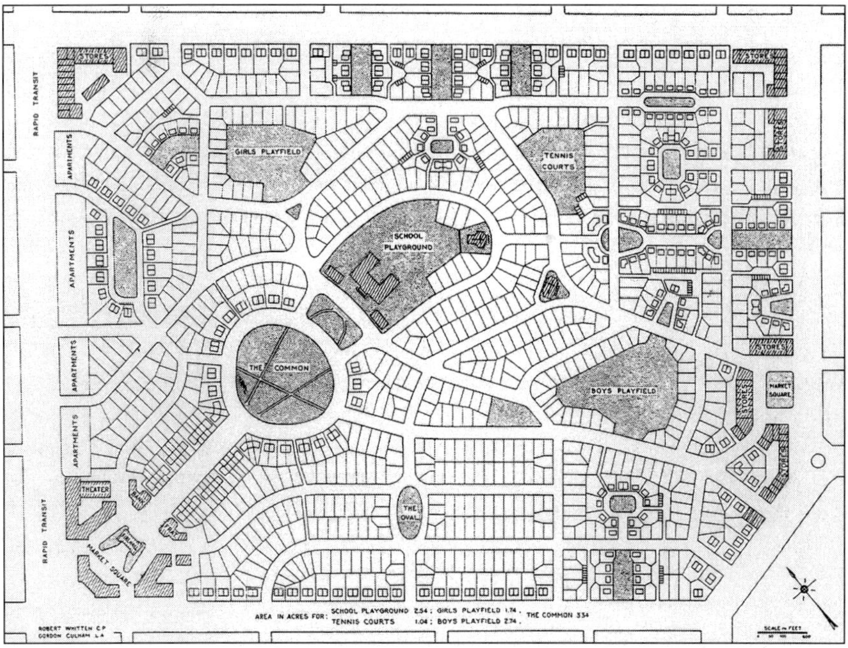

Figure 7.1. Clarence Perry's neighbourhood unit. This version emerged from the Regional Plan of New York and Its Environs (1929). The conventional grid street system is eschewed as the neighbourhood becomes a self-contained suburban subdivision. (*Source*: Perry, 1929)

responsible for the planning and architecture of the estate. Not one element of the Radburn plan was truly new, although Lewis Mumford praised the plan as 'the first major departure in city planning since Venice' (1975, plate 51). It was a (sub)urban model which promoted communal lifestyles and was meant to meet modern demands. The transport concept was the main innovation of Radburn. Car ownership in the United States rose from 1.2 million in 1914 to over 23 million in 1930. By 1926, the number of cars per household was eight times that of Great Britain and 35 times that of Germany (Hass-Klau, 1990). The road system laid out on a grid in suburban areas proved increasingly disadvantageous with mass-motorization. Road accidents and deaths gave rise to the idea that pedestrians should be better protected from car traffic. A hierarchy of roads was pursued by making residential streets culs-de-sac and further segregating the modes of transport. This concept was adopted worldwide in the planning of suburban settlements (Fagence, 1973; Miller, 1969). It marks the Americanization of the Garden City concept by responding to modernization with regard to demands made by the automobile (figure 7.2).

In May 1929, the first owners moved to Radburn; in autumn came the Wall Street crash. Many residents lost their jobs and had to move. Radburn was never completed and became a victim of the global recession (Schaffer, 1981). Three neighbourhood units with about 25,000 inhabitants had been planned, each built around an elementary school and all clustered around a single high school. The neighbourhoods were designed in a way that children could walk to school. It was not possible to attract industries to Radburn and a proposed green belt was not implemented. The houses were sold off, there being no public or co-operative land ownership.

Figure 7.2. Aerial photo of Radburn (1929). A new suburban model segregating modes of traffic and integrating roads and paths into the landscape.
(*Source*: Stein, 1966)

Radburn's planners had high hopes for the new sense of community, but its inhabitants held predominantly conventional and conservative values. The residents were more educated than the average population and were predominantly middle class; over 70 per cent commuted to New York and New Jersey (Christensen, 1986). 'It should be borne in mind that Radburn is essentially a community and a housing development for families of moderate incomes', wrote Louis Brownlow (1930, p. 7), the President of the International City Management Association and Radburn resident. Residents were organized in the Radburn Citizens Association to discuss questions of communal interest. Radburn would became a Mecca for planners, but the daily lives of the 'Radburnites' conformed on the whole with those of other American suburban housing estates (Birch, 1983).

Although Radburn was soon transformed from urban planning ideal to financial disaster, Roosevelt's New Deal in the 1930s created other opportunities for progressive new town design and the further evolution of the neighbourhood idea as a touchstone of community planning. In 1934, the *National Housing Act* was passed and the Federal Housing Administration (FHA) set up. The degree to which the FHA influenced model housing policies during the following decades cannot easily be overestimated. It advanced suburbanization by building large suburban settlements with neighbourhood units. Neighbourhoods, according to the FHA, were to be socially and racially homogenous to promote a sense of community among the residents (McKenzie, 1994). In 1935, an innovative Greenbelt Towns programme was adopted by the Federal government to create employment, make available cheap housing, and to demonstrate new urban planning concepts. Fifty new towns were planned, later reduced to eight, with three actually built: Greendale (Milwaukee), Greenhills (Cincinnati), and Greenbelt (Maryland). The Greenbelt towns could not be compared to the English Garden Cities, as they were small settlements of less than 1000 households without places of work.

What of the existing city? Rexford Tugwell, Director of Roosevelt's Resettlement Administration, had declared 'My idea is to go just outside centres of population, pick up cheap land, build a whole community, and entice people into it. Then go back into the cities and tear down whole slums and make parks of them' (quoted in Buder, 1990, p. 176). The concept was for problem inner city localities to be rebuilt along the lines of neighbourhood unit theory. Many planners linked redevelopment and new housing estates with slum clearance and demolition of old tenements, but the decentralist urge was again decisive. 'The attack [on slum districts] . . . can be indirectly assisted by the development of model home neighbourhoods in the suburbs just as much as by replanning and rebuilding the slum areas themselves. No direct attack on the slum districts will yield completely satisfactory results' (Adams, 1934, p. 265). Those in control of housing policies, the construction industry, and private community builders continued to emphasize peripheral urban expansion, causing the inner city to be neglected.

NEIGHBOURHOODS IN ENGLAND

During the 1930s, an urban design model became dominant which envisioned low densities, decentralization, and the restructuring of the urban conglomerate with neighbourhood units. The re-organization of the nineteenth-century city was a major concern. It was realized that this could no longer be achieved by simply clearing the inner city slums and developing new housing estates. Principles of structuring large cities were a major theme at the international conference of the International Federation of Housing and Town Planning in London in 1935.

Critics in England complained about the absence of social mix in public housing estates. 'The loss of neighbourhood values has its further bearing on socially disorganized areas' (Tylor, 1939, p. 177). Neighbourhoods composed of only flats for lower income groups were seen to promote youth crime and vandalism. For this reason, estates with a uniform social structure were avoided. New estates were to be developed with neighbourhood units containing social facilities such as schools and other communal institutions. They were to help minimize crime and act positively against forms of deviant behaviour. In this context, different models were discussed, based on the situation found in London. In 1940, the Barlow Report (Royal Commission on the Geographical Distribution of the Industrial Population) suggested a new spatial distribution of the industrial population in Britain and new towns with 'mixed neighbourhoods'. Lower densities in inner city areas were suggested, making re-housing operations necessary. The damage caused by bombing in World War Two gave impetus to the arguments for decentralization, lower densities, and neighbourhood organization. The focus was always on London. From a military point of view, 'London was the weakest place on earth . . . the Achilles heel of Britain and the British Empire', a journalist wrote in 1939 (quoted in Lees, 1985, p. 263).

The MARS plan of London (1942), developed by a group of architects and town planners in private practice, was also based on neighbourhood units to structure the metropolis. Borough units and neighbourhood units, with schools and infrastructure based on public transport corridors, formed the basic idea. 'Only by forming clearly defined units, which in turn are part of larger units, can social life be organized' (Korn and Samuely, 1942, p. 143). The East End of London was to be comprehensively remodelled with modern housing estates along neighbourhood lines.

The neighbourhood unit also formed the central planning element in London's official plan, the London County Council Plan (1943). The plan prescribed extensive action, even in areas which had escaped destruction during the war. It foresaw new dimensions of rebuilding in accordance with the neighbourhood ideal. 'Partial solutions are not sufficient', Forshaw and Abercrombie wrote in the foreword to the plan, redevelopment and slum clearance on a big scale were mandatory. The planning goals were demonstrated using a neighbourhood unit in Shoreditch and Bethnal Green in the East End. The community of Eltham, was clearly structured on these principles. Another plan was made for Bermondsey and the South Bank. The redevelopment areas were to be similar in size to New Towns. They were to have 60,000 to 100,000 inhabitants in neighbourhood units of 6000 to 10,000 people each: 'The composite plans which we have prepared provide a proportion of lofty blocks of flats, spaced well enough apart for groups of trees, with terraced houses dispersed in regular but not monotonous form, the whole interspersed with open space and organically related to the smaller neighbourhood centre and finally the centre of the whole community' (Forshaw and Abercrombie, 1943, p. 9).

Shortly after the war began, discussion grew on how postwar England should look. It was accepted unanimously that large-scale redesign of the cities was necessary. It became evident that town planning would not only play an important role, but occupy the central position in creating a new England. Because of his preliminary work for the County of London Plan, Patrick Abercrombie, who worked for the Ministry of Country and Town Planning, was entrusted with the design of a plan for the Greater London area. Whereas the 1943 plan had concentrated on the administrative area of the London County Council, the new plan covered a larger area within a 30 miles radius of the city centre. One element of the plan of 1943 to be developed further was the concept of organic communities. Abercrombie wrote:

> Both the neighbourhood and the town should be given physical definition and unmistakable separateness, and the population should be socially stable. This stability can largely be achieved by the provision within the community of a variety of houses and dwellings to meet the needs of all population groups . . . We have used the community as the basic planning unit . . . Each community would have a life and character of its own, yet its individuality would be in harmony with the complex form, life and character of its region as a whole. (Abercrombie, 1945, pp. 112-113)

The East End of London, again, served as a model for rebuilding according to modern principles of neighbourhood planning. Abercrombie thought that the buildings and dwellings in the slum areas of the East End which were not destroyed by German bombs should also be demolished. Redevelopment areas were established and the relocation of the population necessary for implementing the idea of modern neighbourhood units was planned.

THE 'LOCAL GROUP AS A SETTLEMENT CELL' IN GERMANY

In 1930s Germany, discussion had a much stronger ideological taint. The National Socialists saw a direct connection between urban planning, physical arrangements, and the '*Volk* without space' ideology. Their urban design concepts drew upon the anti-urban critique of large cities and postulated 'de-densification'. 'The city as the seat of Judaism' and 'place of Marxism', in the words of leading National Socialist ideologist Gottfried Feder, was to be thinned out and brought to order. Feder also suggested the *Volksschule* (elementary school) as a basis for creating a new order. 'This urban organism will be composed of a series of cells, which will be grouped in cell associations within different sub-cores around the centre of the city' (Feder, 1939, p. 19). By means of urban development, the 'health of the body of citizens' was to be achieved. Programmatic statements by the National Socialists called for a decrease of urbanization, or even its reversal in back-to-the-land migration. They were connected to ideas of autarky in an agrarian society, 'blood and soil', population policies, and anti-aircraft defence. But by the late 1930s the Nazis had come to see large cities as a necessary evil.

American and English plans for neighbourhood units had been presented at international conferences and sparked discussion among German planners. Articles by Gurlitt (1929) and Lederer (1930) about Radburn were published in German periodicals. Henry Wright (1932) described his vision of the neighbourhood idea to the German planning community in the journal *Die Neue Stadt*. In 1935, Bruno Schwan also published a map and photographs of Radburn (Schwan, 1935). It was not the Anglo-American vision which raised criticism, but, rather, the 'backward world views' of Western planning that were prohibitive to implementation. But how was the principle of bringing orderliness to the city – of redesigning the Führer-cities and 'developing the East' – to be adopted without merely imitating 'decadent' Western democracies?

There were conflicting aspects in the ideological claims of hostility towards the city and the reality of a highly industrialized armament production, between the image of the idyllic homeland and the ideals of economic modernization. Hitler himself was a (secret) follower of Fordism who was in support of motorization and the construction of motorways. The disintegration of large cities would be accelerated by new settlements for car owners (Hass-Klau, 1990). The conversion of cities to accommodate cars and new settlement concepts was an integral part of the link between political power and spatial planning. Hitler had to postpone mass-motorization as a secondary political goal until the war was won.

The political framework was also transposed to the redesign and expansion of existing towns. The idea of the *Die Ortsgruppe als Siedlungszelle* ('local group as a settlement cell') was a consensus which emerged from Nazi Party theory and planning practice. It adapted neighbourhood theory to National Socialistic ends. The emphasis was placed on Germanic-national origins to link community with kinship, neighbourhood and camaraderie. The ideas of the National Socialist German Workers Party were expanded to establish principles of urban design with local groups as settlement cells (Pahl-Weber and Schubert, 1991).

In Hamburg in 1941, Konstanty Gutschow, an architect in private practice, was responsible for town planning. Gutschow declared: 'The anonymity of the city is the result of an amorphic formation. It is necessary to make it more transparent again, to structure and design it to create order. For neighbourhoods to evolve the settlement units must be clearly set apart' (Gutschow, 1941). The organizational principle of the urban landscape was to follow the political structure of the party. According to this principle, units of the Nazi Party were imitated in the planning of new housing estates to represent a cross-section of German society. A mixture of terraced houses (owner occupied), small blocks of flats and tenement buildings was planned. Dwelling and population density were to be kept low and a maximum of three storey buildings permitted including local group units of 6000 to 8000 people. Although neighbourhood unit planning was the product of various international roots, German planners insisted that the idea offered a typically German solution. Different design models were developed: 'organic' ones, like those of architect Hans Bernhard Reichow, or geometric ones, by Walter Hinsch, with a mix of housing densities and a party building on a central axis. But these plans were never wholly implemented during the Nazi period.

Local groups were also to be used as a structuring element in the conquered eastern zones. The 'local group as a settlement cell' was never built in its pure form, but there are many plans which illustrate the desired appearance. Heinrich Himmler, the Reichsführer of the SS, planned to use this principle to 'secure German national tradition in the new east'. He announced:

> In the design of housing areas, schemes on a massive scale should not be allowed to take over. Instead, homely settlements for the promotion of the common good should be created in the interest of urban design . . . The criteria for the structure of housing areas, with a view towards developing the community, can be drawn from the same source which guides the political structure of the *Volksgemeinschaft* [the community of all Germans in the National Socialist sense]. The structure of housing areas must thus, as far as possible, conform with the political organizational structure of the *Volksgemeinschaft*, organized in cells, local groups and districts. The urban form appropriate to the local group would, in this sense, consist of small cells and ultimately in small scale streets, as well as the clear arrangement of squares, residential courtyards and neighbourhood groups. (Quoted in Lehmann, 1944, pp. 13-14)

In the wake of destruction inflicted during the war, a general plan for Hamburg was created by Gutschow in 1944. He was well informed about the London plans by the Foreign Secret Information Service. His plan followed the principle of reducing housing densities with the goal of 'local group as neighbourhood cell' and developed it aggressively:

> Every previous master plan for Hamburg which wanted to avoid the danger of becoming utopian and which attempted to remain realistic had to take the existing physical situation in the central areas more or less for granted. A totally effective renewal, even if implemented gradually, was reserved for a very distant future, especially a reduction of the irresponsible population densities in the areas which had traditionally housed the Communist 'electorate'. The new master plan is based on the reality of destruction and the entirely new possibilities it offered . . . and the new master plan sees it as its task to build a city in which, despite its size, no national comrade (*Volksgenosse*) feels like a mere number, but is the member of a neighbourhood. (Gutschow, 1944)

This concept was to form the basis for spacious rebuilding of residential areas in Hamburg (Pahl-Weber and Schubert, 1991). Bombed areas were divided into neighbourhood units within an ideological organization reflecting the political structure of the Nazi Party.

Neighbourhoods in Postwar Reconstruction

After 1945, in Germany, the term neighbourhood had negative connotations because of the analogy to 'cells' and 'blocks' and the Nazi Party's mechanisms of discipline and control. As the concept of the 'local group as neighbourhood cell' was discredited because of its National Socialist origin, it was variously renamed 'cell', 'node' or 'estate unit'. National Socialist town planning and architecture remained forbidden subjects of study for a long time. Nevertheless, the goals formulated after 1945 appeared similar to the pre-1945 vision, despite the quasi de-nazification of the terminology. The myth of the Stunde Null ('zero hour') of 1945, as a completely new start, was misleading. There was some continuity in the people involved and, of course, in the planning paradigms. The National Socialist 'local groups as neighbourhood cell' idea was transformed into a Western, democratically envisioned neighbourhood unit called the 'estate node'. The depoliticized neighbourhoods were then founded on biological analogies.

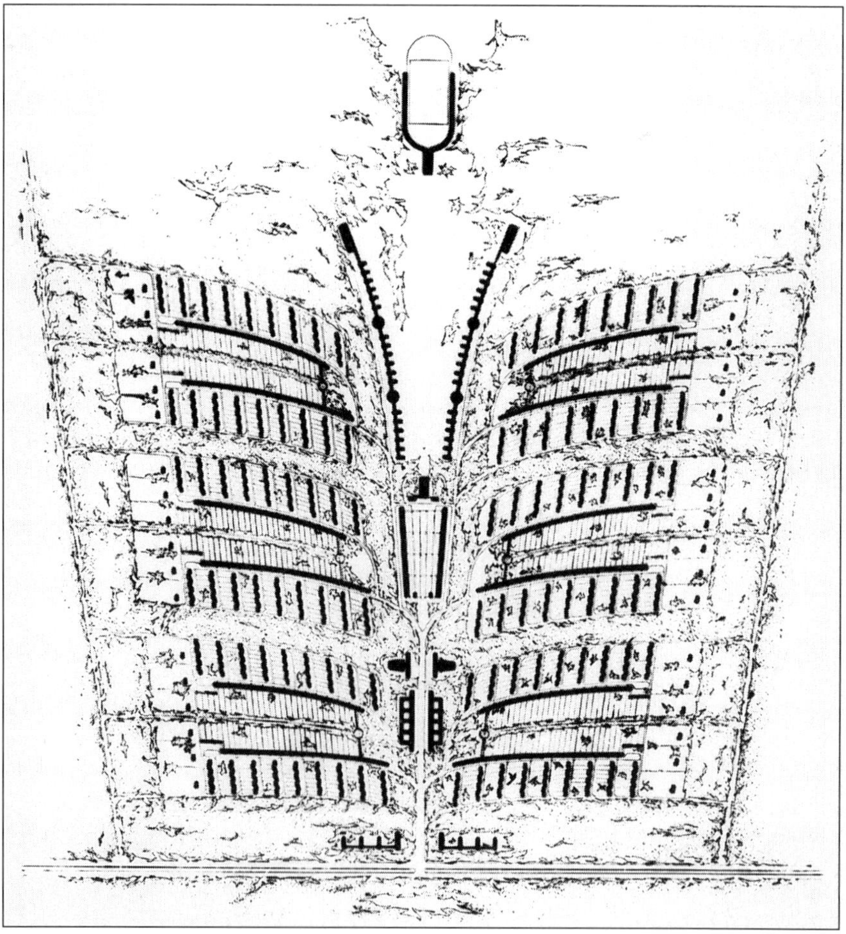

Figure 7.3. Neighbourhood as a settlement cell. A design by German architect H.B. Reichow. The idea was developed in his 1948 book *Organic City Planning*. (*Source*: Reichow, 1948)

H.B. Reichow, having propagated the 'local group as neighbourhood cell' in Germany before 1945, later switched to planning 'organic neighbourhoods' (Reichhow, 1948). Once employed by Gutschow, he had no trouble mutating National Socialist terminology into an apparently technical terminology which derived from examples in nature. He used the concept of branching for creating street patterns in postwar housing estates and borrowed the idea of segregation of modes of transport from Radburn (figure 7.3). His projects are considered some of the most influential in postwar Germany, and his books *Organische Stadtbaukunst* (1948) (Organic City Planning) and *Die Autogerechte Stadt* (1959) (The Car-suitable City), in which he primarily propagates the 'Radburn principle', were best-sellers. Similarly, although more technocratic and lacking biologic analogies, was the influential work *Die Gegliederte und Aufgelockerte Stadt* (The Structured, Low-Density City). Its authors (Göderitz *et al.*, 1957) did not disguise the fact that it had been drafted during the National Socialist era and thus made no attempt to change the terminology. One of the authors, Roland Rainer, pointed to the neighbourhood theory as early as 1948 and published the plans of Clarence Perry's Radburn in German. So there was a continuity of planning models with many similarities between the expansive plans during the war and postwar plans which were all based on the neighbourhood unit.

In postwar Great Britain, lower densities, decentralization, slum clearance, housing construction and spatial re-structuring were fused into an integrated conventional wisdom. The neighbourhood concept was central, backed by some of the most influential planners in the land. Gordon Stephenson had visited Radburn in 1929. He had held various important positions in the Ministry of Town and Country Planning and later succeeded William Holford to the Lever Chair of Town Planning at Liverpool University. Thomas Sharp, planner and publicist of important books on planning, saw neighbourhoods as defining urban expansion in rings around the city (Sharp, 1948). Frederick Gibberd, responsible for the design of number of major urban schemes, stressed the meaning of neighbourhoods for physical structuring, but pointed out that their significance lay in the development of a sense of community:

> The object of arranging the town housing in the form of neighbourhoods is to enable the family unit to combine, if it so wishes, with other families into a community which has definite social contacts and a recognisable physical unity. The neighbourhood is essentially a spontaneous social grouping, and cannot be created by the planner. (Gibberd, 1969, p. 229)

In 1946, the *New Towns Act* was passed. New towns and the dispersed regional city were part of the dream of physical and social reconstruction, of a vision for a new, planned Britain transcending 'the interwar years of social sterility, physical sprawl and architectural banality' (Cherry, 1986, p. 16). The New Towns represented a large-scale experimental field for building the neighbourhood concept (Bracey, 1964). Most used neighbourhood units varying in size from 5000 to 10,000 'but their effectiveness in creating "neighbourhood" consciousness seems to vary' (Osborn and Whittick, 1969, p. 146). A 'balanced social composition' avoiding segmentation was a social planning goal (Osborn and Whittick, 1969). The sociologist Peter Mann (1958) suggested 'socially balanced neighbourhood units', each a small-scale copy of the social structure of the entire city. By the early 1970s, planning according to the principle of neighbourhood units with segregated routes of transport on Radburn lines was practised in most local planning departments in Britain.

In England, the concept of neighbourhood planning found its greatest expression in municipal housing construction and new communities; the private housing market operated along other principles. By contrast, in the United States, where government subsidized housing construction was nearly non-existent, real estate speculators co-opted selective elements of the neighbourhood concept as selling devices. The dominance of market forces was in no way affected by the fact that the neighbourhood concept was propagated by the government. The preamble of the *Housing Act* of 1949 required that the homes of citizens be integrated into neighbourhoods. In 1969, eighteen of the most important planning and building organizations in the United States gave clear support or general advocacy to the concept of neighbourhood units (Banerjee and Baer, 1984, p. 26).

EPILOGUE AND CONCLUSIONS

Rarely has there been such international consensus among planners in the twentieth century as that reached on the neighbourhood concept. The goal of (re)structuring the city was internationally accepted. The method of achieving this aim – using smaller, cellular urban units – was widely accepted. Only the question of what exact form the city should take caused opinions to drift apart. Some of the best-known residential estates worldwide are based on the neighbourhood unit theory, including Vällingby near Stockholm, Sondergaardsparken in Copenhagen, and Linda Vista in San Diego. There is a multitude of other examples in many different cultural settings (e.g. Attia, 1963; Dahir, 1947; Rasmussen, 1957; Ritter, 1960-61; Tarantul, 1962).

The worldwide planning euphoria of the 1960s produced technocratic models which reduced and restricted the neighbourhood theory to technical, organizational norms of infrastructure planning. Even representatives of the Modern Movement, such as Walter Gropius, supported the neighbourhood theory and the goals it denoted. He stated that lower densities and not the complete diffusion of the city, was the goal of organic neighbourhood planning (Gropius, 1956). Jane Jacobs, on the other hand, strongly criticized the myth of the neighbourhood; the 'doctrine of salvation by bricks' was a worn out ideal of planning, she argued (Jacobs, 1969, p. 79).

Many empirical studies since the 1960s have demystified the neighbourhood myth. The anticipated strengthening of community has rarely happened. It progressively became apparent that human behaviour, social integration and political consensus could not, or could only marginally, be induced, steered, changed or affected by spatial concepts. The attempt to produce change in social behaviour, mixing of class and political harmony proved a context of false premises. Now, at the start of a new century, the age of globalization and individualization causes a further dissolution of traditional family ties (Rohr-Zänker, 1998). Increased mobility, pluralization of lifestyles, social isolation, and decreased feelings of solidarity promote spatial dispersal and the dissolving of traditional spatial neighbourhood contacts. At the same time, and paradoxically, we find calls for a greater localization of social life through place-bound environments which can be experienced and strengthen local identity.

In real estate propaganda, the old ideology of neighbourhoods is still an important issue, the concept instrumentalized as an empty formula in advertising brochures. Stripped of its original reform connotations, it is used as a marketing tag, along with terms like 'garden cities'. Two contrasting concepts have been developed in the North American residential property market which culminate in re-interpretations of the neighbourhood concept.

Developers have long offered 'master-planned communities' or 'planned unit developments' (PUDs) which have open spaces and sports facilities for communal use. At the end of the 1980s, 12 million Americans lived in 'Common Interest Developments' (CIDs) and more than 225,000 of such settlements were in the development pipeline for the year 2000 (McKenzie, 1994). CIDs are generally put on the market by developers for groups with special demands, such as elderly people, golfers, and singles. Security, a sense of community and 'neighbourhood' are promised. The homogeneity of the residents and their common interests should give rise to mutuality and community. The purchase of such properties means the contractual acceptance of conditions which govern life in the 'community' in detail. Unwanted neighbours can be excluded and a form of voluntary segregation and 'positive ghettoism' emerges. The current trend is for wealthy Americans to enclose such precincts with high fences. These 'gated communities' not only give their owners a feeling of stronger social control, but also simulate the feeling of neighbourliness. T.C. Boyle describes this paranoia in his novel *The Tortilla Curtain* (1995) in which he describes the contradiction of wishing for a peaceful neighbourhood in a globalized world.

The approach of the 'New Urbanism' – a return to an intact world of compact small towns, mixed land use and ample community facilities – is similar, but with the exclusivity managed more subtly by property prices rather than security systems. The so-called 'Congress for the New Urbanism' sees its role as initiating a renewal of the urbanized world (Kegler, 1998). The 'Charter of the New Urbanism' states:

> We stand for the restoration of existing urban centres and towns within coherent metropolitan regions, the reconfiguration of sprawling suburbs into communities of real neighborhoods and diverse districts, the conservation of natural environments,

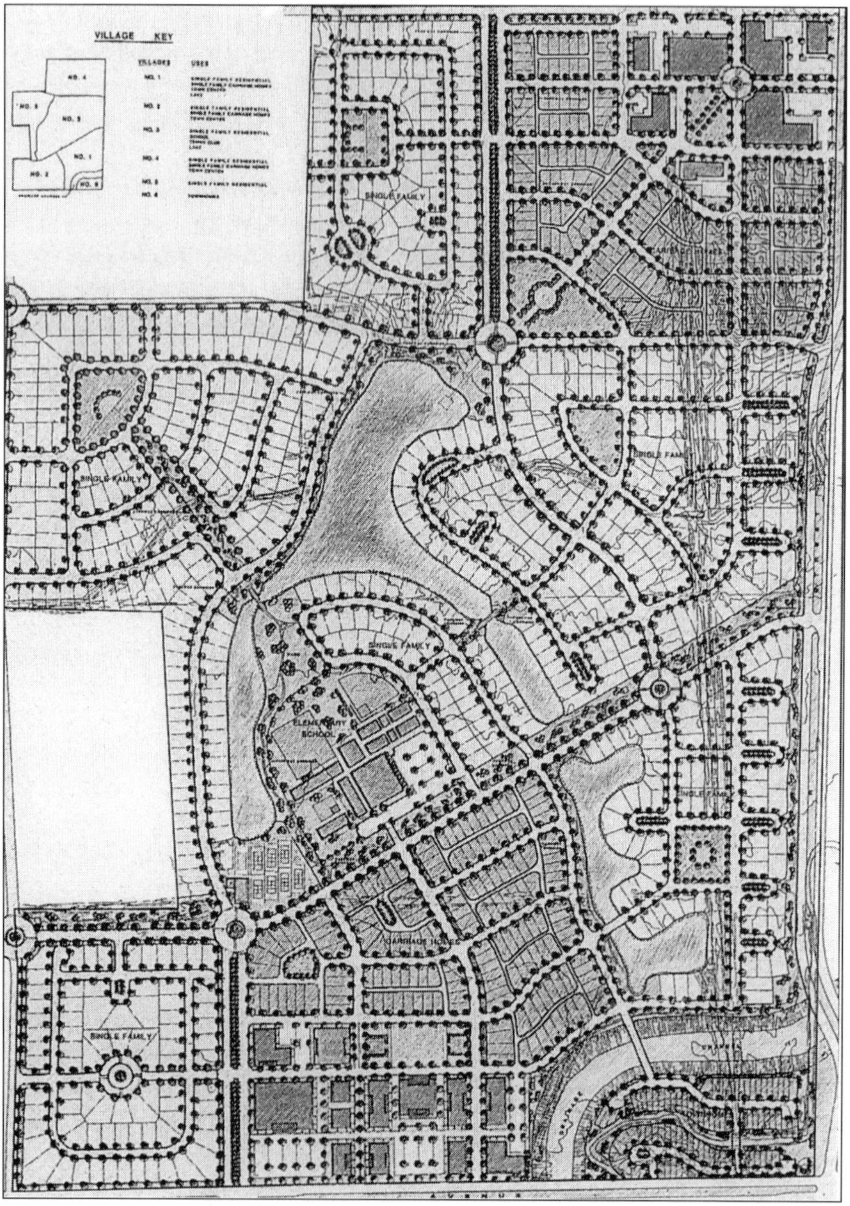

Figure 7.4. Plan of North Park Villages, Merced, California. New neighbourhoods set within a larger 'master-planned-community' in the neo-conservative style of the 'New Urbanism'. (*Source*: Calthorpe, 1993)

and the preservation of our built legacy . . . The neighborhood, the district, and the corridor are essential elements of development and redevelopment of the metropolis. They form identifiable areas that encourage citizens to take responsibility for their maintenance and evolution . . . Within neighborhoods, a

> broad range of housing types and price levels can bring people of diverse ages, races and incomes into daily interaction, strengthening the personal and civic bonds essential to an authentic community.

New urbanists explicitly refer to Perry's neighbourhood theory. Andreas Duany and Elizabeth Plater-Zyberk, for example, nominate several familiar principles of ideal neighbourhood design: giving precincts distinct centres and edges; defining an optimal size based on walking distance; injecting a balanced mix of activities; articulating through and local traffic networks; and prioritizing public space and the location of civic buildings (Katz, 1994; Krieger, 1991). Although the list is dominated by built spatial aspects, it is presumed that such planning will have community forming effects (figure 7.4). Calthorpe (1993) refers explicitly to the Radburn precedent.

There are many examples of model settlements of both movements. Whilst purchase of a property is voluntary, membership in the community and the administrational organization are compulsory. In the United States, most of the settlements are located in the Sunbelt States, especially Florida and California (Mohney and Easterling, 1991). These developments are reactions to criticism of large cities in the guise of encouraging community cohesion. Community-promoting architecture and planning are said to strengthen neighbourhood links. These surrogates promise to save the city via a form of social cleansing and exclusion (Bodenschatz, 1998). Many gated communities are enclaves for the rich based on the exclusion of undesirable outsiders and the abrogation of social responsibilities for 'outsiders' (Aguilar-San Juan, 1997, p. 35). In this sense, the trend is a perversion of the social ideals, however flawed, of the early neighbourhood propagandists.

History has illustrated that the planning theory of neighbourhood units has always had a technical content manifest as a physical design paradigm for traffic segregation, culs-de-sac, progressive housing layouts, and provision of social infrastructure. But it also contained elements of social engineering and political ideology. Hopes were continually raised and dreams nurtured, making the idea so enduring and so successful. This adaptability also proved to be one of the limitations of the theory in the twentieth century.

REFERENCES

Abercrombie, P. (1945) *Greater London Plan 1944*. London: His Majesty's Stationery Office.

Adams, T. (1934) *Design of Residential Areas: Basic Considerations, Principles and Methods*. Cambridge, Mass: Harvard University Press.

Aguilar-San Juan, K. (1997) Gated communities: Where property reigns. *Architecture California*, **2**, pp. 34-39.

Attia, A.A. (1963) *The Neighbourhood as a Basic Unit in Planning New Towns and Town Extensions*. Dielsdorf: Akerets Erben.

Banerjee, T. and Baer, W.C. (1984) *Beyond the Neighborhood Unit: Residential Environments and Public Policy*. New York: Plenum Press.

Birch, E.L. (1983) Radburn and the American planning movement: The persistence of an idea, in Krueckeberg, D.A. (ed.) *Introduction to Planning History in the United States*. New Brunswick: Center for Urban Policy Research, pp. 122-151.

Bliss, W.D. (ed.) (1908) *The New Encyclopaedia of Social Reform*. New York: Funk and Wagnalls.

Blowers, A. (1973) The neighbourhood: Exploration of a concept, in *The Open University: The City as a Social System*. Bletchley: Open University Press.

Bodenschatz, H. (1998) Alte Stadt: Neu gebaut. *Die Alte Stadt*, **4**, pp. 299-317.

Boyle, T. Coraghessan (1995) *The Tortilla Curtain*. New York: Viking.

Bracey, H.E. (1964) *Neighbours on New Estates and Subdivisions in England and U.S.A.* London: Routledge and Kegan Paul.

Brownlow, L. (1930) Radburn: A new town planned for the Motor Age. *International Housing and Town Planning Bulletin*, February, pp. 4-11.

Buder, S. (1990) *Visionaries and Planners: The Garden City Movement and the Modern Community*. New York: Oxford University Press.

Calthorpe, P. (1993) *The Next American Metropolis: Ecology, Community, and the American Dream*. New York: Princeton Architectural Press.

Cherry, G.E. (1986) Settlement planning and the regional city, in Gordon, G.E. (ed.) *Regional Cities in the U.K. 1890-1980*. London: Harper and Row, pp. 11-24.

Christensen, C.A. (1986) *The American Garden City and the New Towns Movement*. Ann Arbor: UMI Research Press.

Coit, S. (1891) *Neighbourhood Guilds: An Instrument of Social Reform*. London: Swan Sonnenschein.

Crawford, M. (1995) *Building the Workingman's Paradise: The Design of American Company Towns*. New York: Verso.

Dahir, J. (1947) *The Neighborhood Unit Plan: Its Spread and Acceptance. A Selected Bibliography with Interpretative Comments*. New York: Russell Sage Foundation.

Fagence, M. (1973) The Radburn idea –1. *Built Environment*, **2**, pp. 467-470.

Feder, G. (1939) *Die neue Stadt: Versuch der Begründung einer neuen Stadtplanungskunst aus der sozialen Struktur der Bevölkerung*. Berlin: Julius Springer.

Forshaw, J.H. and Abercrombie, P. (1943) *County of London Plan*. London: Macmillan.

Fritsch, T. (1896) *Die Stadt der Zukunft*. Leipzig: Hammer.

Gibberd, F. (1969) *Town Design*. London: Architectural Press.

Göderitz, J., Rainer, R. and Hoffman, H. (1957) *Die Gegliederte und Aufgelockerte Stadt*. Tubingen: Wasmuth Verlag.

Gropius, W. (1956) *Architektur: Wege zu einer optischen Kultur*. Frankfurt: Fischer Bücherei.

Gurlitt, C. (1929) New Yorker neue Siedlungen. *Stadtbaukunst*, **2**, pp. 27-31.

Gutschow, K. (1941) Erläuterungsbericht zum Generalbebauungsplan. State Archive Hamburg, AKG, A 125.

Gutschow, K. (1944) Skizze Generalbebauungsplan. State Archive Hamburg, AKG, A 44 D34.

Hass-Klau, C. (1990) *The Pedestrian and City Traffic*. London: Belhaven Press.

Howard, E. (1898) *To-Morrow: A Peaceful Path to Real Reform*. London: Swan Sonnenschein.

Jacobs, J. (1969) *Tod und Leben großer amerikanischer Städte*. Berlin: Bertelsmann Fachverlag (translation of *The Death and Life of Great American Cities*, 1962).

Johnson, D.A. (1996) *Planning the Great Metropolis: The 1929 Regional Plan of New York and its Environs*. London: E & FN Spon.

Johnson, D.L. (1998) Origins of the neighbourhood unit, in Freestone, R. (ed.) *The Twentieth Century Urban Planning Experience, Proceedings of the 8th International*

Planning [History] Society Conference and 4th Australian Planning/Urban History Conference. Sydney: University of New South Wales, pp. 429-430.

Katz, P. (1994) *The New Urbanism: Toward an Architecture of Community*. New York: McGraw Hill.

Kegler, H. (1998) Mehr als Sehnsucht nach der alten Stadt: New Urbanism in den USA. *Die Alte Stadt*, **4**, pp. 335-346.

Korn, A. and Samuely, F.J. (1942) A Master Plan for London, Based on research carried out by the Town Planning Committee of the M.A.R.S. Group. *Architectural Review*, **91**, pp. 143-150.

Krieger, A. (ed.) (1991) *Andreas Duany and Elizabeth Plater-Zyberk: Towns and Town-Making Principles*. Cambridge, Mss: Harvard University Graduate School of Design.

Lederer, R. (1930) Die Stadt Radburn. *Der Städtebau,* **25**, pp. 529-530.

Lees, A. (1985) *Cities Perceived: Urban Society in European and American Thought, 1820-1940*. Manchester: Manchester University Press.

Lehmann, E. (1944) *Volksgemeinschaften aus Nachbarschaften, Eine Volkskunde des deutschen Nachbarschaftswesens*. Berlin: Noebe Verlagsbuchhandlung.

Lubove, R. (1963) *Community Planning in the 1920s: The Contribution of the Regional Planning Association of America*. Pittsburgh: Pittsburgh University Press.

Mann, P.H. (1958) The socially balanced neighbourhood unit. *Town Planning Review*, **29**, pp. 91-98.

McKenzie, E. (1994) *Privatopia: Homeowner Associations and the Rise of Residential Private Government*. New Haven: Yale University Press.

Miller, A. (1969) Radburn and its validity today. *Architectural & Building News*, No. 5, pp. 40-49; No. 6, pp. 30-34.

Mohney, D. and Easterling, K. (eds.) (1991) *Seaside: Making a Town in America.* New York Princeton Architectural Press.

Mumford, L. (1954) The neighbourhood and the neighbourhood unit. *Town Planning Review*, **24**, pp. 256-270.

Mumford, L. (1975) *The City in History*. Harmondsworth: Penguin.

Osborn, F.J. and Whittick, A. (1969) *The New Towns: The Answer to Megalopolis*. London: Hill.

Pahl-Weber, E. and Schubert, D. (1991) Myth and reality in National Socialist town planning and architecture: Housing and urban development in Hamburg 1933-45. *Planning Perspectives*, **6**, pp. 166-181.

Perry, C.A. (1929) The neighborhood unit: A scheme of arrangement for the family-life community, in *Regional Survey of New York and its Environs, Vol. VII, Neighborhood and Community Planning*. New York (Reprint Routledge/Thoemmes Press, London, 1998).

Rasmussen, S.E. (1957) Neighbourhood planning. *Town Planning Review*, **27**, pp. 197-218.

Reichow, H.B. (1948) *Organische Stadtbaukunst: Von der Großstadt zur Stadtlandschaft.* Berlin: Georg Westermann.

Reichow, H. B. (1959) *Die autogerechte Stadt: Ein Weg aus dem Verkehrs-Chaos*. Ravensburg: Otto Maier Verlag.

Ritter, P. (1960-61) Radburn planning: Foreign examples. *The Architects Journal*, No. 10, pp. 680-684; No. 2, pp. 176-182.

Rohr-Zänker, R. (with W. Müller) (1998) *Die Rolle von Nachbarschaften für die zukünftige Entwicklung von Stadtquartieren*. Bonn: Arbeitspapiere Bundesamt für Bauwesen und Raumordnung.

Schaffer, D. (1981) Garden Cities for America: The Radburn Experience. Ph.D Thesis, Rutgers University, New Brunswick.

Schaffer, D. (1992) The American garden city: Lost ideals, in Ward, S. (ed.) *The Garden City, Past, Present and Future*, London: E & FN Spon, pp. 127-145.

Schubert, D. (1982) Theodor Fritsch und die völkische Version der Gartenstadt. *Stadtbauwelt*, **73**, pp. 463-468.

Schubert, D. (1998) *Stadterneuerung in London und Hamburg: Eine Stadtbaugeschichte zwischen Modernisierung und Disziplinierung*. Braunschweig/Wiesbaden: Friedr Vieweg & Sohn Verlagsgesellschaft.

Schwan, B. (ed.) (1935) *Städtebau und Wohnungswesen der Welt*, i.A. des Deutschen Vereins für Wohnungsreform, Berlin.

Scott, M. (1969) *American City Planning since 1890*. Berkeley, CA: University of California Press.

Sharp, T. (1948) *Städtebau in England*. Berlin.

Silver, C. (1985) Neighborhood planning in historical perspective. *Journal of the American Planning Association*, **51**, pp. 161-174.

Stein, C. S. (1966) *Toward New Towns for America*. Cambridge, Mass: MIT Press.

Sussman, C. (1976) *Planning the Fourth Migration: The Neglected Vision of the Regional Planning Association of America*. Cambridge, Mass: MIT Press.

Sutcliffe, A. (1983) Urban planning in Europe and North America before 1914: International aspects of a prophetic movement, in Teuteberg, H.J. (ed.) *Urbanisierung im 19. und 20. Jahrhundert. Historische und geographische Aspekte*, Köln-Wien: Böhlau Verlag.

Tarantul, A.I. (1962) A neighbourhood in the USSR. *Town and Country Planning*, **30**, pp. 264-267.

Toennies, F. (1887) *Gemeinschaft und Gesellschaft* (ed. 1922). Berlin: Verlag Karl Curtius.

Tylor, W.R. (1939) The neighbourhood unit principle in town planning. *Town Planning Review*, **18**, pp. 174-186.

Wright, H. (1932) Planen oder nicht Planen, in *Die Neue Stadt: Internationale Monatszeitschrift für architektonische Planung und städtische Kultur* (Reprint RWTH Aachen, Lehrstuhl für Planungstheorie).

CHAPTER 8

Planning for Social Betterment: From Standard of Living to Quality of Life

Raphaël Fischler

The twentieth century has seen the worst horrors of human history. Planners and designers, unfortunately, have been involved in genocide, if only by glorifying tyrants in monumental architecture and laying out concentration camps. In less murderous regimes, even in democratic ones, they have helped to enforce segregation on the basis of race, ethnicity or class. Yet, at the same time, this century has shown that planning, as a collective effort to shape people's living conditions, can do good. Much of what we know as urban planning today follows from a genuine desire to remedy the evils of the industrial metropolis. If some reformers were mostly interested in increasing the economic efficiency of the congested city, many others were bent on making it, first of all, a better place to live. Social workers, sanitary engineers, landscape architects, public officials and activists of various stripes fought for decades to enact laws and institute programmes to improve people's well-being, in particular to ameliorate their home and work environments. One outcome of this crusade was the creation of professional urban planning.

Reformist efforts did not take on the same form everywhere, nor did they remain identical over time. Different problems call for different solutions, different ideologies for different actions. An important, though not radical, change has occurred over the century in the way progressive planners and policy-makers define their mission, a change that can be captured in a discursive shift. In 1909, the American reformer Herbert Croly defined the aim of good government as the fulfillment of the century's 'promise' to the masses: 'What the wage-earner needs, and what it is to the interest of a democratic state he should obtain, is a constantly higher standard of living' (Croly, 1964, p. 206). Some ninety years later, the Independent Commission on Population and Quality of Life provided an alternative formulation of the policy agenda: 'Governments should aim . . . at constant, equitable, and sustainable improvements in quality of life' (Independent Commission on Population and Quality of Life, 1996, p. 85). The two mission statements have much in common but also much that sets them apart. In both, the state is given the responsibility of securing decent living conditions for all its citizens. Yet the use of the expressions 'standard of living' and 'quality of life' points to a significant evolution in political theory and in planning practice.

In this chapter, I explore what the shift from 'standard of living' to 'quality of life' tells us about urban planning, and about public policy in general, since the early 1900s. In some ways, public intervention has shown a relative constancy over the century; it has ostensibly remained predicated on the objective identification of people's wants and the organization of collective action to meet their demands. But policy has changed, too, in the range of needs targeted, in the priorities that we establish among them, and in the ways in which we pursue their satisfaction. At the world scale, personal well-being and social development have been redefined to include political, cultural and environmental factors, and the role of the state has been recast from that of provider to that of enabler. As an indicator of material conditions, the standard of living has become but one variable among many to assess the quality of life. In Western countries, 'standard of living' and 'quality of life' have in fact become the keystones of radically opposed political discourses based on economic position. Maintaining their standard of living appears to be the goal of people who are losing ground in a post-industrial, high-tech economy, while ensuring their quality of life is the objective of those who are gaining from the societal transformation.

The shift from 'standard' to 'quality' reflects the fact that, by the 1960s, material conditions of life had improved so significantly in industrialized nations that higher-order wants could be placed on the political agenda. Perhaps more important, dissatisfaction with the social and environmental side-effects of modernization provoked a fundamental critique of the modernist agenda of economic growth. In addition, the transformation of capitalism, together with the high-technology revolution that prompted it and the political revolution that it fostered in turn, led to a revision of modes of thinking and ways of doing in governments around the world. It is in this social, political and economic context that the passage from 'standard' to 'quality' must be studied. Yet at the same time, factors of change should not obscure elements of continuity. As concepts, the standard of living and the quality of life do show much overlap in meaning; as tools, both belong to an old tradition of needs-analysis. To grasp the origins of the two expressions, to understand the different approaches to policy and planning that they encapsulate, to point out the continuity that exists between them, and to learn from this analysis where urban planning stands at the dawn of a new century: these are the goals of this chapter.

THE STANDARDIZATION OF NEED

In the early days of the century, the idea of the standard of living was the focal point of reformist discourse, both in Progressive political theory and in labour activism (Glickman, 1997). Edward Devine, a leader of organized philanthropy in New York City, declared: 'It [the standard of living] is at once the point of departure and the object of solicitude. It determines the size and the nature of the tasks to be undertaken' (Devine, 1917, p. 2). Before World War One, ensuring to all an adequate (and rising) standard of living – Herbert Croly's 'Promise of American Life' – was seen increasingly as an economic possibility and a political necessity. After the Great War, it became a sacred duty, 'a debt to the dead which must be paid to the living, in terms of life and health and opportunity' (Rowntree, 1918, p. 144). By 1935, this goal had been enshrined into what Walter Lippmann called 'The New Imperative' of the Welfare State. New Deal policies emerged as the natural expression of the idea that 'government must henceforth hold itself consciously responsible for the maintenance of the standard of life prevailing among the people' (Lippmann, 1935, p. 1).

In the first half of the century, in urban planning as in many other fields, professionals came to share a new vision of modern society, a new language to describe its ills, a new set of measures to cure them. From Canada to Australia, their mandate was to translate social norms into policy standards and to meet 'the higher expectations of what [is] a reasonable standard for a modern advanced society' (Troy, 1995, p. 270). At any time, what is seen as 'decent' housing, 'adequate' nutrition, 'normal' access to recreation, and so forth, is indicative of a society's level of development, of its wealth and welfare. The role of the state is to help its citizens meet the increasingly demanding standards of food, clothing, housing, recreation and education that the community, local or national, sets as its own. The task of professionals – urban planners and others – is to determine what these standards are, that is, to mediate between social need and economic interest and set thresholds that are at once useful and affordable (Fischler, 1998). This approach to public policy is inherent to the regulation of industrial capitalism in the first half of the twentieth century.

The problem of the nineteenth century was misery, especially in the industrial city. But when the popular malaise with the slum translated into political action, the target of public intervention was not really poverty or disease. It was waste. The goal, then, was greater efficiency. In order to reduce waste and increase efficiency, reformers concurred, cities and dwellings had to be planned rationally. In particular, housing construction and habitation, sanitary expert Ellen Richards wrote, had to be 'regulated by the scientific laws of right living'

(Richards, 1910, p. 83). What was needed to guide urban development was not only a city plan that made the city more efficient (and more beautiful), but also legislation that brought all dwellings up to a 'standard of healthful housing' (p. 53). Plans and regulations went together. Both were needed to give workers the facilities that they generally lacked in the slums. The questions of housing reform, of town extension, of urban infrastructure, of economic development and of moral reform – these questions were to a large extent one question: how to meet, as two Belgian proponents of the garden city put it, 'the primordial necessities' of the population in an economical manner (De Heem and De Ridder, 1914, p. 386; author's translation). In this broad policy of social betterment, the paradigmatic instrument of control was the quantitative specification standard.

For the economist A.C. Pigou, speaking in 1914, public policy in the twentieth century was to be predicated on a simple axiom:

> It is the duty of a civilized State to lay down certain minimum conditions in every department of life, below which it refuses to allow any of its free citizens to fall. There must be a minimum standard of conditions in factories, a minimum standard . . . of leisure, a minimum standard of dwelling accommodation, a minimum standard of education, of medical treatment in case of illness, and of wholesome food and clothing . . . The standards must be upheld all along the line, and any man or family which fails to attain independently to any one of them must be regarded as proper subject for State action. (Pigou, 1914, p. 36)

In their battle against poverty and the slum, reformers would indeed rely first and foremost on 'legislative action to establish and maintain fair standards of wages, hours, and housing' (Bremner, 1956, p. 138) or, in the words of Benjamin Marsh, on 'the effort to standardize living conditions for every citizen' (Marsh, 1909, p. 39).

The reference to 'standardization' is not accidental. By the 1910s, technical normalization had become the credo of professionals, industrialists, and politicians alike, including those who sought to improve workers' housing (Wright, 1981). Planners were eager to apply the lessons that businessmen had learned in their quest for greater efficiency. As Raymond Unwin put it: 'The age which has to an unrivaled extent applied foresight and organization to the development of its industries and other activities, has suddenly realised that the same principles are applicable to the development of towns' (Unwin, 1914, p. 30). In planning, as in many other fields, standardization, the enunciation of minimum or maximum requirements, and the regulation of production according to these norms, became an instrument of progress, a symbol of modernity. Municipalities throughout the world adopted all manner of construction, housing and planning codes.

Of course, building regulations were nothing new. But in the second half of the nineteenth century and especially in the beginning of the twentieth, regulation became more systematic, based more firmly on empirical research, and applied in a more bureaucratic fashion. The modern standard is a threshold between good and bad enunciated by experts and applied by officials in order to prevent the production of goods deemed inadequate. It translates a political will into a legal norm by means of scientific analysis and judicious compromise. In effect, as the profession of urban planning appeared on the public scene, the whole social question, of which town planning was an integral part, was being rewritten in quantitative terms. Poverty itself was being redefined in this manner. The work of reformers such as Seebohm Rowntree in England and Robert Hunter in the United States ushered in a new era in social welfare policy, one in which the lives of citizens would be evaluated according to one supreme standard, the synthesis of all norms, the standard of life.

In their influential books, Hunter and Rowntree describe with great accuracy the poverty that prevailed in their respective countries at the turn of the century, and they denounce it with great vehemence (Hunter, 1965; Rowntree, 1901). They, too, work from the premise that the state owes all citizens the maintenance of a certain level of well-being, in particular in matters of public health. To quantify this minimum standard of welfare, they rely on a simple proxy, the income required to meet 'the human needs of labour' (Rowntree, 1918). 'The world rests upon their shoulders; it moves by their muscle', Hunter says of industrial workers, and yet 'they are not able to obtain *those necessaries which will permit them to maintain a state of physical efficiency*' (Hunter, 1965, pp. 4-5; emphasis in text). These basic needs are food, clothing and shelter, and their necessity is a direct function of the industrial system of production. Ensuring that every worker gets what it takes to 'keep the body in working order', that is all Hunter demands from society; the norm he proposes is 'a purely physical standard', the same standard 'that a man would demand for his horses or slaves' (p. 7). That norm, the 'standard of life', is the financial basis of productive citzenship. It is the generic standard behind all specific standards, the economic condition for the reproduction of labour power and the maintenance of social order.

THE PARADOX OF THE STANDARD OF LIVING

Political efficacy demanded that reformers generate clear standards and straightforward indicators. Yet observation told them that the identification of basic needs and the calculation of minimum revenues could not suffice to assess human welfare in all its complexity and variety. If '[d]ecent housing was at the core of what to . . . the social reformer constituted a reasonable standard of living' (Petrow, 1995, p. 6), that core never really hardened into a strict, consensual definition of housing or neighborhood quality. Even Rowntree, the scientific investigator of family budgets, remained sensitive to the subjective value of consumption goods and living conditions. In a discussion of the housing problem in Great Britain before the Great War, he lamented the fact that 'half [of the country's] workers reside in an environment which is without variety, without beauty, and but slenderly supplied with light and air – an environment which devitalises both body and soul' (Rowntree, 1914*a*, p. 89; see also Rowntree, 1914*b*). The target of Rowntree's opprobrium was in large part 'Bye-Law' housing, housing produced along the guidelines of the *Public Health Act* 1875. By the 1910s, the residential projects that had given numerous English working-class families more spacious and more sanitary homes were being condemned on aesthetic grounds (Burnett, 1978).

This type of critique was not particular to housing and its design. The very ideas of quantification and standardization were being put into question. Many reformers, among them Edward Devine, were reluctant to follow Hunter and Rowntree in defining need in monetary terms. Devine doubted that the standard of living could be captured by means of 'a formula' or a list of 'physical necessities and comforts' (Devine, 1917, pp. 2, 4). Quantification has its limits, critics argued, precisely because there is a subjective dimension to the standard of living. 'Each family', a colleague of Devine noted early on, 'has its own standard of living, by which I understand . . . its opinion as expressed in practice, as to the minimum provision for itself, that it regards as consistent with health, self-respect, and the general well being of all its members' (Folks, 1901, pp. 415-416). In addition, human needs are not all physical in nature. 'Food, rent, and household sundries are the material items in the standard of living', another New York social worker commented, 'but one must remember that every family demands something else in the way of education, amusement, and social intercourse, for which its members will go hungry, if necessary' (Ayres, 1902, p. 216). Many planners agreed: proponents of the City Beautiful addressed the psychic needs of the urban population in their campaigns to beautify, clean and green the city (Wilson, 1989). Hunter and Rowntree did not disagree either with the arguments of Folks, Ayres and Devine. But they understood that simple quantitative measures could have great practical virtues in politics, as they had in industry, by establishing unambiguous thresholds between what was acceptable and what was not. Their pecuniary standard of living was perhaps controversial, but it was strategically useful and politically effective.

The debate between Hunter and Devine, though, reflects a deeper problem: as a sociological notion, the standard of living suffers from an internal contradiction. For historian Daniel Boorstin, the conflict between quantitative and qualitative assessment was inscribed from the start within the heart of the idea, in the juxtaposition of the two terms 'standard' and 'life'. When, in 1798, Sir John Sinclair produced a statistical profile of Scotland, he did so 'for the purpose of ascertaining the quantum of happiness enjoyed by its inhabitants, and the means of its future improvement' (quoted by Boorstin, 1973, p. 167). The conundrum of Sinclair's 'quantum of happiness' has remained complete in Rowntree's and Hunter's 'standard of life'. The standard of living expresses a monetary objective of increased means but really serves a non-monetary goal of higher well-being. Early statisticians and reformers perhaps 'did not use the term "quality of life"', another historian notes, 'but [they] came close to it' in their descriptions of human welfare (Jordan, 1997, p. 40). In fact, they did not only come close to it, but they did use it once in a while.

In 1918, the American architect Charles Harris Whitaker, an active participant in his country's war-housing effort, wondered aloud 'why [designers] were . . . willing to accept advances in sanitation, comfort, convenience, with so little thought of the preservation of those other qualities of charm and picturesque attractiveness which we so much admire when we visit Europe' (Whitaker, 1918, p. 5). To his colleague Frederick Ackerman, another participant in the US war-housing effort, the British experience in workers' housing taught an important lesson: 'The houses in a town, the arrangement of a town, *the quality of life* made possible in a town, serve as an infallible barometer whereby we may read the state of social, physical, economic, and moral development of a people' (Ackerman, 1918, p. 18; emphasis added).

Indeed, Ackerman and his contemporaries understood that the standard of life was a poor representation of human welfare, and that good environmental conditions added a qualitative dimension to human well-being. But that insight did not translate for a long time into a new political discourse. It would take half a century of modernist planning and the reactions it provoked, together with a radical transformation of systems of production, for the notion of quality of life to become a new habit of thought.

The Quality-of-Life Movement

Two developments were instrumental in putting the qualitative dimension of well-being explicitly on the political agenda. The first was strong economic growth; higher productivity helped to feed the coffers of an expanding Welfare State and to increase per capita income. This translated, for instance, into higher spending on dwellings and infrastructure. Public and private building gave Western urbanites significantly improved housing conditions, both in terms of construction and in terms of crowding. For example, between 1921 and 1981, the average number of persons per unit in England and Wales declined by 43 per cent, from 4.75 to 2.71, after being reduced only by 7 per cent between 1861 and 1921 (Mitchell, 1988, pp. 9, 389). Expectations concerning the wider home environment changed as well, and neighbourhood quality, no longer a prerogative of the rich, became a major concern of planners in the inter-war years (Adams, 1934). For Le Corbusier, providing the masses with good housing required that one adhere to a coherent set of quantitative standards, pertaining to various scales of planning and design (Le Corbusier, 1982). Planners in other countries followed the same reasoning and generated 'scientific' neighbourhood schemes and standards. In the United States, these can be found in the Regional Survey of New York and Its Environs (Perry, 1929) and in the publications of the

Committee on the Hygiene of Housing of the American Public Health Association (American Public Health Association, 1948). In France, the United States and other countries, such normative systems became a reference to governments and developers who participated in the building boom after World War Two. In effect, a higher standard of living turned yesterday's targets into accepted parameters of a decent urban life.

The other development that drew attention to the qualitative aspects of the environment, and of well-being in general, was the very fact of modernization, or rather the reaction that it elicited. The triumph of technocracy, of professional normalization, and of material consumption in the 1950s and 1960s spelled the end of the ideology that informed them. As the effects of standardization and economic growth became clear, the goals of public and corporate action were no longer taken for granted. Among others, the concept of standard of living lost its legitimacy and, hence, its efficacy. To reflect the new political consciousness, a new generation of reformers embraced the alternative notion of 'quality of life' as their guiding concept of public policy.

With massive government intervention in urban development after World War Two, modernist planners and designers set out to create the new Radiant City that they had envisaged in plan before the war. The result of thirty years of modernist urban renewal was a radical transformation of the urban landscape in many cities. But what was put forward as a constructive response to the housing crisis was increasingly perceived as a tragic mistake on the part of arrogant experts. By the 1970s, urban renewal and public housing projects were being condemned, both literally and figuratively speaking. The physical demolition of Pruitt-Igoe in St. Louis, Missouri on 15 July 1972 materialized the intellectual destruction of modernism that was taking place through the pen of social and architectural critics (Jencks, 1981). Denouncing the 'incompetence' of contemporary planners and 'the pseudoscience of city rebuilding and planning', Jane Jacobs saw in most contemporary schemes 'a celebration, in art, of the potency of statistics and the triumph of the mathematical average' (Jacobs, 1961, pp. 7, 13, 436).

Jacobs's attack on modern city building was part of a larger critique, coming from different directions, of the political and cultural mores of the era. The counter-cultural movements of the 1960s made Westerners aware of the fact that greater affluence had been accompanied by alienation more than by liberation (Bloomberg and Schmandt, 1969). Resistance to urban redevelopment grew, and so did efforts to preserve historic areas and revitalize declining ones. Environmental activism and neighborhood organizing gave many urban planners a new sense of mission and put the profession in search of 'high levels of environmental quality and public services', even at the cost of 'modest sacrifices of private consumption' (Baumol and Oates, 1969, p. 2). The emphasis was now on livability, on qualitative rather than quantitative features of life. In various social-science disciplines, researchers tried to grasp 'the subjective reactions of men and women to economic and social processes, both affective and cognitive' (Hankiss, 1983, p. 15). In medicine, perceived quality of life became an object of scrutiny, alongside diseases of modern life (Parmenter, 1994).

A similar revision of beliefs and practices occurred in the field of economic development planning. Soon after World War Two, international organizations called on experts to devise international indicators of well-being that went beyond monetary factors and that reflected the multiplicity and variability of wants (United Nations, 1954). Accordingly, the Gross National Product was deemed inadequate as a measurement of national welfare (Miles, 1985); the notion of development, understood as growth, was itself put into question (Routledge, 1995); and the concept of need became the object of widespread criticism for its failure to represent the complexity of human existence (Doyal and Gough, 1991). The standard of living, too, was declared to be of limited value. For Amartya Sen, winner of the 1998 Nobel Prize in Economics, it is but one way, and a highly problematic one at that, of describing the quality of life (Sen, 1986). Assessments of people's well-being, Sen and a colleague tell us, ought to take into account 'not only . . . the money they do or do not have, but a great deal about how they are able to conduct their lives . . . [and even] how people are enabled by the society in question to imagine, to wonder, to feel emotions such as love and gratitude, that presuppose that life is more than a set of commercial relations' (Nussbaum and Sen, 1993, p. 1).

As fundamental as these intellectual revisions may be, they can take us only so far in the realm of policy. Rowntree and Hunter acknowledged that there was more to human well-being than the satisfaction of basic material wants. But as men of action, they understood that objective measures were needed in order to frame and to solve the problem of the unskilled wage-earner. In the redefinition of our social problems, at the start of the twenty-first century, pragmatism again must temper reformist zeal. It is nice to recognize the multiplicity and subjectivity of personal aspirations, but it remains necessary to rationalize the evaluation of living conditions and the distribution of resources. Collective action calls for commonly accepted definitions of what is wrong and of what would be right. Hence quality-of-life policies still require some sort of objective measures of need. That, in turn, calls for some agreement on what is to be measured. 'While the nature of the good life is ultimately an individual matter', David Smith maintains, 'the extent to which human beings have common needs suggests some consensus as to quality of life criteria' (Smith, 1977, pp. 30-31).

What these needs are, whose satisfaction determines our quality of life, remain an open question. What indicators must be selected to assess well-being? When doing inter-city comparisons, what factors must be taken into account? When setting standards of adequacy for municipal services, what benchmarks must be used? Papers published since 1974 in the journal *Social Indicators Research* ('An International and Interdisciplinary Journal for Quality-of-Life Measurement') have not settled any of these questions. Nor did papers presented at the First International Conference on Quality of Life in Cities, held in Singapore in March 1998. As one speaker put it, 'there are no standard measures of QOL [quality of life] at either the individual or the city level. Thus, there is no consensus about the objective components that make up QOL' (Leitmann, 1998, p. 333). The times are past when experts, alone, could make edicts on public standards and technocrats could impose universal solutions. To be relevant, the same World Bank planner explained, 'urban QOL indicators should be developed at the level where they will be applied (e.g. neighborhood, district, city, metropolitan area, or region) and by representatives of the stakeholder groups that are both knowledgeable and concerned about QOL issues' (p. 337). Indeed, the very process of consensus-building in socially and culturally fragmented cities and regions has emerged as a major issue in planning practice and planning theory (Healey, 1997). The strength of local democracy is itself considered an important quality-of-life indicator, for 'social inclusion and civic participation . . . the ethic of citizenship and democracy, are [not only] as important as some of the more immediate material aspects of individual well-being [but] probably better predictors of national well-being in the long term' (Burke and Salvaris, 1998, p. 61).

In the 1950s, psychologists told us that individuals have a hierarchy of needs, higher as well as lower ones, including the need for love and respect, the need for learning and beauty, and the need for self-actualization (Maslow, 1954). In the 1960s, social activists impressed on us the fact that communities, too, have needs, the need for active citizenship and for the expression of cultural identity. For all these elements of human well-being, a quality of life index includes objective indicators. But what distinguishes the quality of life from the standard of living is not only the fact it accounts for non-material wants. What sets the two concepts apart, most of all, is the fact that the quality of life, as understood today, pertains not to the average individual or household, but to the community. And when it comes to community, that is, to cultural and political identity, there is no average, only difference.

Quality of Life in Post-Industrial Society

The notion of quality of life has two main components: the idea that people live by more than bread and water and the idea that what gives meaning to life is also rooted in the collective: 'In human happiness, creative activity and a sense of community count at least as much and maybe more than material standard of living' (Peattie, 1998, p. 248). Individual well-being, today's progressive planners tell us, demands also the satisfaction of cultural and political needs, in particular the need for social identity and democratic voice (Friedmann and Douglass, 1998). We must therefore institute collaborative planning practices that include all people, irrespective of class, race or gender. These practices, in turn, will help to rehabilitate 'the city of memory, of desire, of spirit; the importance of place and the art of place-making; the local knowledges written into the stones and memories of communities' (Sandercock, 1998, p. 4). Ironically, business actors in capitalist urban development can readily agree with a good part of this programme. Post-modern urbanism, which is to post-industrial capitalism what modernist planning was to industrial capitalism, also translates a desire to evoke memories, create image, make places rather than spaces (Ellin, 1996).

Since the late 1980s, the quality-of-life enterprise is no longer a monopoly of social reformers. Developers have not remained insensitive to the new spirit of the age, and the discourses of quality of life, sustainability and identity can also be heard in the arena of growth politics and city marketing (Kearns and Philo, 1993). Builders have become much more sensitive to demands for urban livability and environmental preservation. Not only have they learned to anticipate the resistance of local associations and advocacy groups, they have recognized that many of their best-paying customers, precisely those baby-boomers who were part of the protest generation of the 1960s and 1970s, demand high-quality amenities and local character (Bookout, 1994). Hence, designing the healthy city and the city of memory, which had become morally imperative to some, has become financially interesting to others.

Likewise, mayors seek voter support by promising to improve the city's quality of life, and municipal agencies legitimate their activities in those terms as well. (In Montreal, the Public Works Department notifies residents of upcoming disruptions by means of a leaflet that reads in part: 'To improve your quality of life, work crews will soon begin repairs on your street'.) With their expertise in how to improve the attractiveness of places, urban designers have started to work in tandem with economic development planners to enhance local competitiveness (Kotler, Haider and Rein, 1993). Place-identity, heralded at first as an antidote to modernist planning, has evolved into a sales gimmick (Erickson and Roberts, 1997). As David Ley notes, the search for urban quality of life, which originated in a grass-roots protest against private enterprise and public policy, has been taken up and redefined by those people who were the target of critique; the discourse of the counter-culture and of social movements has been assimilated in part by the members of urban growth machines (Ley, 1996).

If the citizen in industrial society was a subject of need (Laufer and Paradeise, 1990), the citizen in post-industrial society is a subject of desire. An increasing share of household consumption is related to lifestyle choices; even basic necessities are sold as sources of pleasure and bought as means of self-fashioning. Consumption and entertainment are more and more intertwined; design and the arts, tourism and recreation are growing in importance as engines of urban development (Whitt, 1987). At the same time, rootlessness and multiculturalism have put a premium on community identity. Advocates of community empowerment reject not only technocratic modernism but also the global consumer culture that deprives people of their traditions and robs their emotional life of spontaneity. Leonie Sandercock expresses both concerns in her call for culturally sensitive planning:

> One symptom of the narrowness of modernist planners' horizons is the fact that they find it very hard to focus on desires rather than needs. A need is supposedly an objectifiable entity, identified in 'needs surveys' . . . A desire, by contrast, involves the subconscious, a personal engagement, dreams and feelings, an ability to intuit the atmosphere and feeling of a place. How does the city of desire translate into planning? Perhaps by giving more attention to places of encounter, specifically those which are not commercialized – the street, the square – and which are not placed under the gaze of surveillance technologies. Perhaps also by recognizing that some places of encounter must necessarily be appropriated, and not trying to regulate the uses of all public spaces. (Sandercock, 1998, p. 212)

Following Jane Jacobs, Sandercock wants to see less control of urban space. Like the free-market advocate, she deplores the inability of residents to live their lives as they see fit and to organize themselves on the basis of negotiated agreements. Sandercock and Jacobs may not agree on much, politically speaking, but they do concur in their condemnation of modern planning. And that verdict is, to a large extent, also an indictment of the Welfare State 'as we know it' (to use the expression of Clinton Democrats), that is, as a system of 'standard operating procedures, . . . vertical chains of command, and . . . standardized services' (Gore, 1993, p. xxxiii).

The Welfare State, of which urban planning was a significant vector in the first decades of the profession, is undergoing a severe fiscal, ideological and political crisis. It cannot sustain past levels of expenditure, its bureaucratic institutions have lost public confidence, and the social contract on which it was based – the solidarity of people exposed to the dangers of industrial production, its accidents, its seasonal and cyclical fluctuations – has unravelled (Rosanvallon, 1995). The role of the state as provider, indeed the very idea of collective consumption, has been undermined. In the face of growing diversity, 'one-size-fits-all' approaches will not do: 'When consumers accustomed to choices confront public institutions that offer standardized services, they increasingly go elsewhere' (Osborne and Gaebler, 1993, p. 168). That 'elsewhere' is most likely the private market, the arena where desire – not only need – is so readily satisfied. The demise of the Welfare State is largely an effect of the transformation of capitalism in the last third of the twentieth century. The increasing flexibility of production methods and mobility of production inputs has intensified competition between firms and between geographic areas. The order of the day for elected officials is to make their territory globally competitive and to advertise it worldwide to holders of financial or intellectual capital.

Policy-makers have found in quality-of-life assessment an important marketing device to attract the right people to their city, region or nation. The right people, in the new economy, are knowledge workers, not manual labourers. Hence urban planning focuses on what makes life easier for the educated individuals who run the service economy, on what makes the city more attractive to them. It is their needs – or rather, their desires – that matter now, not those of unskilled workers. Because knowledge has replaced physical strength as the worker's greatest asset, standards of education rather than standards of nutrition and housing have become objects of attention on the part of business leaders (Committee on Economic Development, 1985). The residential environment of lower-income families and their facilities for collective consumption matter less to us than they did to our parents and grand-parents. At the same time, the environment of higher-income households is the object of continued and even increasing care on the part of agencies and corporations. The public and private sectors collaborate, in increasingly nebulous configurations, to create the post-modern region, a region where those who have clout support initiatives that tend 'to enhance the quality of life for themselves and other symbolic analysts' (Reich, 1992, p. 279). While they rely on the private sector for most goods and services, the upper classes still resort to public consumption for specific wants, namely those that involve the public realm as such. Hence government is still relatively useful to them in order to preserve the beauty of natural or urban landscapes and to ensure the safety of public spaces. Even here, though, owners' associations (in business improvement districts or gated communities) and privately funded non-governmental organizations are playing an increasing role.

Today's calls for urban beautification and natural preservation echo those heard in the early 1900s. Likewise, contemporary attention to workers' mobility replicates the interest expressed around World War One. If, as we saw above, the concept of quality of life appears in 1918 in the writings of a war-housing specialist, it is because his observations confirmed what the developers of suburban estates had learned at the end of the nineteenth century and what all manner of companies and governments learned again at the close of the twentieth, namely that quality matters when there is competition. Frederick Ackerman noted that the provision of amenities in new English developments helped to secure a stable work force by reducing 'the very serious daily labor "turnover"' in factories (Ackerman, 1918, p. 31). Lawrence Veiller, the father of modern housing regulations in the United States, stated that American industrialists were sensitive to the issue as well. During the war, he explained, his work and that of his colleagues in the Bureau of Industrial Housing and Transportation had been guided by a key principle: 'That in order to attract and hold the right kind of *skilled* worker [the government] must build houses of an attractive type; houses that will not only provide the essentials of light and air, shelter, warmth and convenience of living, but also be reasonably attractive' (Veiller, 1918, p. 346; emphasis added). Thus, if unskilled workers, who have little to spend and of whom there is an ample supply, can or should only be given the basics, skilled workers, who have greater spending power and who are more in demand, must get something extra. That added value, which can influence the residential decisions of mobile spenders – skilled industrial workers then, symbolic analysts now – is a certain quality of life.

While no economy can function without an abundant supply of low-wage, low-skill workers, these people's well-being is no longer an economic requirement; nor is it, therefore, a political priority. In its weakened condition, the Welfare State cannot compensate for the decline of the labourer's utility. Decades ago, the idea of a rising standard of living served to promote a sense of common expectations in terms of material and social achievement. The 'average man' could make it into the middle class, and urban planning was one tool that the state used to facilitate his integration. Today, the idea of local quality of life serves all too often to assess whether taxpayers are getting their tax dollars' worth and whether the city is competitive on the national and international stage. Because the idea of quality of life is still vague and open-ended among researchers, because the notion has been hijacked to serve business interests, and because the material living conditions of the poor have returned to abysmal levels in many countries, the standard of living has remained the term of reference for critics of current government policies (Robins, 1997). As one proponent of urban quality-of-life assessment admitted, 'there is a limit below which no one dares mention the word "quality"' (Fradier, 1976, p. 7).

Conclusion

The constructs of 'standard of living' and of 'quality of life' exemplify our continued attempts to represent human reality in objective terms and to rationalize collective decision-making. But the passage from one to the other shows a shift in public discourse from individual health to community health and even to global health. 'Quality of life' is to our post-industrial and multicultural society what 'standard of living' was to the industrial and assimilationist society of the past: a shorthand for people's conception of what is required to live a worthy life. As our collective understanding of human worth has changed, both in the political and in the economic realm, so has our representation of goals and remedies.

In principle, then, the shift from 'standard' to 'quality' denotes a widening of our perspective on human welfare, a more holistic and democratic approach to planning. The phrase 'quality of life' gives voice to our concern with the non-material dimension of well-being, beyond the basic needs of the individual. It reflects a more capacious notion of human productivity and a heightened recognition that happiness is also a function of cultural vitality and political autonomy. In the first decades of the twentieth century, a time of government expansion, the policy agenda revolved to a large extent around the maintenance of physical productivity and, therefore, around the improvement of workers' material living conditions. This priority translated in particular in an attempt to provide labourers with housing units that met certain standards pertaining to health and privacy. As affluence increased, urban planning helped to meet the demand of a growing middle-class for good neighbourhoods. In the last decades of the century, the planning agenda came to focus on improving quality of life, primarily creating better conditions of life in the public sphere.

In practice, however, in a time of government retrenchment, those who seem to be gaining the most from quality-of-life policies are not the unskilled workers of the new tenements, but the knowledge-workers of the renovated lofts, those who represent economic, social and cultural capital for the city. Devised at the beginning of the twentieth century as a synthesis of basic requirements, the standard of living soon became a general norm of social achievement. Adopted at the end of the century as an expression of social and environmental considerations, the quality of life has quickly become a marketing tool in the global economy. Still, the quality-of-life movement has also effected positive changes in public policy. It has helped to diffuse an ethos of sustainable human development, a sensitivity both to 'The Carrying Capacity of the Earth' and to 'The Caring Capacity of Humankind' (Independent Commission on Population and Quality of Life, 1996, p. xiii). More concretely, the movement has fostered the creation of new indicators to be used by governments, international agencies and non-governmental organizations. Researchers have been working steadily to generate objective measurements of urban quality of life in terms of housing, leisure and recreation, health, safety, education, employment, neighbourhood environment and public services, as well as political and civic life (Maclaren, 1995). But, like their predecessors many decades ago, reformers today will have to translate research findings into binding norms of human well-being. Without official standards, written into law, quality-of-life measures – the last word means both 'interventions' and 'assessments' – will remain ineffective as instruments of betterment for all.

The twentieth century planning experience is in part a story of benevolent reform, a continued attempt to improve the lot of the citizen by means of standards-based policies, and also a story of responsiveness to the demands of capitalism. At the same time, as a story of environmental improvement, it shows how our conception of the environment itself has broadened to include the community's political institutions and the planet's global ecology. Because physical efficiency matters less, while capital, knowledge and nature matter more, the notion of standard of living has largely given way to that of quality of life in our mental universe. There is both promise and danger in this shift. The standard of living represents the danger of modernity, the quixotic project of objectifying the subjective and of normalizing the individual. The quality of life, on the other hand, symbolizes the problem of post-modernity: we are obsessed with difference, yet we must somehow find collective rules to act in concert and define common norms to fight inequality. Averting the pitfalls of policy-making based on identity and desire and at the same fulfilling the promises of individual and community empowerment, that is the challenge for planning in the twenty-first century.

ACKNOWLEDGEMENTS

The author wishes to thank Eugenie Birch, Annick Germain, Sherry Olson and Jeanne Wolfe for their comments on an earlier draft of this essay. Part of the research for this work was supported by a grant from the Social Science and Humanities Research Council of Canada.

REFERENCES

Ackerman, F.L. (1918) What is a house? IV, in Whitaker, C.H., Ackerman, F.L., Childs, R.S. and Wood, E.E.(eds.) *The Housing Problem in War and in Peace*. Washington DC: American Institute of Architects, pp. 17-40.

Adams, T. (1934) *The Design of Residential Areas: Basic Considerations, Principles, and Methods*. Cambridge: Harvard University Press.

American Public Health Association, Committee on the Hygiene of Housing (1948) *Planning the Neighborhood.* Chicago: Public Administration Service.

Ayres, P. W. (1902) The standard of living. *Charities.* **IX**, pp. 216-220.

Baumol, W.J. and Oates, W.E. (1979) *Economics, Environmental Policy, and the Quality of Life*. Englewood Cliffs: Prentice-Hall.

Bloomberg, W., Jr. and Schmandt, H. J. (1969) Introduction: The problem of contemplating quality of life, in Schmandt, H.J. and Bloomberg, W. Jr. (eds.) *The Quality of Urban Life*. Urban Affairs Annual Reviews, Vol. 3. Beverly Hills: Sage Publications, pp. 13-23.

Bookout, L.W. (1994) Valuing landscape, site planning, and amenities. *Urban Land*, **53**, pp. 39-43.

Boorstin, D.J. (1973) *The Americans: The Democratic Experience*. New York: Random House.

Bremner, R.H. (1956) *From the Depths: The Discovery of Poverty in the United States.* New York: New York University Press.

Burke, T. and Salvaris, M. (1998) Citizenship, Progress and the Measurement of Well-Being. *Proceedings of the First International Conference on Quality of Life in Cities: Issues and Perspectives, Singapore, 4-6 March 1998*. Singapore: School of Building and Real Estate, National University of Singapore, pp. 44-66.

Burnett, J. (1978) *A Social History of Housing, 1815-1970.* London: Methuen.

Committee for Economic Development (1985) *Investing In Our Children: Business and the Public Schools*. Washington, DC: Committee for Economic Development.

Croly, H. (1964) *The Promise of American Life*. New York: Capricorn Books (originally published 1909).

De Heem, P. and De Ridder, A. (1914) La cité-jardin et l'hygiène urbaine, in W. Van Kuyck, W. (ed.) *Compte-Rendu des travaux du Quatrième Congrès international [Assainissement et salubrité de l'habitation] tenu à Anvers du 31 août au 7 septembre 1913*. Antwerp: De Vos-van Kleef, pp. 386-395.

Devine, E.T. (1917) *The Normal Life*. New York: Douglas C. McMurtrie.

Doyal, L. and Gough, I. (1991) *A Theory of Human Need.* New York: The Guilford Press.

Ellin, N. (1996) *Postmodern Urbanism*. Oxford: Blackwell.

Erickson, B. and Roberts, M. (1997) Marketing local identity. *Journal of Urban Design* **2**, pp. 35-59.

Fischler, R. (1998) For a genealogy of planning: Zoning and the Welfare State. *Planning Perspectives*, **13**, pp. 389-410.

Folks, H. (1901) The care of needy families in their homes. *Charities.* **VII**, pp. 413-419.

Fradier, G. (1976) *About the Quality of Life*. Paris: UNESCO.

Friedmann, J. and Douglass, M. (1998). Editors' Introduction, in Douglass, M. and Friedmann, J. (eds.) *Cities for Citizens: Planning and the Rise of Civil Society in a Global Age*. Chichester: John Wiley, pp. 1-6.

Glickman, L.B. (1997) *A Living Wage: American Workers and the Making of Consumer Society*. Ithaca: Cornell University Press.

Gore, A. (1993) *Creating a Government That Works Better and Costs Less: The Report of the National Performance Review*. New York: Plume Books, Penguin.

Hankiss, E. (1981) Cross-cultural quality of life research: An outline for conceptual framework, in *Quality of Life: Problems of Assessment and Measurement*. Paris: UNESCO, pp. 9-48.

Healey, P. (1997) *Collaborative Planning: Shaping Places in Fragmented Societies*. Houndsmills: Macmillan Press.

Hunter, R. (1965) *Poverty: Social Conscience in the Progressive Era*. New York: Harper and Row (originally published 1904).

Independent Commission on Population and Quality of Life (1996) *Caring for the Future: Making the Next Decades Provide a Life Worth Living*. Oxford: Oxford University Press.

Jacobs, J. (1961) *The Death and Life of Great American Cities*. New York: Vintage Books.

Jencks, C.A. (1981) *The Language of Post-Modern Architecture*. London: Academy Editions.

Jordan, T.E. (1997). *Ireland and the Quality of Life: The Famine Era*. Lampeter: The Edwin Mellen Press.

Kearns, G. and Philo, C. (eds.) (1993) *Selling Places: The City as Cultural Capital, Past and Present*. Oxford: Pergamon Press.

Kotler, P., Haider, D.H. and Rein, I. (1993) *Marketing Places: Attracting Investment, Industry, and Tourism to Cities, States, and Nations*. New York: The Free Press.

Laufer, R., and Paradeise, C. (1990) *Marketing Democracy: Public Opinion and Media Formation in Democratic Societies*. New Brunswick, New Jersey: Transactions Publishers.

Le Corbusier (1982) *Manière de penser l'urbanisme: Soigner la ville malade*. Paris: Denoël Gonthier (originally published 1946).

Leitmann, J. (1998) Can City QOL Indicators Be Objective and Relevant? Towards a Tool for Sustaining Urban Development, in *Proceedings of the First International Conference on Quality of Life in Cities: Issues and Perspectives, Singapore, 4-6 March 1998*. Singapore: School of Building and Real Estate, National University of Singapore, pp. 331-342.

Ley, D. (1996) The new middle class in Canadian central cities, in Caulfield, J. and Peake, L. (eds.) *City Lives and City Forms: Critical Research and Canadian Urbanism*. Toronto: University of Toronto Press, pp. 15-32.

Lippmann, W. (1935) *The New Imperative*. New York: Macmillan.

Maclaren, V.W. (1995) *Developing Indicators of Urban Sustainability: The Canadian Experience*. Ottawa: Environment Canada, Canada Mortgage and Housing Corporation, and Intergovernmental Committee on Urban and Regional Research.

Marsh, B.C. (1909) *An Introduction to City Planning. Democracy's Challenge to the American City*. New York: The author.

Maslow, A.H. (1954) *Motivation and Personality*. New York: Harper and Row.

Miles, I. (1985) *Social Indicators for Human Development*. London: Frances Pinter (Publishers) for the United Nations University.

Mitchell, B.R. (1988) *British Historical Statistics*. Cambridge: Cambridge University Press.

Nussbaum, M. and Sen, A. (1993) Introduction, in Nussbaum, M. and Sen, A. (eds.) *The Quality of Life*. Oxford: Clarendon Press, pp. 1-6.

Osborne, D. and Gaebler, T. (1993) *Reinventing Government: How the Entrepreneurial Spirit is Transforming the Public Sector*. New York: Plume Books, Penguin.

Parmenter, T.R. (1994) Quality of life as a concept and measurable entity. *Social Indicators Research*, **33**(1-3), pp. 9-46.

Peattie, L. (1998) Convivial cities, in Douglass, M. and Friedmann, J. (eds.) *Cities for Citizens: Planning and the Rise of Civil Society in a Global Age*. Chichester: John Wiley, pp. 247-553.

Perry, C.A. (1929) The Neighborhood Unit: A Scheme of Arrangement for the Family-Life Community. *Regional Survey of New York and Its Environs. Vol. VII: Neighborhood and Community Planning*. New York: Regional Plan of New York and Its Environs, pp. 21-140.

Petrow, S. (1995) John Daniel Fitzgerald: Planning as progressive reform. *Planning History*, **17**(1), pp. 4-7.

Pigou, A.C. (1914) Some Aspects of the Housing Problem. *Lectures on Housing: The Warburton Lectures for 1914*. Manchester: Manchester University Press, pp. 35-66.

Reich, R.B. (1992) *The Work of Nations: Preparing Ourselves for 21st-Century Capitalism*. New York: Vintage Books, Random House.

Richards, E.H. (1910) *Euthenics: The Science of Controllable Environment*. Boston: Whitcomb and Barrows.

Robins, H. (1997) The test of the city's recovery: Is life better for the average New Yorker? *Social Policy* **27**(4), pp. 4-29.

Rosanvallon, P. (1995) *La nouvelle question sociale*. Paris: Éditions du Seuil.

Routledge, P. (1995) Resisting and reshaping the modern: Social movements and the development process, in Johnston, R.J., Taylor, P.J., and Watts, M.J. (eds.) *Geographies of Global Change: Remapping the World in the Late Twentieth Century*. Oxford: Blackwell, pp. 263-279.

Rowntree, B.S. (1901) *Poverty: A Study of Town Life*. London: Macmillan.

Rowntree, B.S. (1914*a*) Housing reform in Great Britain, in *Report of Proceedings of the Imperial Health Conference, London, May 18th to 21st, 1914*. London: Victoria League, pp. 89-99.

Rowntree, B.S. (1914*b*) How far it is possible to provide satisfactory houses for the working classes at rents which they can afford to pay, in *Lectures on Housing: The Warburton Lectures for 1914*. Manchester: Manchester University Press, pp. 3-31.

Rowntree, B.S. (1918) *The Human Needs of Labour*. London: Thomas Nelson and Sons.

Sandercock, L. (1998) *Towards Cosmopolis: Planning for Multicultural Cities*. Chichester: John Wiley.

Sen, A. (1986) The standard of living, in McMurrin, S.M. (ed.) *The Tanner Lectures on Human Values VII*. Cambridge: Cambridge University Press, pp. 1-51.

Smith, D.M. (1977) *Human Geography: A Welfare Approach*. London: Edward Arnold.

Troy, P. (1995) Cooperation Between Governments, in Troy, P. (ed.) *Australian Cities: Issues, Strategies and Policies for Urban Australia in the 1990s*. Cambridge: Cambridge University Press, pp. 265-282.

United Nations (1954) *Report on International Definition and Measurement of Standards and Levels of Living*. New York: United Nations.

Unwin, R. (1914) Town Planning, in *Report of Proceedings of the Imperial Health Conference, London, May 18th to 21st, 1914*. London: Victoria League, pp. 29-46.

Veiller, L. (1918) Industrial developments in America. Part II: The Government's standards for war housing. *Architectural Record*. **43**, pp. 344-59.

Weiss, D. (1992) Changing paradigms of development in an evolutionary perspective. *Social Indicators Research*, **26**, pp. 367-389.

Whitaker, C.H. (1918) What is a house? in Whitaker, C.H., Ackerman, F.L., Childs, R.S. and Wood, E.E. (eds.) *the Housing Problem in War and Peace*. Washington, DC: American Institute of Architects,pp. 3-8.

Whitt, J. A. (1987) Mozart in the metropolis: The arts coalition and the urban growth machine. *Urban Affairs Quarterly*, **23**, pp. 15-36.

Wilson, W. H. (1989) *The City Beautiful Movement*. Baltimore: The Johns Hopkins University Press.

Wright, G. (1981) *Building the Dream: A Social History of Housing in America*. Cambridge, Mass: MIT Press.

CHAPTER 9

The Paradoxes of Anti-Sprawl Reform

Robert Bruegmann

One of the most conspicuous developments in the realm of planning thought during the last decades of the twentieth century has been a concerted attack on urban 'sprawl'. Although this word was first commonly used in something like its present sense in Britain in the interwar years and was imported to the United States in the 1950s, it has been primarily in the 1990s that it has served as the focus of a broad-based reform movement. According to many reformers today, sprawl, which has typically been described as low-density and unplanned development at the periphery of large cities, is ugly, inefficient, inequitable, and environmentally damaging. To combat sprawl, a large, if loose, coalition of activist planners, architects, environmentalists, historic preservationists and other reformers, has come together to promote policies encouraging what they believe will be more carefully planned, more efficient, compact and environmentally friendly cities.

In this chapter, which is primarily focused on relatively affluent cities in the English-speaking world, I will examine the evolution of the term 'sprawl' and the reasons that the current anti-sprawl crusade appeals, at least in theory, to a large number of diverse interest groups in cities worldwide. I will also argue that because these ideas were never clearly defined, are biased by class attitudes, and have been extrapolated from insufficient and out-of-date data, the results are likely to be irrelevant, ineffectual or even counter-productive. Anti-sprawl efforts will almost certainly not stop the development of the new multi-centred and highly dispersed urban settlements that are appearing in cities throughout the world.

FROM 'BLIGHT' TO 'SPRAWL'

The origins of contemporary concern about sprawl, as paradoxical as it might seem at first glance, are deeply rooted in the long-standing concern of urban reformers over excessive density. This is, on reflection, not surprising, however, since the forces of centralization and decentralization have been acting simultaneously on cities since the beginning of urban history – particularly since the beginning of the Industrial Revolution. In a period when cities were growing with unprecedented speed, increasing densities at the core were usually accompanied by the movement of affluent citizens out of the centre to new, less dense neighbourhoods at the urban periphery. In both sides of this process new technologies played a central, if always ambivalent, role. The railroad and streetcar, the steam engine and electric power, for example, aided decentralization not only by making industry less dependent on specific locations for power supply and transportation, but also by allowing workers to live further from their jobs. These same new technologies, however, intensified the competitive advantage of certain parts of the city, notably the central core, and made it possible to build larger, more tightly packed and profitable districts. In the biggest cities, like Paris and New York, these densities could reach figures of well over 100,000 people per square mile (Harris, 1956). There was a widespread feeling that this excessive density caused not only disease but also social pathologies such as crime, juvenile delinquency and civil disorders.

By the early twentieth century, the term 'blight', earlier used by farmers to describe the effects of microscopic insects on plants, had been taken over to describe the pathological results of overcrowding. The medical imagery was very useful to reformers because it gave them a role in the city analogous to that of the doctor in the treatment of the human body. They would plan the surgical operations that would remove the diseased portions. Widespread demolition of old, overly dense quarters started in Europe in the mid-nineteenth century and reached its apogee in Europe and the United States in the aftermath of World War Two when massive urban renewal schemes were envisioned as a way to eliminate entire districts and rebuild them as healthy new neighbourhoods at much lower densities.

From the beginning of this decentralization effort, reformers realized that a corollary to the rebuilding of central cities was the creation of mechanisms to guarantee a satisfactory distribution of population on the periphery. A long line of reformers descended from Ebenezer Howard codified the available tools including new satellite cities, limitations on settlement size, green belts and other measures to make the process tidy and preserve the countryside. These ideas became the basis of accepted wisdom among planners about how to develop the periphery of cities.

The residential population of industrial cities in the Western world did, indeed, thin out, often dramatically. This process appears to have started in the central core of the most mature cities. The City of London, for example, counted 120,000 inhabitants per square mile in 1801 but its population was probably already in the process of declining. This was caused by both push and pull factors. As more space was needed for commercial purposes the cost of land went up, driving many of the inhabitants out to adjoining neighbourhoods. At the same time, many of those able to afford the choice had, at least since the seventeenth century, been moving to the comparative tranquillity of elegant new districts like the West End. In newer cities, particularly those in the United States, the peak of density in central neighbourhoods was not reached until the very end of the nineteenth century. The Lower East Side of New York, for example, registered over 500,000 people per square mile at the turn of the century, certainly one of the highest densities on record (Hall, 1966). Here, too, densities started to fall precipitously shortly thereafter.

In virtually every city in the Western world, the density gradient – the rate at which density falls as one gets further from the centre – has been declining for over a century as population density in the centre has declined and the population has spread itself over a constantly larger area. By about 1950, many of the large metropolitan areas of Europe and America exhibited similar overall densities of about 4000-6000 people per square mile (Harris, 1956). This was also the density of many middle class American postwar automobile suburbs. Levittown on Long Island, for example, counted about 6000 people per square mile. Since that time, most Western cities, and especially American ones, have continued to disperse. Except in a few central core areas, American urban areas now average around 3000 people per square mile, and many affluent suburban areas – for example those with two-acre lots and small families – well under 1000 people per square mile (Jackson, 1985; Morrill, 1991).

When densities fell, they usually did so with very little regard to the dictates of the reformers and planners. Already in the 1920s, the low-density suburban landscape that was increasingly visible throughout England and America horrified some of the very reformers who had most fervently advocated decentralization. By the end of the interwar years, a campaign against the newly discovered evil of 'sprawl' was underway. In this chapter, I will follow four tracks of argument against the evils of sprawl.

Aesthetic Objections

The first and most persistent strain of attack on sprawl has been based on aesthetic objections. This strand of argument was already conspicuous in England by the time of World War Two. In the nineteenth century, according to the *Oxford English Dictionary*, the old verb 'to sprawl', meaning to spread out in straggling fashion, had come into use as a noun. It meant a display or array of objects spread out in straggling fashion, for example a 'sprawl of people'. It appears that between the wars this term was taken over to describe parts of the quickly expanding suburban landscape around British towns, particularly the 'ribbon development' that bordered the roads running from the suburban fringe into the surrounding countryside. The objections were almost always made in the context of laments about the despoiling of the traditional British countryside.

An early example of this kind of criticism can be found in an acerbic little book published by the architect Clough Williams-Ellis: 'As the Joneses fly from the town, so does the country fly from the pink bungalow that they have perched so hopefully on its eligible site. The true countryman will know that the area is infected – the Jones have brought the blight of their town or suburb with them' (Williams-Ellis, 1928, p. 40).

This passage is notable both for the way the word 'blight' is still used for what would soon become 'sprawl' and for the obvious class bias that has marked so much of the sprawl reform literature from that day to this. Obviously, for the true 'countryman', the proper kind of building in the countryside would be the great country house and the agricultural village that depended on it, in other words the component parts of the feudal landscape. From this point of view, the strivings of the middle classes to obtain for themselves what had been the privilege of the landed gentry could only result in disorder and ugliness. Although this might be called an aesthetic idea, notice that this, like any important aesthetic idea, was far from being just a matter of taste. It was deeply rooted in very basic notions about the nature of the natural social order. These very deep but often deliberately obscured underpinnings give aesthetic belief an unmatched power and tenacity.

Perhaps the most powerful and revealing spokesman for the countryside as unspoiled preserve for aristocratic pleasures was Thomas Sharp. 'The new suburbs and town extensions sprawl out in a sloppy diffuseness all over the countryside,' he wrote in 1932, using the word sprawl as a verb (Sharp, 1932, p. 146). 'Tradition has broken down. Taste is utterly debased. There is no enlightened guidance or correction from authority. The town, long since degraded, is now being annihilated by a flabby, shoddy romantic nature,' he wrote in an earlier part of the same book (Sharp, 1932, p. 6). For Sharp, the end of this 'negative semi-suburbia' and a return to sane building would only come with a return to what he imagined were eternally valid ordering principles: 'The crying need of the moment is the re-establishment of the ancient antithesis: the town is town: the country is country; black and white; male and female. Only in the preservation of these distinctions is there any salvation' (Sharp, 1932, p. 11).

Both urban and rural beauty in the past had come from 'some degree of enlightened autocratic control,' Sharp wrote (Sharp, 1932, p. 218). He suggested the way back to a satisfactory path could only be obtained through a 'stringent but enlightened control from authority' that would be able to 'preserve for dull democracy that rural beauty which was created by enlightened autocracy' (Sharp, 1932, pp. 220, 224). For many observers of Sharp and his allies in the Council for the Preservation of Rural England, this process of 'preservation' was actually nothing more than a way for the gentry to preserve privilege (King, 1980).

By the end of World War Two, the term sprawl as a noun meaning all that was so disliked by Thomas Sharp was in such common usage in Britain that it could be employed without any explanation in the famous *Greater London Plan* of

1944. In this volume not only did the word appear a number of times in the text but it figured graphically as an alarming band of colour on a map featured on the volume's dust jacket (Abercombie, 1945). For Abercrombie the continuing, and still most important, fight against blight had to be balanced by the new fight against sprawl.

During the postwar years, as many cities continued to decentralize rapidly, the fight against blight gradually receded and the fight against sprawl came to the fore. In Britain the most conspicuous war cry was the set of special issues of the British journal *Architectural Review*. In the first number, entitled 'Outrage,' Ian Nairn described in nearly apocalyptic terms the urban area as spreading across vast tracts of farm and open space in the form of urban sprawl. He called it a 'thing of terror,' and something that had the power to 'get you up sweating at night' (Nairn, 1955, p. 865).

Nairn appears to have been the chief link between the British anti-sprawl crusade and what would soon become its American counterpart. Nairn's work in the *Architectural Review* quickly came to the attention of William H. Whyte, an editor of *Fortune* magazine. In the 1950s, Whyte organized what was probably the first conference on sprawl. This conference resulted in a widely read little book entitled *The Exploding Metropolis* (Fortune Magazine Editors, 1958). In his own contribution, an essay simply entitled 'Urban Sprawl', Whyte's chief criticisms were, as usual, aesthetic, but he also remarked, almost parenthetically, that sprawl was not only bad aesthetics but bad economics (Whyte, 1958). This was a theme that would soon be taken up by others in a serious way. Closely associated with Whyte in the magazine and the conference was Jane Jacobs. Although she did not use the term sprawl in any conspicuous way in her enormously influential 1962 work, *Death and Life of Great American Cities*, she did provide a classic defence of high-density inner city living and critique of contemporary low-density suburban settlement patterns. Jacobs, like many others before and since, used Los Angeles as an example of all that was wrong with American cities, curiously attributing to that city exactly the kinds of high crime levels and urban pathology that observers at the beginning of the century had associated with the overcrowded centres of the great industrial cities (Jacobs, 1962, p.64).

A clear indication that 'sprawl' had arrived as a standard term by the beginning of the 1960s can be found in Lewis Mumford's magisterial work *The City in History*. Mumford, who had been one of the strongest supporters of urban decentralization for much of his life, entitled one section of his book 'Railroad Line, Greenbelt, Motor Sprawl'. This triad for Mumford encapsulated a century of history. First the railroad helped to spawn the dense industrial city. Then came the efforts of which he was a part to reform and decentralize this city. Finally, however, these efforts were undermined by the appearance of the private automobile, the agent that fuelled sprawl and spelled the end of the coherent urban and suburban landscapes Mumford was advocating. In his concise epigrammatic history, Mumford revealed the powerful class-bound aesthetic biases that he probably inherited from his British planning mentors:

> Whilst the suburb served only a favored minority it neither spoiled the countryside nor threatened the city. But now that the drift to the outer ring has become a mass movement, it tends to destroy the value of both environments without producing anything but a dreary substitute, devoid of form and even more devoid of the original suburban values. We are faced by a curious paradox: The new suburban form has now produced an anti-urban pattern . . . Unwin's salutary demonstration, 'Nothing Gained by Overcrowding' must now be countered with a qualifying admonition: 'Something Lost by Overspacing' (Mumford, 1961, p. 510)

The postwar American discussion of sprawl was closely connected with attacks on 'mass culture', 'middle-brow' taste, and on suburbia generally, which was often described as socially stifling and culturally barren. Many observers, particularly academics in the Northeast of the United States, continued to take a special delight in castigating Los Angeles, which they viewed as the apotheosis both of middle class life and sprawl. William Whyte, for example, observed that the view as seen by an airline passenger flying over the stretch of land between Los Angeles and San Bernardino offered 'an unnerving lesson in man's infinite capacity to mess up his environment' (Whyte, 1958, pp. 115-116). Low-density Los Angeles had come to stand for the crisis of the contemporary city in much the same way that the high-density East End of London had served for an earlier generation of reformers (Peel, 1995).

During the 1960s and 1970s, the debate over sprawl and the suburbs raged. 'This book is not written in anger. It is written in fury', wrote architect Peter Blake on the first page of his book *God's Own Junkyard* (Blake, 1964). He went on to say that Americans were about to turn their beautiful country into the 'biggest slum on the face of the earth'. The anti-suburban and anti-sprawl crusade did find its own vigorous critics. Sociologist Herbert Gans mounted an extremely effective defence of suburbia by showing the wide variety of people that lived there and the ways in which their life was quite similar to that of their counterparts in the city. On the matter of sprawl he wrote, 'I have never seen any persuasive evidence that sprawl has bad effects or high-density development significant virtues' (Gans *et al.*, 1982, p. 176). The eminent geographer Jean Gottmann, whose book *Megalopolis* became one of the most important urban texts of the 1960s, wrote 'There is no doubt . . . that the urban sprawl will go on; that it will accelerate in most countries; and that, like it or not, the great majority of mankind is praying for it to come, to develop and satisfy them' (Gottmann and Harper, 1967, p. 6). Despite these rebuttals, the criticisms of suburbia have continued unabated (Kunstler, 1993; Sorkin, 1992).

Efficiency Arguments

A second main track of argument about sprawl, more measured in tone, has been about efficiency. By the mid-1960s, most of the elements of the 'costs of sprawl' debate had been put in place (Clawson, 1962; Harvey and Clark, 1965; Lessinger, 1962). Although sprawl might be very efficient for the new arrival buying cheap land at the city edge, these critics argued, it was actually inefficient for society at large because it spreads people out, induces longer commuting, eats up valuable farmland, and requires more infrastructure. Although for various of these writers sprawl might mean low-density scattered settlement patterns, ribbon development along roads leading out of towns, and leapfrog development – that is development that skipped over areas of land that remained undeveloped – there were, from the very start, major disagreements on which elements constituted sprawl and which were harmful.

The 'costs of sprawl' literature, moreover, almost always uses the typical social science format in which a survey of the literature is followed by a series of empirical studies which test previous hypotheses and then a conclusion which is derived from the new evidence. Although this format suggests a scientific neutrality, in fact the assumptions of many social scientists – for example, a strong belief in environmental determinism and an equally strong belief that government should play a major regulatory role in land use – are so deeply engrained that they virtually guarantee that only a relatively narrow range of conclusions will be reached and that evidence to the contrary will be discounted.

The landmark study was a report entitled *The Costs of Sprawl* prepared for several American government agencies in the early 1970s (Real Estate Research Corporation, 1974). In this study the authors devised a theoretical model to analyse six different kinds of neighbourhood prototype, ranging from low-density unplanned development (sprawl) to high-density planned (meaning increased clustering and compactness) development. They concluded that planned communities were less costly to build and operate than unplanned communities and that high and moderately high densities (for example walk-up apartments) were less expensive than low densities. Altogether, the authors found, a moderately high-density planned community would cost 41 per cent less than a low-density sprawl community. In addition, the report found that unplanned low-density development produced undesirable environmental effects, led to longer commuting, higher fuel consumption and pollution, and meant an increased cost burden on municipalities.

These findings drew immediate rebuttals. Reviews and subsequent studies fatally undermined many of the key points. The report failed to control for the size of the housing units, which meant that much of the reported increase in costs for lower density settlement could be simply explained not by density but by construction costs. The report also neglected many of the most important benefits of sprawl, some critics maintained (Altshuler, 1977, Windsor, 1979). In actual practice, argued Richard Peiser, planned development did not present anything like the savings the report predicted, and Helen Ladd has shown that increasing suburban density does not necessarily lower municipal service costs (Ladd, 1992; Peiser, 1989). Over the years the objections to the findings of the

Costs of Sprawl report have continued to pile up, particularly in the work of Peter Gordon and Harry Richardson, who have devoted a good part of their careers to refuting the economic arguments of the sprawl reformers (Gordon and Richardson, 1997*a*, 1997*b*).

In the latest and most careful American summary of the costs-of-sprawl to date, a report created for the Transportation Research Board and National Research Council, authors Robert Burchell, Anthony Downs and their colleagues tried to take these critical views into consideration. They conceded that sprawl, although considered beneficial by much of the public, still would need to be stopped:

> Sprawl is so endemic to the culture of the United States that it is almost impossible to change. Americans like its outcome. It provides safe and economically heterogeneous neighbourhoods that are removed from the problems of the central city. In low-density, middle-class environments, life is lived with relative ease, and when residents wish to relocate, they typically leave in better financial condition – the result of housing appreciation.
>
> The public services available to residents in sprawl locations are more than adequate – and their cost, until recently, has been relatively inexpensive. But costs are beginning to increase. Americans are looking, albeit halfheartedly, for an alternative to current development patterns. There is a general sentiment that communities and individuals specifically, and society as a whole, cannot continue to pay for the costs of sprawl. Costs have been held at a manageable level only because existing infrastructure is ignored and developmental infrastructure is not repaired or replaced adequately. (Burchell *et al.*, 1998, p. 36)

Perhaps the most startling aspect of this conclusion was the way it undermined the statistical basis of almost all of the costs of sprawl literature to date. Sprawl had not yet proved excessively expensive, the authors concluded. They then offered two different arguments. Either it would soon become too expensive or it was already too expensive but was masked by deferred maintenance. But since no real argument was made in the report for either of these two propositions, the document perhaps succeeded better in demolishing accepted wisdom than in establishing a new basis for argument.

EQUITY ARGUMENTS

The third strain of anti-sprawl argument, one that has been particularly important in the United States, has been based on issues of equity. These arguments were clearly modelled directly on the anti-suburban literature of the 1960s and 1970s. Following the logic of this literature, the contemporary urban pattern could be explained by the flight of affluent white citizens to the suburbs which in turn caused an increasing concentration of poor non-white residents in the centre city and, at the same time, siphoned tax funding away from the central city (which needed it most) to the suburbs.

Among the many difficulties in sustaining this argument was the question, if one believed that race was a primary cause of suburban growth, of why metropolitan areas with small minority populations have decentralized just like those with large minority populations. Likewise, if a desire to avoid taxes to help poor city residents was important, why have places where the central city covers all or most of the metropolitan area decentralized much like those with a great deal of political fragmentation and without any greater redistribution of income? How can one prove that low-density settlement at the edge caused deterioration at the core, particularly when some of the cities with the most robust centres also have a great deal of low-density growth at the edge? In the hands of the sprawl reformers, these connections did not have to be proved. They were self-evident. Consider this example from the late 1960s:

> In hindsight we can see how the affluence of the great majority of Los Angeles residents led them to the development of low-density residential areas and to an elaborate automobile freeway network, the combined effect of which eroded mass transportation service to the point where important parts of the metropolitan area became inaccessible to those who could not afford to own an automobile or could not drive. The riots in the Negro suburb, Watts, were in large measure a manifestation of the growing inaccessibility of centralized job opportunity. (Thompson, 1967, p.177)

Here, in a single paragraph, the low-density suburb leads directly to inner city violence.

One perennial remedy for perceived inequities has been the creation of metropolitan governments. According to this logic, such a governmental structure would level the playing field, allowing all parts of the metropolitan area to compete on an even footing and solve many urban problems (Rusk, 1995). Of course, as others have pointed out, the result might be quite the opposite as bureaucracy would grow and minority groups would have their voices greatly diminished in the larger governments. In the matter of sprawl, moreover, no one has proved that metropolitan areas with larger central cities have been more effective in controlling sprawl or in fostering equity than those with more fragmented government. In some ways the equity arguments have been undercut by new work by Myron Orfield who argues that the inner suburbs now share many of the same problems as the central city (Orfield and Rusk, 1998). Although it was Orfield's stated intention to create a coalition based on the coincidence of interests of central city and inner suburb, it can easily be argued that, given scarce resources, the inner suburbs deserve attention much more than the inner cities because they have the problems but not the major resources of the central cities. The shift of focus away from the problem of the central cities will undoubtedly be very unappealing to the entrenched academic and institutional interests in the central cities who have traditionally been the main standard bearers for the anti-sprawl crusade.

The debate about the effects of sprawl on equity in the United States has chiefly been waged between reformers in the liberal tradition of the New Deal and the 1960s, on the one hand, and, on the other, fiscal conservatives and libertarians. The former are likely to argue for governmental controls to soften the effects of the private market, constrain sprawl, redistribute income, and increase equity. The latter usually argue either that sprawl has little effect on equity issues or that it has been the interference of government itself, notably regulations that drive up housing costs and policies that subsidize the concentration of poor people in the city, that has been the chief cause of the problem.

ENVIRONMENTAL ARGUMENTS

The most recent, and certainly most contentious, aspect of the sprawl debate relates to environmental issues and to 'sustainability' (Breheny, 1992, 1995, 1996; Jenks *et al.*, 1996). Like many of the other strands of this debate, this is a reformulation of issues that were hotly debated in the 1960s. In this instance, it has been the concept of global warming that has revivified earlier environmental reform efforts that had lost much of their urgency as urban air and water got noticeably cleaner in almost all Western cities since World War Two and the energy crises of the early 1970s failed to have their predicted dire impact on urban patterns.

The factual basis for the current environmental attack on sprawl is an observation that low-density cities use more energy than high-density cities (Newman and Kenworthy, 1989). The conclusions usually drawn are that this is inequitable because small numbers of people are using a disproportionate share of a fixed amount of energy and that it is unsustainable because mankind will run out of energy or greenhouse gases will damage the Earth's ecosystem. On the basis of these conclusions, a number of agencies worldwide have put forward plans of action to change development patterns (European Commission, 1990; World Commission on Environment and Development, 1987). For critics of sprawl, this line of argument has great advantages. It serves the same function for reformers today as did the correlation of disease incidence with urban density for reformers in the early 1900s. It appears to suggest that there is a crisis and, with it, conclusive statistical evidence for a specific programme of reform as well as immunity from the charge that the prescriptions are based on subjective personal taste or class-bound notions.

A major problem both one hundred years ago and today is the shaky causal link between the perceived problem and observed phenomena. In fact, we know today, it was micro-organisms and not high density itself that caused disease at the turn of the century. Likewise, it is almost certainly true that low-density sprawl itself does not cause high energy use or pollution today, as even a quick glimpse at many low-density farming communities with low energy consumption would suggest. What makes the current environmental arguments more difficult to sustain is the unresolved debate over whether mankind is in any danger of running out of energy or whether there is potentially unlimited energy that can be tapped using new technology. There is also, even if one assumes that global warming is a long-term condition and not a statistical blip, a fundamental lack of agreement on whether it could be reversed by changing land use patterns, or, indeed, whether the effects of global warming might not be good rather than bad for at least some parts of the world (Moore, 1998; Philander, 1998).

In the end, the environmental attack on sprawl is clearly predicated on some very basic notions about the relationship of people and nature, whether there is a self-regulating, purposeful nature and whether humankind is part of it or not. In short, environmental arguments that have attempted to lift the sprawl debate out of the realm of personal taste and subjective opinion, may, in fact, have actually raised it into the realm of basic science and metaphysics. Although the debate is far from played out, and scientists may, indeed, prove conclusively that it will be necessary to alter urban development patterns in ways advocated by anti-sprawl critics, it seems more likely that with the latest environmental objections, the arguments about sprawl have merely come back full circle to the highly subjective grounds on which they were initially founded, that is to the realm where answers to the question of proper city form flow more or less directly out of deep personal belief and conviction rather than from any empirical evidence.

The Debate becomes International

Historically concerns about sprawl first became prominent as a response to a rapid rise in affluence and a resulting democratization of the fringes of Anglo-American cities, and the sprawl debate has always been most active in England and the United States. The argument has also been extended without much alteration in its terms into other wealthy, predominately English-speaking countries, notably Canada and Australia. In Australia, however, where planning controls are similar to those in Britain, strong opposition to anti-sprawl controls has come from the political left rather than primarily from the political right as has been the case in the United States. A number of individuals, following Patrick Troy, have argued that compact development policies actually help foster inequalities because they continue to privilege the city centre rather than the places where most Australians of modest means live (Troy, 1996). Michael Poulton has made similar arguments from the Canadian context (Poulton, 1995). Poulton has argued that the real beneficiaries of sprawl control are an 'incumbents' club', those who already enjoy great benefits from their situation in the city and wish to halt further change. These are the individuals most ready to support the compaction of the less privileged part of the population into higher-density neighbourhoods, a process for which the derogatory term 'town cramming' has been applied.

The debate has also extended itself to places outside the Anglo-American world, notably in continental Europe. This has happened, because, despite obvious differences between American and European cities, particularly cities in the south of Europe, it has become clear that many Europeans, given a choice, prefer low-density living of the kind so much in evidence in the United States (Bauer and Roux, 1976; Dezert, Metton and Steinberg, 1991; Monclús, 1998). Although stopping scattered low-density development is now official policy in Europe, the very presence of the debate suggests that there is considerable question whether the highly dispersed settlement patterns increasingly evident in Europe can be stopped or whether they are the all but inevitable result of increasing affluence and democratization (Bruegmann, 1996; Commission for the European Community, 1990). Debates about sprawl, often in quite different contexts and using very different definitions, finally, have become common in other parts of the world, notably in Japan and Israel (Hebbert, 1994; Razin, 1998). It might be suggested that criticism of sprawl, like sprawl itself, is the inevitable complement of increasing affluence and democratization.

The Anti-Sprawl Coalition

It is in the United States that the anti-sprawl rhetoric is loudest and the anti-sprawl coalition most developed. The current coalition consists of a large group of academics, architects and planners, historic preservationists, land conservationists, environmentalists as well as advocates for farmland protection, the central city and public transportation. Each of these groups believes the anti-sprawl fight can be used effectively to further its mission (Moe and Wilke, 1997; *Newsweek*, 1995; Sierra Club, 1998). The coalition unites individuals on the political and artistic right who would like to return to an earlier, more orderly, even authoritarian world, with individuals who consider themselves liberals in the 1960s tradition who would like to use government powers to promote what they view as a more efficient, equitable society. Among the assumptions shared by many individuals on both sides is an upper middle-class disdain for the messy reality of the middle-class fringes of large cities and a common desire to move control upward in the governmental hierarchy away from ordinary citizens and into the hands of professional experts.

Much of the current anti-sprawl popular literature, like the writing of Thomas Sharp and his contemporaries in the 1930s, is less interested in defining sprawl than in using it as a polemical weapon to argue for a return to some presumably golden age of the past. In fact the definitions of the term have been so numerous and have changed dramatically since the 1930s, or even the 1960s, that it is increasingly necessary to discuss which definition of sprawl one is using. This became strikingly apparent in the confusion over the meaning of the phrase 'Los Angeles Style Sprawl' in a recent important debate on the topic in the pages of the *Journal of the American Planning Association* (Ewing, 1997; Gordon and Richardson, 1997*a*, 1997*b*). Although the term and the object of attack may have changed, however, the rhetorical tradition has not. Most of the same criticisms made of sprawl today, that it is disordered, privatized, congested and ugly, are the same ones used by critics of Chicago in the late nineteenth century and of

London in the early nineteenth century. Thus the very places that are often held up as the golden ages of the past were themselves criticized for many of the same things in their own day. In fact these criticisms seem to be common in almost every place that is growing quickly.

CONCLUSION

If history is any guide, it is unlikely that the current anti-sprawl coalition will survive for any length of time given the very different, even warring, assumptions that lie behind the agendas of the constituent groups. History suggests, moreover, that even if the coalition were successful, the measures might well backfire, at least as perceived by reformers of the next generation. Here the medical analogy is suggestive. Many doctors, while admitting that certain measures of previous eras – bloodletting, for example – could have been counter-productive, rarely admit that their current prescriptions are just as liable to be condemned in a subsequent generation. So, too, the dams and atomic energy plants, championed by many environmentalists of a previous era to cut down on fossil fuel use and pollution, came to be the objects of emotional attacks by the same constituency a generation later.

The same kind of reversal of opinion has characterized many if not most of the large scale attempts in modern urban planning. Urban renewal, public housing, urban expressways and pedestrian main streets, all advocated by some planners as a way for the central city to compete effectively with the suburbs in the postwar years, came, by the 1960s, under attack by the very reform groups that had pushed hardest for them in the first place. The same is true with planning for the urban periphery. Although many of the planning devices to control growth at the metropolitan fringe, notably England's green belts and satellite cities, did have some of the effects intended, it is quite likely that they also had side effects like increased housing prices and driving growth out beyond the greenbelt, that could, for significant parts of the population, be considered worse than the initial problem (Hall *et al.,* 1973).

The same might be said of the growth management, slow-growth and anti-growth efforts of the 1970s and 1980s in the United States that are the direct precursors for the 'Smart Growth' and 'Sustainable Growth' mantras of today. In these decades, the mechanisms to stop, slow or channel growth included building moratoria, growth boundaries, large-lot zoning, and 'concurrency' regulations that made building contingent on having sufficient infrastructure in place before development could proceed. Although these did do some of the things they were designed to do, they also tended to drive up the price of land, thus hurting those least able to afford housing, and they were mostly unsuccessful on a larger scale since growth usually just moved to areas beyond the regulations, meaning that it might easily spread even further out than it would have without them (Porter, 1997).

Although current reformers often decry important aspects of these previous policies, the popular appeal of the current wave of anti-sprawl efforts appears to be based on exactly the same desire to stop change where it will be most visible to current inhabitants (Audirac, Shermyen and Smith, 1990). Instead of stopping peripheral growth entirely, which proved impossible, the new ideal is to make it invisible by channelling it as much as possible into existing areas. However, just as 'sprawl' is always almost used to refer to places other than where the speaker lives, so, too, do most people using the term appear to imagine that densification will happen in neighbourhoods other than their own. So far most of the anti-sprawl measures have not had much bite, even in places like Portland, Oregon, where anti-sprawl has been public policy for decades (Abbott, Howe and Adler, 1994; Knapp and Nelson, 1992; O'Toole, 1999). The moment that they do and the abstract principles become reality, when apartment buildings replace single family homes in established neighbourhoods, and when automobile congestion inevitably increases because of the greater density and the refusal to build new roads, the enthusiasm for these measures will almost certainly decline.

These historical reversals and policy backfires underline one of the critical weaknesses of the current anti-sprawl crusade: the flimsy evidence on which it is based (Breheny, 1996). Our urban systems are simply too large and complicated for us to understand, let alone to 'fix'. The vast literature on the key question of the land use-transportation connection, for example, has been mostly based on home-to-work commuting patterns of single household members at a given moment in time in cities that are understood as having a single dominant centre. None of this even remotely reflects current urban reality (Morrill, 1991). Almost all of the projections into the future, moreover, tend to accept a steady state of technology, rather than the probability that our urban world will have quite different technology and functions in 50 years.

Does the record of paradoxes, backfires and unintended consequences suggest that professional planning can accomplish nothing, that there is no way to solve any urban problem? Definitely not: planners and reformers have had important places in the management of the city, if only in their efforts to make understandable the impacts of changes on various parts of the society. They have also undoubtedly helped in creating courses of action that have minimized some of the negative impacts on those hurt by urban change. It is mostly a matter of perspective and stance. If reforms are seen as temporary and makeshift solutions to specific conditions that cannot be defined very precisely and will probably have changed by the time the reform takes place, it will be easier to resist the temptation to rush to broad-brush solutions. It will also be easier to realize that many of our frustrations are not the result of things getting worse, at least in cities in developed countries. It is, in fact, because conditions have got so much better for such a large part of the urban population so quickly and such a large part of the population has been able to enjoy the privacy, mobility and choice that were once the privilege of only the wealthy that expectations on the part of citizens often run ahead of current realities. The agitation against sprawl is a clear indication of the dynamism and resiliency of urban systems.

REFERENCES

Abbott, C, Howe, D. and Adler, S. (1994) *Planning the Oregon Way: A Twenty Year Evaluation*. Corvallis, Oregon: Oregon State University Press.

Abercrombie, P. (1945) *Greater London Plan 1944*. London: HMSO.

Altshuler, A. (1977) Review of the costs of sprawl. *Journal of the American Institute of Planners*, **43**, pp. 207-209.

Audirac, I., Shermyen, A. and Smith, M. (1990) Ideal urban form and visions of the good Life: Florida's growth management dilemma. *Journal of the American Planning Association*, **56**, pp. 470-482.

Bauér G. and Roux, J. M. (1976) *La rurbanisation, ou la ville éparpillée*. Paris: Editions du Seuil.

Blake, P. (1964) *God's Own Junkyard*. New York: Holt Rinehart and Winston.

Breheny, M. (ed.), (1992) *Sustainable Development and Urban Form*. London: Pion.

Breheny, M. (1995) Counter-urbanisation and sustainable urban forms, in Brotchie, J., Batty, M., Blakely, E., Hall, P. and Newton, P. (eds.) *Cities in Competition: Productive and Sustainable Cities for the 21st Century*. Melbourne: Longman Australia, pp. 402-429.

Breheny, M. (1996) Centrists, decentrists and compromisers: Views of the future of urban form, in Jenks, M. *et al.* (eds.) *The Compact City: A Sustainable Urban Form?* London: E & FN Spon, pp. 13-35.

Bruegmann, R. (1996) The American city: Urban aberration or glimpse of the future, in Cohen, M., *et al.* (eds.) *Preparing for the Urban Future: Global Pressures and Local Forces*. Baltimore: Johns Hopkins University Press, pp. 336-367.

Burchell, R., Listokin, C. Shad, N., Phillips, H., Downs, A., Seskin, S. Davis, J., Moore, T., Helton, D. and Gall, M. (1998) *Costs of Sprawl Revisited*. Washington: Transportation Research Board, National Research Council.

Clawson, M. (1962) Urban Sprawl and Speculation in Suburban Land. *Land Economics*, **32**, pp. 99-111.

Commission for the European Community (1990) *Green Paper on the Urban Environment*. Brussels: European Commission.

Dezert, B., Metton, A. and Steinberg, J. (1991). *La périurbanisation en France*. Paris: Sedes.

Ewing, R. (1997) Is Los Angeles-style sprawl desirable? *Journal of the American Planning Association*, **63**, pp. 107-126.

Fortune Magazine Editors (1958) *The Exploding Metropolis*. New York: Doubleday.

Gans, H., Kasarda, J. and Molotch, H. (1982) Symposium: The state of the nation's cities. *Urban Affairs Quarterly*, **18**, pp. 163-86.

Gordon, P. and Richardson H. (1997*a*) Are compact cities a desirable planning goal? *Journal of the American Planning Association*, **63**, pp. 95-106.

Gordon, P. and Richardson, H. (1997*b*) Where's the sprawl? *Journal of the American Planning Association*, **63**, pp. 275-278.

Gottmann, J. (1961) *Megalopolis, The Urbanized Northeastern Seaboard of the United States*. New York: The Twentieth Century Fund.

Gottmann, J. and Harper, R. (1967) *Metropolis on the Move: Geographers Look at Urban Sprawl*. New York: John Wiley.

Hall, P. (1966) *The World Cities*. New York: McGraw Hill.

Hall, P., Gracey, H., Drewett, R. and Thomas, R. (1973) *The Containment of Urban England* (2 volumes). London: George Allen and Unwin..

Hall, P. (1988) *Cities of Tomorrow: An Intellectual History of Urban Planning and Design in the Twentieth Century.* London: Blackwell.
Harris, C. (1956) The pressure of residential-industrial land use, in Thomas, W. (ed.) *Man's Role in Changing the Face of the Earth.* Chicago: University of Chicago Press, pp. 881-895.
Harvey, R. and Clark, W. (1965) The nature and economics of urban sprawl. *Land Economics,* **61**, pp. 1-9.
Hebbert, M. (1994) Urban sprawl and urban planning in Japan, in Shapira, P., *et al.* (eds.) *Planning for Cities and Regions in Japan.* Liverpool: Liverpool University Press, pp. 141-158.
Jackson, K. (1985) *The Crabgrass Frontier: The Suburbanization of the United States.* New York: Oxford University Press.
Jacobs, J. (1962) *The Death and Life of Great American Cities.* New York: Vintage Books.
Jenks, M., Burton, E. and Williams, K. (1996) *The Compact City: A Sustainable Urban Form?* London: E & FN Spon.
King, A. (1980) Historical patterns of reaction to urbanism: The case of Britain 1880-1939. *International Journal of Urban and Regional Research*, **4**, pp. 453-69.
Knapp, G. and Nelson, A. (1992) *The Regulated Landscape: Lessons on State Land Use Planning from Oregon.* Cambridge, Mass: Lincoln Institute of Land Policy.
Kunstler, J. (1993) *The Geography of Nowhere*. New York: Simon & Schuster.
Ladd, H. (1992) Population growth, density, and the costs of providing public services. *Urban Studies*, **29**, pp. 273-296.
Lessinger, J. (1962) The case for scatteration. *Journal of the American Institute of Planners*, **28**, pp. 159-69.
Moe, R. and Wilke, C. (1997) *Changing Places: Rebuilding Communities in the Age of Sprawl.* New York: Henry Holt & Co.
Monclús, F. (ed.) (1998) *La ciudad dispersa: Suburbanizacion y nuevas periferias.* Barcelona: Centre de Cultura Contemporània de Barcelona.
Moore, T. (1998) *Climate of Fear: Why We Shouldn't Worry About Global Warming.* Washington DC: Cato Institute.
Morrill, R. (1991) Myths about metropolis, in Hart, J.F. (ed.) *Our Changing Cities.* Baltimore: Johns Hopkins University Press, pp. 1-11.
Mumford, L. (1961) *The City in History: Its Origins, Its Transformations, and Its Prospects*. New York: Harcourt, Brace and World.
Nairn, I. (1955) *Outrage*. London: The Architectural Press.
Newman, P. and Kenworthy, J. (1989) *Cities and Automobile Dependence: An International Sourcebook*. Aldershot: Gower Publishing.
Newsweek (1995) Bye, bye suburban dream: Fifteen ways to fix the suburbs. 15 May 1995, pp. 40-53.
Orfield, M. and Rusk, D. (1998) *Metropolitics: A Regional Agenda for Community and Stability*. Washington: Brookings Institution Press.
O'Toole, R. (1999) Dense thinkers. *Reason*, **30**(8), pp. 45-52.
Peel, M. (1995) The urban debate: From 'Los Angeles' to the urban village, in P. Troy, (ed.) *Australian Cities: Issues, Strategies and Policies for Urban Australia in the 1990s*. London: Cambridge University Press, pp. 39-64.
Peiser, R. (1989) Density and urban sprawl. *Land Economics*, **65**, pp. 193-204.
Philander, S. (1998) *Is the Temperature Rising? The Uncertain Science of Global Warming*. Princeton: Princeton University Press.
Porter, D. (1997) *Managing Growth in America's Communities.* Washington: Island Press.
Poulton, M. (1995) Affordable homes at an affordable (social) price, in Fallis, G., *et al.* (eds.) *Home Remedies: Rethinking Canadian Housing Policy*. Toronto: C.D. Howe Institute, pp. 50-122.

Razin, E. (1998) Policies to control urban sprawl: Planning regulations or changes in the 'rules of the game'? *Journal of the American Planning Association*, **35**, pp. 321-340.
Real Estate Research Corporation (1974) *The Costs of Sprawl: Environmental and Economic Costs of Alternative Residential Development Patterns at the Urban Fringe*. Washington DC: United States Government Printing Office.
Rusk, D. (1995) *Cities Without Suburbs*, 2nd ed. Washington, DC: Woodrow Wilson Center Press.
Sharp, T. (1932) *Town and Countryside: Some Rural Aspects of Urban and Rural Development.* London: Oxford University Press.
Sierra Club (1998) *The Dark Side of the American Dream: The Costs and Consequences of Suburban Sprawl*. San Francisco: Sierra Club.
Sorkin, M. (1992) See you in Disneyland, in Sorkin, M. (ed.) *Variations on a Theme Park*. New York: Noonday Press, pp. 205-232.
Thompson, W. (1967) Urban economics, in Eldredge, H. (ed.) *Taming Megalopolis*, Vol. 1. New York: Praeger, pp. 156-190.
Troy, P. (1996) *The Perils of Urban Consolidation*. Sydney: Federation Press.
Whyte, W. (1958) Urban sprawl, in Fortune Magazine Editors, *The Exploding Metropolis*. New York: Doubleday., pp. 115-139.
Williams-Ellis, C. (1928) *England and the Octopus*. Port Meiron: Penrhydeudraeth.
Windsor, D. (1979) A Critique of the costs of sprawl. *Journal of the American Planning Association*, **45**, pp. 279-292.
World Commission on Environment and Development (1987) *Our Common Future*. Oxford: Oxford University Press.

CHAPTER 10

Motor Vehicles and the Inner City

Jeffry M. Diefendorf

Without doubt one of the most important themes for urban planning in this century has been traffic planning. Over the centuries, most of the old cities of Europe and the oldest cities in America had evolved with complex, irregular street patterns, determined by the interaction of topography and urban growth, and densely populated areas where housing was mixed with commerce and industry. These were cities of pedestrians and horse-drawn carts and carriages. In contrast, those few cities that had enjoyed the fruits of baroque planning were privileged with a more regular pattern of broader avenues, though these often served as boundaries for densely built-up blocks, and the avenues had been conceived more for display than as carriers of large volumes of traffic. The nineteenth century brought industrialization, explosive population growth, and a new intrusion in the urban fabric – railroads. Whether elevated or at ground level, they cut swaths through cities, disrupting ancient flows of pedestrians and horse-drawn vehicles. The railroad was then adapted for urban use in the form of trams, drawn by horses but electrified toward the end of the century, that occupied those streets wide enough and straight enough to accept them. Since town planners commonly lacked control over state-owned or state-supported railroads, they could at most seek to master the competition between tramways, pedestrians, and horse-drawn carriages by putting trams underground in new subways, although even here high costs and weak planning legislation meant that this solution was seldom implemented (McKay, 1976).

By reaching into the central city, railroads and tramways facilitated the sprawl of the population into the suburbs, but they also made possible that great planning innovation at the turn of the century, the garden city movement. Rail transport seemed ideal for moving people and goods into the central city from surrounding garden cities or, as was more often the case, garden suburbs. However, by the 1920s the mass production of motor vehicles for private use transformed thinking about traffic planning in cities. The low cost of cars and fuel, rapid suburbanization, and ideologies that equated car ownership with individual freedom, rapid mobility, and masculine potency brought rapid increases in the use of private automobiles (Flink, 1988; McShane, 1994). Though best suited for long-distance travel, motor vehicles quickly began to compete with other modes of transit on city streets. The challenge to planners was two-fold. How could long-distance through-traffic be kept out of the central city, and how could traffic within the central city best be accommodated? New circumference highways might answer the first challenge, but the second was more formidable, especially in older preindustrial cities with dense networks of narrow streets. However great the challenges they posed, motor vehicles and broad highways as prime examples of modern technology fascinated engineers and the public, and traffic engineering developed into a new profession. Planning for motorized traffic was a perfect juncture of modern technology, engineering, architectural design, and industry. Consequently, accommodating the volatile growth in urban motorized traffic has ever since been central to nearly all planning paradigms, utopian and pragmatic, regardless of whether the polity has been socialist, capitalist, democratic, or authoritarian.

For example, in 1922, Le Corbusier composed his plan for an ideal modern city of three million people, where high-rise apartments towered over express roads. The centre of the city would be the crossing point of great superhighways, a main rail terminal, and even an airport. His 1925 plan for Paris called for radical demolitions on the Right Bank and the construction of a grid of roadways and skyscrapers (Fishman, 1977, pp. 190-191, 211). Those plans remained utopian dreams, but in the United States, Robert Moses built urban parkways to channel automobiles into New York City and give access to his new parks and beaches (Caro, 1974). The 1933 Charter of Athens, that great manifesto of architectural modernism, demanded the dissolution of the dense old cities through functional zoning with distinct areas for housing, work, and recreation, with these zones integrated into a whole by rapid motorways (Gideon, 1941). In the 1930s and 1940s, Nazi planners dreamed of gigantic avenues in the hearts of German cities, and autobahns and Volkswagens were to facilitate a new dimension of mass mobility (Durth and Gutschow, 1988; Dülffer, Thies, and Henke, 1978).

Although town planners were aware of the potential impact of motor vehicles on cities, the actual consequences of this mode of transportation became most visible in the 1950s. The Great Depression and World War Two had postponed mass purchase of motor vehicles by private individuals, but peace, European reconstruction, and spreading prosperity brought accelerated use of the private car and a decline in the use of public transit on both sides of the Atlantic. In 1953, a leading West German town planner, Herbert Jensen, acknowledged this change when he proclaimed: 'We live in an age of traffic. Traffic is trump; we live in fact in an intoxication, an ecstasy of traffic, and of motorized traffic at that' (Jensen, 1953, pp. 16-18).

In the late 1950s and 1960s, even as urban highway projects were still being built, urban reformers denounced the negative consequences brought by the domination of cities by the automobile. In spite of urban highways, motor vehicles still clogged city streets, killed drivers and pedestrians in accidents, fouled the air, and claimed precious space for new roads and parking. Lewis Mumford and Jane Jacobs in America, Hans Bernhard Reichow, Edgar Salin and Hans Paul Bahrdt in Germany called for the revival of mixed-use neighbourhoods where the street served the social interaction of all kinds of people, not automobiles, and where a new urbanity could take root. Thus Mumford compared urban highway construction to a tornado or atom bomb, noting that new roads, interchanges, parking garages and parking lots 'butcher up precious urban space in exactly the same way that freight yards and marshalling yards did when the railroads dumped their passengers and freight inside the city.' He challenged car manufacturers to 'abandon their fantastic commitment to the indecently tumescent organs they have been putting on the market' (Mumford, 1963, pp. 238, 242). Only mass transportation, he argued, serving dense central areas where pedestrians, not cars ruled, could solve the problem of great cities, and he complained that 'in the act of making the core of the metropolis accessible, the planners of congestion have already almost made it uninhabitable' (Mumford, 1961, 548; see also Bahrdt, 1968; Jacobs, 1958; Jacobs, 1961; Reichow, 1959; Salin, 1960). These critics, in other words, demanded that town planners rethink what they were doing and stop giving motor vehicles top priority.

Here then we find two conflicting motifs of twentieth century urban planning: the desire to serve the automobile, and the demand that their use be subordinate to other, more important needs. How has the tension between the two worked in practice, particularly in those old cities whose street pattern long predated the modern age? The primary concern in this chapter will be traffic planning for the central areas rather than highways on outskirts of the cities. Citizens in most cities, and especially in Europe, still identify the city in terms of its central core, not the suburbs, and it has been to these core areas that motor vehicles have posed the greatest threat.

This chapter will attempt to assess urban traffic planning in both Europe and America by sketching the experiences of Cologne, Basel, and Boston, three large, old metropolises with dense central areas containing street patterns that evolved over time in relation to the local topography. Certain commonalities, and differences, make a comparison of the three cities illuminating. In all three, the historic core developed on a waterfront. Basel and Cologne had evolved as semi-circles along the Rhine's left bank, with much smaller sister towns across the river. Boston's growth was shaped both by rivers and the Atlantic harbour front. Boston is, of course, a much younger city than Basel and Cologne, but for purpose of comparison, it is perhaps the most 'old-world' of America's large cities, with distinctive architectural traditions and significant numbers of historic buildings. Initially important commercial as well as cultural centres, since the mid-nineteenth century these cities grew from the expansion of industry. In all three cities, large industrial sites developed on the edges of town, and, as is typical of such cities, the growing flow of people and goods in and out of the central core caused increasing congestion, especially with the advent of motor vehicles. Moreover, even long-distance traffic not destined for the core tended to take the straightest route through the crowded central areas. Because citizens of all three cities closely identified with the historic centre, changing street patterns to facilitate traffic flow provoked considerable resistance.

As will be evident, there was a remarkable similarity in the approaches to traffic planning in these three cities on two continents, even though the outcomes of those efforts were quite different, which in turn suggests how local planning fits into the larger story of modern traffic planning. In all three cities examined here, planners sought for decades to construct major traffic arteries through the hearts of their historic cities in order to relieve the congestion of narrow streets. They were seeking pragmatic solutions to problems specific to their cities. The fact that Cologne was badly bombed during World War Two and extensively rebuilt, whereas Basel and Boston escaped any war damage, proved surprisingly unimportant. Eventually, planners had to listen to the jeremiads of the urban reformers and the complaints of citizens about the negative impact of arteries in the inner cities, and plans were modified accordingly. At the start of the twenty-first century, the need to plan for motorized traffic remains critically important, but town planners in Cologne, Basel, and Boston, like their colleagues elsewhere, no longer share Herbert Jensen's 'ecstasy of traffic.'

COLOGNE: THE PERSISTENCE OF ARTERIAL PLANNING

Urban planning in Cologne has been associated with some of the most famous names in German planning: Joseph Stübben, Fritz Schumacher, Paul Bonatz, and Rudolf Schwarz (Diefendorf, 1991). Once the Prussian state agreed to relinquish its fortifications and defence zone around the city in 1877, Cologne expanded its territory – almost entirely on the Left, or West Bank of the Rhine – to accommodate a growing population. In 1880, Joseph Stübben, one of the founders of modern planning, developed a plan for the expansion of the city. Of interest here are two features of Stübben's plan. First, he enclosed the historic old city with a ring boulevard constructed on the site of the inner fortification wall. Influenced by Parisian post-1670 *grands boulevards* and Haussmann's spectacular work in that city, Stübben envisaged an elegant street serving pedestrians and coaches, but traffic was not necessarily its primary feature. Stübben's ring included green 'islands' in the centre for pleasant strolling. He also laid out new blocks for housing development just outside the ring. Stübben recognized that the city's railroad line constituted a major disruption in the city fabric. The main railroad station had been built next to the cathedral, and the tracks cut through the heart of the city from west to east. Stübben preferred removing the train station and tracks from the central city to a peripheral location but, failing that, his plan called for elevating the tracks to allow for uninterrupted passage underneath. In fact, removing the rail lines altogether was beyond the resources of the city, and the railroad tracks, whether elevated or at ground level, formed majors rifts in the urban fabric.

In the early 1920s, Cologne's dynamic and ambitious *Oberbürgermeister* Konrad Adenauer brought in a consulting planner, Fritz Schumacher from Hamburg, to prepare a new plan. Under Adenauer, Cologne expanded further. A green belt was constructed around the suburbs planned by Stübben, and a new outer ring road, constructed on the sites of the middle and outer fortification rings, helped carry traffic around the city. Like Stübben, Schumacher recognized both the desirability and impracticality of moving the railroad lines and train station, but Schumacher also saw the need to improve traffic flow around and through the old city. He called for widening existing streets and cutting new arteries through the *Altstadt*, pointing out that playing 'the role of the destroyer' was not really *his* fault, since the city had declined to make such changes earlier (Diefendorf, 1991, p. 254). Schumacher planned a major east-west thoroughfare that would carry traffic through the heart of the central city over the Aachener Straße, past Neumarkt and Heumarkt, to an enlarged bridge over the Rhine.

While Schumacher's plans for the old city were not realized, the ideas by no means disappeared. They popped up again in Bonatz's suggestions of 1930, leading Adenauer to complain that too much 'nonsense' (*Unfug*) was being advocated in the name of 'traffic'. Cologne and its old streets, Adenauer said, 'are not there primarily for autos,' and Cologne's town planner Wilhelm Arntz observed that 'the historic old city cannot be saved if we sacrifice it for traffic' (Diefendorf, 1991, p. 255). True enough, but this attitude did not solve the problem of growing congestion in the inner city, and little progress was made.

For all the anti-urbanism in Nazi propaganda, the Nazis were in practice committed to certain kinds of modernization of their large cities, and they were also clearly fascinated with technology, including automobiles. The building of the famous autobahn network and the launching of the 'people's car' had, to be sure, much to do with job creation and politics, but they also marked an unmistakable commitment to motor vehicles. (The main models for autobahn construction were the new autostradas of Mussolini's Italy and the parkways built by Robert Moses in New York City (Seely, 1998).) The autobahns were to link Germany's cities; new arteries, widened streets, and major axes through historic cities were to smooth the flow of traffic. In the case of Cologne, the plans for its redesign as a city representative of Nazi power and values included completion of the east-west artery advocated by Schumacher, plus the construction of a north-south artery from the Neusser Straße to the Bonner Straße. These two arteries would bisect the historic core. The Nazis also wanted to close the old train station and replace it with two stations, one on the western edge of the city, the other across the Rhine in Deutz, the location of a proposed Forum. Finally, much of the historic pattern of small, crooked streets was to be replaced with a grid of wide streets lined with large buildings. Some of these innovations were supposed to be completed by 1940 in time for a proposed international exhibition on traffic, but the war precluded both staging the exhibition and realizing these plans, though considerable demolition and street widening along the route of the east-west artery did take place.

If we leave aside the hardships suffered by people whose homes would have been (or were) demolished for widened streets or brand new arteries as a result of these Nazi-era plans, and if we ignore the brutal architectural style which new buildings along those streets would have taken, the very continuity of ideas for traffic planning in the first half of the century shows that those ideas were logical, technical solutions for a real problem, the need to do something about motor vehicle traffic flowing into and through an ancient city designed for pedestrians and horses. For 50 years, Cologne's planners had focused on relocating the railway lines and main station and building new thoroughfares for automobiles. It can come as no surprise, then, that some of the same ideas emerged again during postwar reconstruction. Indeed, the great devastation of the city during the war seemed to invite the building of new modern roadways.

In postwar Cologne there were many advocates for retaining the basic street system in the inner city. With most of the buildings reduced to rubble, only a few monuments and the street pattern could evoke historic Cologne. *Oberbürgermeister* Hermann Pünder called the main train station a 'thorn in the side' of the city and sought its relocation. Like his predecessors, in this he was unsuccessful; once again the costs were too great, the logistics too complex, and the need to get the railroads functioning again too pressing. Regarding motor vehicles, Pünder promised that 'the monster traffic' would not be allowed to determine the character of the historic metropolis, but ultimately this did not preclude construction of east-west and north-south axes and upgrading the capacity of the street along the Rhine (Pünder, 1946). The reconstruction plan produced by Rudolf Schwarz, Cologne's most important postwar planner, called not only for such axes but also for a grid of medium capacity streets that would outline neighbourhoods, within which the small streets of the past would prevail. Since major crossroads remained, these axes were not to be true urban autobahns, like that built by Friedrich Tamms in nearby Düsseldorf on the model of Los Angeles, but they were supposed to carry a high volume of traffic (Schmucki, 1998).

The east-west artery, well under way under the Nazis, moved ahead, but the final decision for a north-south artery did not come until a general traffic plan was drawn up in 1956 (Wehner, 1957). That plan argued that this artery was simply essential. Motorists were currently clogging both the Ring and the Rhine shore drive to the point that drivers were seeking alternative routes through small residential streets. Planners argued that the number of work-hours lost through traffic delays was such that it cost the economy around 10 million DM a year, compared to the 15 million DM needed to build the artery. The north-south artery was completed in September 1967 (figure 10.1). Because a key section was put underground, it was possible for the city to open a long, 'L' shaped pedestrian shopping zone running from the train station to Neumarkt. Since a complete autobahn ring was finished in mid-1965 and the first section of an inner-city subway opened in 1968, Cologne's planners might have hoped that these problems had been solved (Jatho, 1966).

Figure 10.1. Cologne, north-south artery (1963). This high-capacity artery, part of the postwar reconstruction plan of Rudolph Schwarz, runs from top to bottom on the left side of the photograph, The entire area in the photo lies within Stübben's ring road around the historic centre of the city; Cologne's cathedral and central business district are toward the top. An underpass (left centre) takes the artery beneath a pedestrian shopping zone, but the road and cloverleaf constitute a major rupture in the fabric of the southern part of the central city. (*Source*: Historisches Archiv der Stadt Köln, ZSB 4/66; photo Walter Moog)

That proved an illusion. The north-south and east-west thoroughfares successfully channelled much automobile traffic into and out of the central city, and subway use grew. However, the number of cars continued to grow as West Germany grew ever more prosperous, and congestion on the streets and the demand for parking grew apace. On narrow streets, drivers competed with pedestrians by parking their cars with the right-hand wheels on the sidewalks. Both the shore drive and the inner Ring continued to carry a very high volume of automobiles and trams. Indeed, the Ring lost its character as a pedestrian boulevard. Heavy traffic on the shore drive created a barrier between the city's inhabitants and the Rhine. In fact, automobile traffic proved to be the monster that Adenauer and Pünder had feared would destroy the city.

What was to be done? Some relief came from new parking garages in the central city and electronic signs on streets leading into that area instructed motorists which garages had spaces available, but Cologne also took the path chosen by many German cities, namely, deciding that automobiles could no longer enjoy the absolute top priority in planning. For example, to solve the problem of traffic along the Rhine shore, the city chose the costly but attractive solution of housing the shore drive in a tunnel, which opened in 1982. Once again pedestrians have easy access from the central city to the river. In many places streets have been narrowed and obstacles placed to discourage the use of side streets for through traffic. Barriers keep cars off the sidewalks. New areas have been given over to pedestrian zones, with motor vehicles prohibited except for early morning deliveries. And motorists have had to accept the fact that driving through the central city, even on the east-west or north-south arteries, is simply not the quickest way to get from one side of Cologne to the other. Instead, they use the exterior autobahn ring. Much of the time it too is now clogged, but at least the worst of the traffic planning problem has been relocated outside the city centre.

BASEL: THE REWARDS OF FRUSTRATED TRAFFIC PLANNING

Like Cologne, Basel was also constructed mostly on one side of the river. The canton of Basel-city, containing the central city, lies on the south bank of the Rhine; Kleinbasel, a smaller urban area, lies just across the Rhine to the north. Straddling the Rhine, the city also borders both France and Germany, and those countries retain sovereign rights over parts of the train stations, located on the edges of Basel-city and Kleinbasel, where trains from France and Germany end.

Being a border city complicated Basel's town planning, as did the cantonal governmental system, a parliamentary democracy with strong safeguards for the rights of citizens. Always seeking to avoid the expense and domination of a large governmental bureaucracy, important planning activities were farmed out to temporary commissions and to institutions and individuals outside the official governmental structure. Furthermore, important legislation could, and often was, either referred to public referenda or subjected to revocation by initiatives started by the citizenry. This meant that neither city officials nor politicians could ever be truly confident that their decisions would not be promptly overturned by the voters. Cologne's political leaders and town planners, in contrast, usually assumed that their plans would be realized, even if that did not always happen. Town planning in Basel proceeded more cautiously than in Cologne.

In 1892 Basel created its first planning agency, and in 1896, the city parliament, the Great Council, provided funds for a city planning office and also commissioned the famous Cologne planner Joseph Stübben to draw up a plan for a city street network. Basel had already demolished its city walls in 1852, and it did not have large tracts of newly available land that lent themselves to the kind of monumental ring road built by Stübben in Cologne. For three decades after that, the chief local planner, Eduard Riggenbach, drew up plans for individual streets in Basel and Kleinbasel. In 1927, motivated by the fact that a growing volume of motorized traffic was threatening to clog the city streets, the governing council asked Riggenbach to draw up a new general plan. In the same way that Cologne's proposals for new inner city axes continued to have currency from decade to decade, Basel's plan, presented in early 1930, contained features which continued to inform planning discussions until the 1960s. Traffic planning lay at its heart. One proposal was for an underground tramway under the central core (Musfeld-Imhof, 1929). Another was the construction of a major artery through the central area that would carry motorized traffic over a new bridge just downstream from the historic Mittlere Brücke. Construction of the artery would have required demolishing many ancient buildings, which led planners to attach to their proposal a major renewal project for the run-down block near the town hall. Transforming the historic centre, planners argued, was a price worth paying for improvement in traffic flow through old Basel (Festsetzung eines generellen Bebauungsplanes für die innere Stadt, 1930).

The Great Council referred the plan to a special parliamentary commission for study. The commission's recommendation that the council reject the plan was an indication of a changing way of thinking about the city. The commission supported the general idea of a traffic artery through the city and simultaneous renewal of decrepit areas, but it also voiced concern about the possible destruction of the character of the *Altstadt*, the old city core. At the same time, the commission opposed the construction of an underground tramway; that project would be much too expensive, for one thing, and there was scepticism that Basel's size would justify such a project. Since the plan's primary author, Riggenbach, had just died, the commission suggested that now was the time to create a permanent city planning office within the city government and hire a professional planner to draw up new plans (Festsetzung eines generellen Bebauungsplanes für die innere Stadt, 1931). The new planner, Adolf Schumacher, submitted a new general plan in August 1933 that contained many of the same features as the Riggenbach plan but in a less ambitious way (figure 10.2). The proposal to the Great Council asked not for approval of all details but a general endorsement of the overall concept. In that way the planners could proceed in small steps as conditions warranted. The decision to build a new

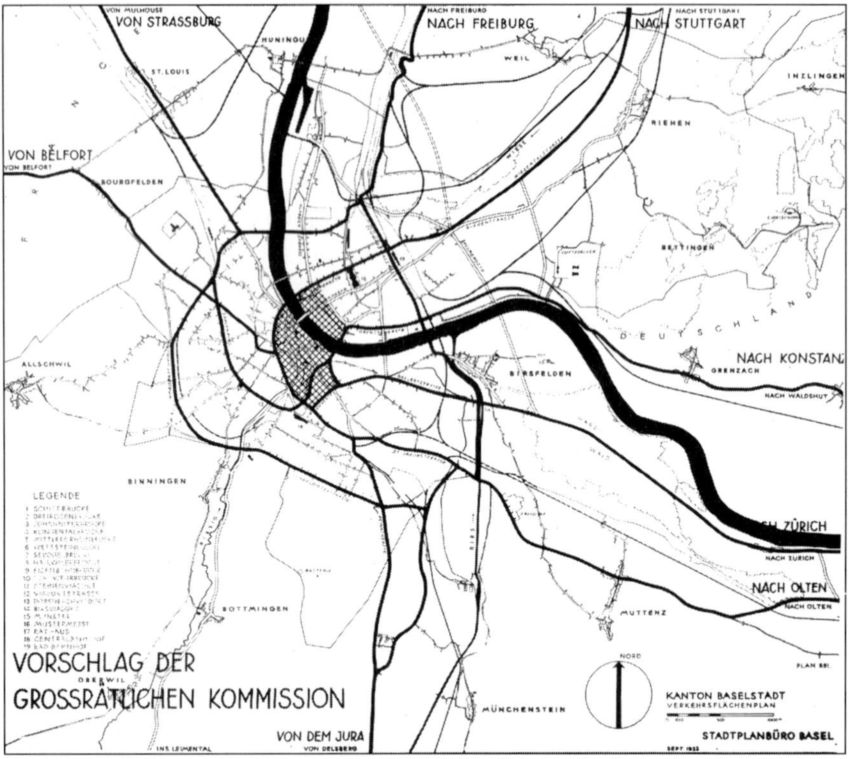

Figure 10.2. Basel, 'Schumacher Plan' (1933-34). Cross-hatched area indicates the historic city, with the cathedral, town hall and central business district on the south side of the Rhine (heaviest line). Medium dark lines show new arteries and new or enlarged bridges to carry motor vehicles through and around the old city. Proposed by Adolf Schumacher, the plan was approved but not implemented because of high costs, the onset of the Depression, and World War Two. (*Source*: Staatsarchiv Basel, BD-Reg. 1A/10-4 (1938-67), Baukommission)

bridge, the traffic artery, and even an underground tramway could be left open (Festsetzung eines allgemeinen Korrektionsplanes für die innere Stadt, 1933).

Once again a commission examined the plan. After thirty-eight sessions, visits to the proposed building sites, and discussions with the local architects' associations, the commission endorsed a revised version of the plan but noted that most of it would be realized only in the distant future. For the moment it was just too expensive. More urgent – and feasible – were smaller projects: urban renewal in the area of the Imbergässlein and Andreasplatz and repairs to old buildings with dangerously unhygienic housing units (Bericht der Großratskommission über den Ratschlag 3370, 1934). Furthermore, even more modest street widening and city-mandated renewal projects required new laws to regulate placement and height of buildings and to facilitate the redrawing of property lines.

The Schumacher plan was approved by the Great Council in July 1934, but in fact conditions were not conducive to the sort of changes in the city's fabric contained in that plan. The Great Depression seriously damaged the city's economy and made many features of the plan prohibitively expensive. Many local architects, including the noted architect/politician Hans Bernoulli, remained critical of parts of the Schumacher plan and announced their intention to prepare alternatives in spite of the council's approval of it.

The central proposals in the 1930 and 1934 plans – the underground tramway, the traffic artery through the central city, the new bridge, and massive urban renewal – were pushed into the background by a large number of much more modest construction projects, most of which were dedicated to renovating rather than demolishing the historic substance of the city, and to creating jobs for unemployed architects and construction workers. Here the conservatives interested in historic preservation and the socialists interested in work creation found a common ground. When a new building law was submitted to the Great Council in 1939, it contained not only technical provisions regulating building heights and redrawing property lines but, even more important, a general zoning plan for the city. While certain areas were designated for housing, industry, and future development, large parts of the historic core were given special protection from development by being including in a special *Altstadt* district. (This became known as the 'violet zone' for the colour used on the plan.) The law stipulated that the building department now had the general responsibility to see that any new construction, demolition of existing buildings, and street alterations in the violet zone did not harm the overall artistic and historic character of the area (Bericht der Großratskommission für die Revision des Hochbautengesetzes, 1939). Whereas earlier planning, motivated primarily by traffic considerations, was frustrated by pragmatic fiscal restraints, henceforth considerations of historic preservation constituted a major, legally sanctioned obstacle to traffic planning.

The outbreak of World War Two meant, quite naturally, that planning remained a matter for discussion only while the city girded itself for a possible invasion. After the war's end, there was new growth of the economies of both the city and the hinterland and a renewal of ties with both Germany and France, all of which produced growing demands for better through traffic for motor vehicles and better access to the central city. Furthermore, the rapid growth of large chemical firms in the western suburbs led to an increased volume of workers commuting to those plants. The planning office and the building department focused on the options of constructing a ring and tangent system of high capacity roads to carry vehicles around the city centre and, once again, constructing some sort of artery into or through the centre. In 1946 a revision of the earlier plans was submitted to the Great Council, and after prolonged study by still another special commission, it was approved in 1949 by both the Great Council and a popular referendum (Festsetzung eines Allgemeinen

Korrektionsplanes für das Großbasel, 1946). This plan called for a partial ring with feeder streets tangential to the ring, some street widening in the city core, and a programme of block renewal that would free the area of the old market (in front of the town hall) for traffic flow.

While some streets were indeed widened and progress made on a segment of a ring road on the south side of the city near the Swiss Federal Railway station, the planners and citizens fought prolonged battles over other aspects of the plan. Preservationists protested the loss of historic buildings, and an unsuccessful public initiative tried to block widening of the Aeschengraben in 1954. Hans Bernoulli and others revived a call for an underground tram system, now coupled with a large pedestrian zone. Meanwhile, the boom of the postwar economic recovery resulted in ever growing pressures on the city's street system.

In an effort to break the logjam, the city sought outside help, as it had in the past. In 1955 it commissioned Professor Kurt Leibbrand of the Technical University of Zürich, a specialist in rail and highway planning, to draw up a new general traffic plan. Completed in 1958, the Leibbrand plan called for building a system of broad roads running tangential to the central area which would divert traffic around the centre, constructing large parking garages on the edge of the centre, placing key stretches of the tram system underground, and connecting Basel to both the Swiss and international autobahn network. Recognizing that all this would be expensive, Leibbrand suggested a 25-year period for realization (Leibbrand, 1958). If adopted fully, this plan would have cost some 220 million Swiss francs between 1960 and 1980, a figure that shocked the frugal Baslers.

It is not surprising, then, that the Leibbrand plan was not implemented. The local associations of architects and engineers joined together to produce a thorough critique of the Leibbrand plan and to make alternative recommendations (Basler Ingenieur- und Architektenverein, 1963). Even this plan sought to retain the idea of placing at least some segments of the tram system below ground, but this received a lower priority than completion of a system of roads tangential to the centre. This would be a so-called 'city-ring' of higher capacity city streets rather than an actual autobahn ring. The major traffic artery through the historic core was finally abandoned in favour of placing greater emphasis on the rights of pedestrians in that area. And with the abandonment of the artery idea, the way was cleared for further restoration of historic buildings in the centre instead of demolition. The central core had to be made liveable for residents and usable for business without being dominated by motor vehicles. Indeed, during the next decade the violet *Altstadt* zone was enlarged to include groups of buildings previously unprotected by the legal authority of the historic preservation law. Today Basel makes do with a surface tram system and very limited automobile access to the central city. Traffic flows on the 'city-ring' around the city and to the chemical plants, allowing historic Basel to function as a charming core of a small metropolis.

Boston: Constructing and Deconstructing a Central Artery

We have seen that both Cologne and Basel drafted dramatic plans to cope with the growth in motor traffic. Cologne's was largely realized, if in fits and starts, but in the end it proved in several ways unsatisfactory, and that city has reverted to restrictions on the place of automobiles in the central city. Basel's plan was only very partially implemented, and the failure to build a major artery through the central area has proved a blessing. What about Boston in the United States?

To be sure, Boston's geographic situation was more complicated than that of either Cologne or Basel. Boston is essentially a peninsula surrounded by the Charles River on the north and a bay opening to the Atlantic on the east and south, which precluded a true circular highway system like Cologne's autobahn ring. Rail lines entered Boston from the north, west, and south, with two major terminals on the edge of the central business district. Like Cologne and Basel, historic Boston's streets were narrow and crooked, unsuitable to moving much vehicular traffic, though the grid avenues in the Back Bay and South Boston – both developments of the mid-nineteenth century – were much more modern.

The chronology of Boston's attempts to come to grips with the demands of modern traffic planning resembles those of Cologne and Basel. It was shaped by the city's complicated fiscal system as well as by bitter and deep-seated political rivalries, based on the one hand on ethnic heritage and neighbourhood rootedness and on the other on the difficulties caused by the fact that Boston is also the capital of the state of Massachusetts (O'Connor, 1993). In 1894 Boston established a Transit Commission, which built its pioneering subway system, but otherwise, for the first half of the twentieth century, it simply proved impossible to develop and implement much of a coherent vision of the city (Cheape, 1980). Whenever major ideas were put forward, the city turned to the state to obtain relief from its traditional dependency on property taxes, and the state consistently failed to help.

There was certainly planning expertise available, with nearby Harvard University and the Massachusetts Institute of Technology both training town planners, and in 1930 the Boston Planning Board, with Robert Whitten consulting, issued a 'Report on a Thoroughfare Plan for Boston' that set forth a programme of seventy-four projects, the most important being twin tunnels between Boston and East Boston, an elevated Central Artery cutting from north to south along the eastern edge of the business district to carry traffic from South Boston (and points south) over the Mystic River to points north, and radial roads to carry traffic out of the central city (City Planning Board, 1930). Like European planners, Boston's planners calculated the monetary cost of current delays transiting the city and estimated that the major projects would be amortized in terms of savings in ten years.

Boston's politics, and the emergencies of the Depression and war, meant that no major projects were begun. The same basic proposals put forward in 1930 reappeared in *The Master Highway Plan for the Boston Metropolitan Area* prepared for Governor Robert Bradford in 1948 and the *General Plan for Boston* of 1950 (Joint Board for the Metropolitan Master Highway Plan, 1948, and City Planning Board, 1950). At about the same time, Professor Martin Wagner and a group of his students at Harvard's Graduate School of Design advocated a truly radical redesign of Boston (Wagner, 1948). Wagner had been Berlin's planner in the late 1920s. He fled the Nazis and followed Walter Gropius to Harvard. Wagner made no distinction between bombed-out cities in Germany and what he considered blighted cities like Boston; in both cases radical surgery was needed. Wagner proposed basically gutting the entire area between the Boston Common and the harbour, replacing it with a wide, two-level circular road lined with two-level parking garages to handle the anticipated growing influx of cars. Most of Boston's historic architecture and street pattern would disappear in favour of massive, modern commercial and residential buildings.

Not surprisingly, Wagner's radical ideas received a more sympathetic audience in Germany – where they were published and discussed by German planners speculating about radical forms of postwar reconstruction – than in Boston, where no one took seriously the thought of demolishing the whole central city. Instead, Boston plodded ahead with the Central Artery. Some contracts for the Artery as proposed in the 1930 Thoroughfare Plan and 1950 Master Highway Plan were in fact let in 1951, but construction proceeded slowly because of legal disputes, opposition from residents in the path of the new arteries, and the lack of funding. The elevated Central Artery – a project also much admired by German planners who advocated urban autobahns – finally opened in the Fall of 1959, after having caused much bitterness among residents and merchants in the Chinatown area and the North End, a now largely Italian neighbourhood cut off from the rest of the city by the new elevated highway. Meanwhile, in contrast to the slow pace of planning and construction in Boston, in the mid-1950s the Massachusetts Turnpike Authority rapidly built a high capacity toll road connecting the western and eastern parts of the state. This highway was extended into Boston in the 1960s, in effect dividing the city in half, though part of the highway passes underground.

It was only with the mayoralty of John Collins and his appointment of Edward Logue as planning consultant in 1960 that Boston really set out to transform itself into a more dynamic and modern metropolis – the 'New Boston' as it was called (O'Connor, 1993). An underground parking garage under the Boston Common, something debated for more than a dozen years, was finally built in 1960-61 to alleviate some of the parking problems. The construction of the Prudential Center on abandoned rail yards along Huntington Avenue led to the redesign of roads in that area. The same was true when a new Government Center, with a new City Hall and Federal Building, replaced the run-down area of Scollay Square in the mid-1960s. High-tech industry, fed by innovations coming from Boston-area universities, began to flourish on Route 128, a multi-lane highway built as a ring to carry long-distance traffic around Boston and its inner suburbs. And, as in Basel, historic preservation issues gained ever greater importance in defining the character of the city (Holleran, 1997). What appeared to be the crowning achievement of Boston's emergence as an attractive, modern city was the renovation of the 1740 Faneuil Hall between 1973 and 1976 and the subsequent revitalization of the waterfront area.

Or was it? No sooner had it 'modernized' its physical infrastructure than Boston, not unlike Cologne, realized that some of the sacrifices made to automobile traffic had been too great. Not only did the anticipated 'savings in time' from the Central Artery and harbour tunnels not materialize – traffic simply increased too fast, constantly clogging these roads – the revitalized business district and waterfront proved incompatible with the brutal presence of the elevated Central Artery and the thundering traffic it carried. Faneuil Hall became a mecca for tourists and residents, but it was much too close to the elevated motorway. Moreover, when the Central Artery was constructed, it was designed to accommodate traffic as it was projected to grow up to 1970. By that date it was clearly already inadequate! Just as Cologne decided to bury its shore drive to reunite the city with the Rhine River, so Boston decided to bury the Central Artery and relieve traffic by replacing the main bridge to the north and

(*a*) (*b*)

Figure 10.3. Boston, 1999 aerial photo (figure 10.3*a*) showing central artery and construction site for the 'Big Dig' project to bury the artery, with arched entry to Rowes Wharf on right. Photographed at noon rather than rush hour, the congested artery now carries more than 200,000 vehicles daily, three times the volume when it opened. Moored boats on the right of Rowes Wharf indicate the proximity of Boston's waterfront. The eight- to ten-lane artery is being put below ground. The artist's drawing (figure 10.3*b*) shows part of the 40 acre park that will replace the surface area now used by the artery. Greenery, pools, pedestrian paths, and a vastly reduced number of motor vehicles are expected to cut noise, improve air quality, and create a pedestrian-friendly environment for residents, visitors and office workers. (*Source*: Central Artery Public Information Office, Boston; photo Louis Martin)

constructing still another harbour tunnel to divert more traffic away from downtown Boston (figures 10.3). Planning is underway to redevelop large areas along the waterfront, where designers hope to mix residential, commercial, and recreational buildings. Frederick Salvucci, former Massachusetts transportation secretary and a force behind this great project, has claimed that 'the most important characteristic of the post-Artery city should be the creation of pleasant – even exciting – pedestrian connections that have been missing for half a century' (Salvucci, 1999). When all of this is completed, the city's residents will regain easy access to the waterfront and North End neighbourhood. How well this huge project – the largest such highway engineering project currently underway in the United States – will work out remains to be seen, but it is an extraordinarily expensive admission that earlier solutions to Boston's traffic problems had not worked.

CONCLUSION

It is instructive to see how similar ideas about traffic planning evolved in quite different circumstances. The challenge was common to all three cities discussed here: how to respond to a growing flow of motorized traffic in and out of cities. The language of traffic planning was everywhere similar, drawn from the marriage of technology and speed (autobahns, expressways), analogies to living organisms (arteries, circulation), and geometry (axes, grids, rings, and tangents). When models of traffic planning were actually put into practice, different outcomes resulted.

In Cologne, new road construction took place first in the context of Nazi planning and then during the postwar reconstruction of the badly bombed city. In both instances, planners believed they had a great opportunity to transform the old city and facilitate the flow of motor vehicles. Neither Basel nor Boston was bombed, but planners nonetheless sought to transform those cities in similar ways. Major arteries were indeed built in Cologne and Boston after World War Two, but it is clear today that these highways have enjoyed mixed success. Boston is in the process of tearing down a major urban freeway and putting it underground – at enormous expense. Cologne has already put part of its inner-city artery system underground. In spite of decades of planning, Basel constructed a medium capacity tangent/ring system but never built its inner-city artery, and the city and its inhabitants are now grateful that there is no mistake there to be rectified. Neither Cologne, nor Basel, nor Boston would claim today that their century-long attempts to come to grips with the growth of motor-vehicle traffic have been unqualified successes. Indeed, the number of cars using the roads and needing parking places continues to grow in spite of the best efforts of their planners. The experiences of these three cities – one German, one Swiss, and one North American – reveal similarities in the conception of and proposed solutions to those problems that demonstrate the universal challenges for traffic planners in the twentieth century and, most likely, in the early decades of the twenty-first century.

Thus at the dawn of a new century, we can see traffic planning in all its glories and its failings. In America and Europe, freeway systems link cities with each other and with suburbs, rivalling or replacing rail services. Motor vehicles have greatly facilitated the development of the megalopolis, bands of contiguous urban areas stretching in some cases hundreds of miles. Cities depend on heavy trucks for delivery and shipment of all kinds of goods. Both mass and individual tourism depend on motor vehicles. For most European and American consumers, private automobiles remain a convenient and comfortable means of transportation, even when moving about within a city.

At the same time, urban sprawl and the dynamism of modern life, driven by industry, mechanization, consumerism, and new technology, mean a continuously high volume of people and goods moving rapidly in and out of cities, much of it carried by a stream of motor vehicles that often appears to embody an irresistible force overwhelming other concerns. Cities everywhere are plagued by traffic jams, accidents and fatalities, air and noise pollution. The fundamental conflict between the imperatives of motor vehicle traffic and other needs of city residents in the areas of housing, culture, work, and recreation remains, and planners must address this conflict. As Colin Buchanan argued in his widely noted study *Traffic in Towns*, the more motorized traffic is allowed to grow in response to the demand of maximized accessibility to individualized transportation, the greater the environmental and cultural damage in the absence of proportional investment in physical alterations to the urban structure. If sufficient investment is made in such things as underground highways, large parking garages, and the like, then cities can accommodate more motor vehicles (Buchanan, 1963). This is the path taken by Boston and, to some extent, by Cologne, in their decisions to place traffic arteries underground. The alternative is a lower level of investment and reduced individual automobile use brought about by such devices as pedestrian zones, highly restricted parking, and 'traffic-calming' strategies such as speed bumps and narrowed streets. Basel and Cologne have taken this path.

Traffic planning, by necessity, constitutes a central part of town planning, but no urban planner would now advocate making cities serve the automobile. After 1960, international criticism of modernist planning models and growing interest in vital urban neighbourhoods increasingly focused attention more on people and less on motor vehicles. Densely inhabited neighbourhoods, with residents and visitors shopping on foot and enjoying architectural or natural vistas unbroken by freeways or expressways filled with speeding cars, has become a common goal of urban planners. At the turn of the century, freeways or autobahns through central cities have lost their irresistible attraction to planners, and residents mobilize against further destruction. As they did at mid-century, planners ask if automobile traffic and the roads they occupy are monsters? Can the traffic engineers, intoxicated by their special subject, be the saviours of cities, or will they be destroyers of cities, as such critics as Lewis Mumford and Jane Jacobs asked? Urban planners are caught in a matrix of strident demands, demonstrable but often conflicting needs, and hostile criticism. Today's urban planners are not to be envied.

REFERENCES

Bahrdt, H.P. (1968) *Humaner Städtebau.* Hamburg: Christian Wegner.

Basler Ingenieur- und Architektenverein. (1963) Gesamtplan der Stadt Basel: Zusammenfassender Bericht über die Planungsarbeit der Fachverbände. *Schweizerbauzeitung*, **81**, no. 16, pp. 253-73.

Bericht der Großratskommission über den Ratschlag 3370. (1934) Ratschlag 3433, 14 Juni 1934, with plan 1395, in Staatsarchiv Basel-Stadt.

Bericht der Großratskommission für die Revision des Hochbautengesetzes zur zweiten Lesung des Gesetzentwurfs. (1939) Ratschlag 3765, presented to Grosser Rat des Kantons Basel-Stadt, am 16 March 1939, in Staatsarchiv Basel-Stadt.

Buchanan, C. (1963) *Traffic in Towns.* London: HMSO.

Caro, R. (1974) *The Power Broker: Robert Moses and the Fall of New York.* New York: Knopf.

Cheape, C. (1980) *Moving the Masses: Urban Public Transit in New York, Boston, and Philadelphia, 1880-1912.* Cambridge, Mass: Harvard University Press.

City Planning Board (1930) *Report on a thoroughfare Plan for Boston.* Boston: City of Boston.

City Planning Board (1950) *General Plan for Boston.* Boston: City of Boston.

Diefendorf, J.M. (1991) Städtebauliche Traditionen und der Wiederaufbau von Köln vornehmlich nach 1945. *Rheinische Vierteljahrsblätter*, **55**, pp. 252-273.

Dülffer, J., Thies, J. and Henke, J. (1978) *Hitlers Städte. Baupolitik im Dritten Reich: Eine Dokumentation.* Cologne and Vienna: Bohlau.

Durth, W. and Gutschow, N. (1988) *Träume in Trümmern: Planungen zum Wiederaufbau zerstörter Städte im Westen Deutschlands 1940-1950.* Braunschweig: Vieweg.

Festsetzung eines Allgemeinen Korrektionsplanes für das Großbasel (1946) Ratschlag 4224, presented to Grosser Rat des Kantons Basel-Stadt, 23 Mai 1946, in Staatsarchiv Basel-Stadt.

Festsetzung eines generellen Bebauungsplanes für die innere Stadt (1930) Ratschlag 2994, presented to Grosser Rat des Kantons Basel-Stadt, 23 January 1930, in Staatsarchiv Basel-Stadt.

Festsetzung eines generellen Bebauungsplanes für die innere Stadt. (1931) Ratschlag 3131, Bericht der Großratskommission für den Ratschlag No. 2994, presented to Grosser Rat des Kantons Basel-Stadt, 7 May 1931, in Staatsarchiv Basel-Stadt.

Festsetzung eines allgemeinen Korrektionsplanes für die innere Stadt. (1933) Ratschlag 3370, presented to Grosser Rat des Kantons Basel-Stadt, 31 August 1933, in Staatsarchiv Basel-Stadt.

Fishman, R. (1977) *Urban Utopias in the Twentieth Century: Ebenezer Howard, Frank Lloyd Wright, and Le Corbusier.* Cambridge, Mass: MIT Press.

Flink, J.J. (1988) *The Automobile Age.* Cambridge, Mass: MIT Press.

Giedion, S. (1941) *Space, Time and Architecture.* Cambridge, Mass: Harvard University Press.

Holleran, M. (1997) *Boston's 'Changeful Times': Origins of Preservation and Planning in America.* Baltimore: Johns Hopkins University Press.

Jacobs, J. (1958) Downtown is for people, in Fortune Magazine (ed.) *The Exploding Metropolis*. New York: Doubleday, pp. 157-184.

Jacobs, J. (1961) *The Life and Death of Great American Cities*. New York, Random House.

Jatho, K. (1966) Neue Straßen im Altstandbereich Kölns. Speech to Stadtplanungsausschuß, 15 September 1966, in Historisches Archiv der Stadt Köln.

Jensen, H. (1953) Der Wettlauf zwischen Verkehr und Städtebau. *Deutsche Akademie für Städtebau und Landesplanung: Jahresversammlung*. Munich, Wasmuth, pp. 16-18.

Joint Board for the Metropolitan Master Highway Plan (1948) *The Master Highway Plan for the Boston Metropolitan Area*. Boston, Commonwealth of Massachusetts.

Le Corbusier (1925) *Urbanisme*. Paris: G. Cres.

Le Corbusier (1973) *The Athens Charter*. New York: Grossman.

Leibbrand, K. (1958) *Gesamtverkehrsplan Basel*. 2 vols., in Staatsarchiv Kanton Basel.

McKay, J.P. (1976) *Tramways and Trolleys: The Rise of Urban Mass Transport in Europe*. Princeton: Princeton University Press.

McShane, C. (1994) *Down the Asphalt Path: the Automobile and the American City*. New York: Columbia University Press.

Mumford, L. (1961) *The City in History: Its Origins, Its Transformation, and Its Prospects*. New York: Harcourt, Brace and World.

Mumford, L. (1963) *The Highway and the City*. New York: Harcourt, Brace and World.

Musfeld-Imhof, F. (1929) *Der Tram im Birsig Tunnel*. Basel, National Zeitung.

O'Connor, T.H. (1993) *Building a New Boston: Politics and Urban Renewal 1950-1970*. Boston: Northeastern University Press.

Pünder, H. (1946) Historisches Archiv der Stadt Köln, Acc. 2 Oberbürgermeister/614/24-27, speech of 23 June 1946.

Reichow, H.B. (1959) *Die Autogerechte Stadt. Ein Weg aus dem Verkehrs-Chaos*. Ravensburg: Otto Maier.

Salin, E. (1960) Urbanität. *Erneuerung unserer Städte* (Neue Schriften des deutschen Städtetages), no. 6. Stuttgart and Cologne: Kohlhammer, pp. 9-34.

Salvucci, F. (1999) Midway through Big Dig, time to check the map. *Boston Globe*, 4 April, pp. C1-2.

Schmucki, B. (1998) Schneisen durch die Stadt – Sinnbild der 'modernen' Stadt. Stadtautobahnen und amerikanisches Vorbild in Ost- und Westdeutschland, 1925-1975. *Werkstatt Geschichte*, **21**, pp. 43-63.

Seely, B.E. (1998) Der amerikanische Blick auf die deutschen Autobahnen. Deutsche und amerikanische Autobahnbauer 1930-1965. *Werkstatt Geschichte*, **21**, pp. 11-28.

Wagner, M. (1948) Der Neubau der City. *Baurundschau*, **38**, pp. 129-160.

Wehner, B. (1957) Stellungnahme zum Generalverkehrsplan der Stadt Köln 1956. Typescript, 2 vols. Berlin, copy in Historisches Archiv der Stadt Köln.

CHAPTER 11

Planning and Heritage: Towards Integration

David Hamer

When viewed in retrospect, the history of the relationship between planning and heritage in the twentieth century appears to have a certain symmetry. The detailed planning of the protection and management of urban heritage in France and Germany that was emerging before 1914, in particular, bears very striking similarities to the growth of historic districts and their integration into urban planning since 1970. Writers such as Patrick Geddes and Camillo Sitte, who were sympathetic to heritage and were very influential at the beginning of the century but then went into a sort of limbo reserved for discarded planning theorists, seem to have much that was relevant to say once again in the final decades of the century. On the other hand, the decades from the 1920s through to the late 1960s were dominated by influences which were highly antagonistic to giving heritage any substantial influence over urban planning. These included detestation of many of the legacies of the Victorian city, the influence of modernism on architecture and urban design, the impacts of the world wars, and the frequently hostile attitudes manifested towards heritage by dictators and many revolutionary and post-colonial regimes. The purpose of this chapter will be to explain why these tendencies were so strong and why and in what ways a reaction against them set in from the 1960s.

THE HAUSSMANNITE TRADITION

A century never provides a neat starting-point for the study of any historical phenomenon, with a *tabula rasa* existing from year one. This is certainly true of the relationship between planning and heritage: much that is relevant to understanding its development in the new century was under way well before 1900. Many cities in late nineteenth century Europe had already undergone large-scale modernization with dramatic and often devastating effects on their built heritage. A notable example is Vienna and the removal of the old city walls for the construction of the Ringstrasse. Overshadowing much of the ensuing century was the legacy of Baron Haussmann, the master-planner of Paris in mid-century (Jordan, 1995). What he stood for in terms of urban reconstruction and destruction of heritage became known as 'Haussmannization' and recurred repeatedly in the discourses on twentieth century urban renewal. Haussmannization also injected into consideration of what should be done about old city centres a political dimension which it never thereafter entirely lost. It was widely believed that a principal reason why Haussmann and his political master, Napoleon III, wanted to get rid of narrow streets and old quarters was to make the suppression of insurrection easier by preventing the installation of barricades and by opening up lines of fire.

One of the most enduring and controversial aspects of the Haussmannite tradition proved to be the clearing of the spaces adjacent to monumental historic buildings. This practice – known as 'disencumbering' – had become standard for locating major new edifices such as town halls and courthouses by the nineteenth century. It was also applied to redesigning the surrounds of historic monuments (Kostof, 1992). Some powerful voices were heard in reaction to this practice. Camillo Sitte, the Austrian author of *Der Städte-Bau nach seinen künstlerischen Grundsätzen* (1889), protested against 'the delusion that everything must be seen at once, that around a thing only empty space is in order' (Collins and Collins, 1965, p.29). Patrick Geddes, the Scottish biologist and town planning theorist, deplored the clearing away of the narrow streets surrounding cathedrals and the buildings clustered and crowded around their walls. For a while into the new century these critics seemed to have an effect. German cities began to think of ways of reconstructing old cathedral places which had been subjected to 'disencumbering'. Ulm, for instance, held competitions in 1906 and 1924 for plans to build up the area around its cathedral all over again (Geddes, 1915; Kostof, 1992).

Haussmann's view was that what had developed adjacent to a building of historic significance or monumentality was 'detritus' that had not been there at the time the building was designed and erected. It was not part of the building's meaning, and clearing it away would enable that meaning to be more easily discerned. Geddes, by contrast, focusing on medieval towns, argued that much of what was being removed was there originally. And, of course, Geddes was particularly concerned over what was being put in the place of the 'detritus'. He referred to 'the incongruous and transient utilitarian ugliness of the present' (Geddes, 1915, p. 202).

If at one pole the nineteenth century legacy of town planning theory seemed to be dominated by Haussmann, at the other pole there was the influence of Camillo Sitte. Sitte analysed medieval urban forms in Germany and deplored the destruction of that heritage. He appreciated the urban aesthetic and the human scale of the design which he saw in medieval towns and deplored the straight streets which contributed so much to the soullessness of emerging modern city planning. His influence was great early in the century (Koshar, 1993) and then died away as modernism gained its ascendancy, only to be recently revived – as is seen, for instance, in the admiration of him expressed by New Urbanists in the United States. As far as heritage is concerned, Sitte's planning ideas involved a considerable idealization of the past. He looked to the aesthetic which underlay medieval planning. Nevertheless, his book did cause planning and historic preservation to be allied for a time, especially in Germany. When his impacts are linked with the reactions in France to the ravages of Haussmann's reconstruction programme on the heritage of Paris, one finds a brief period before the First World War in which in both countries significant protection for heritage and regulation of new construction in historic cities began to be put in place. The question remains why the era following became dominated by influences much more hostile to heritage.

Rejection of the Victorian City

Heritage that is in vogue and is receiving attention and protection is seldom the heritage of the era that has just passed. Patrick Geddes was typical of influential theorists in the early part of the century in going back to the pre-Victorian eras of architecture and town planning for inspiration. He had no time for the Victorian city and condemned the 'modern town' of the nineteenth century. 'For tradition of towns worth living in,' he wrote, 'our few medieval or early renascence houses and monuments, as in Chester, or York, or Old Edinburgh; and our eighteenth-century dwellings, as in Bloomsbury or New Edinburgh, are, each in their way, far more inspiring and serviceable, more enduring probably also, than is the vast mass of nineteenth-century growth' (Geddes, 1915, pp. 209-210).

Many commentators have observed the strength of the reaction against the city created by industrial capitalism among intellectuals in the early twentieth century. The image of 'Coketown', derived from Charles Dickens's *Hard Times*, survived well into the century and was given a fresh lease of life by Lewis Mumford in *The City in History* (1961), with the result that 'post-war planners, already more than willing to prove themselves totally modern, felt justified in their blanket disregard for the nineteenth-century built environment' (Meller, 1997, pp. 11-12). Rejection of the nineteenth century urban inheritance was often associated with a more general anti-urbanism which boded ill for the retention of urban heritage (Cherry, 1988; White and White, 1977; Wiener, 1981; Williams, 1973).

One aspect of the Victorian legacy that drew particular condemnation was the slum. Slum clearance continued to be a central feature of urban 'improvement'. Public health scares periodically set back conservation in the older districts of Paris well into the twentieth century (Sutcliffe, 1970). This priority remained prominent in one of the key documents of architectural modernism, the Athens

Charter, Principle 67 of which stated that preservation must not 'entail the sacrifice of keeping people in unhealthy conditions':

> Certain people, more concerned for aestheticism than social solidarity, militate for the preservation of certain picturesque old districts unmindful of the poverty, promiscuity, and diseases that these districts harbor. They assume a grave responsibility . . . under no circumstances should the cult of the picturesque and the historical take precedence over the healthfulness of the dwelling, upon which the well-being and the moral health of the individual so closely depend. (Le Corbusier, 1973, p. 87)

In the nineteenth century physical disease had been closely linked with moral disease: there was a highly influential perception of the older districts of cities as breeding grounds not just for disease but also for sedition, insurrection, crime, and vice. That they were therefore a danger to the survival of the race became a predominating view as imperial and social-Darwinian ideas gained an ascendancy. An exception was Raymond Unwin whose half-way attitude foreshadowed some of the changed perceptions of the later part of the century. Unlike so many of his contemporaries and successors, Unwin recognized that slums could possess an inner life and vitality of their own. However, according to Walter Creese, he 'did not believe that the social unity of the slum, although valuable, necessitated its artificial maintenance' (Creese, 1966, p. 288).

PLANNERS AND THE PAST: MODELS OF EVOLUTION AND CHANGE

Another legacy from the previous century that greatly affected attitudes towards heritage in the early and mid-twentieth century was evolutionary models of social development. Two of these were inherited from earlier approaches to predicting the course of social development. Both in their different ways incorporated a role for heritage in the management of the urban environment. On the one hand, there was the linear model, emphasizing and predicting continual progress through adaptation. This was a characteristically Victorian approach to understanding and predicting change in society. Evolutionary theorists such as Geddes believed in studying and appreciating the past in order to understand the present and where society was heading. More than, say, Sitte, Geddes saw this as requiring the actual retention of physical fabric and was prepared to intervene to secure or recommend the preservation of historic buildings (Meller, 1990, 1997).

An alternative model, drawing on historical and particularly classical precedent, saw as inevitable the eventual decline and fall of all civilizations (Hamer, 1990). Many planning theorists thought in terms of cycles rather than a regular and endless linear process (Hill, 1993). The classic decline-and-fall model had itself been urban-based: the onset of the decline phase in civilizations had been seen as closely connected with the growth of cities to the point at which, like ancient Rome, they became corrupt and decadent. The state of cities in the twentieth century appeared to give substantial evidence that such a phase had once again set in. The challenge was to find ways to arrest or reverse it. This is where preservation ultimately had a part to play. Since decayed buildings were symbolic of fall, saving and restoring them would help to break the cycle.

However, a third school of thought has been identified that shared characteristics of the two just mentioned but was decidedly less sympathetic to the retention of heritage as well as being more in accord with the approaches to planning that predominated by mid-century. This tendency is much more characteristic of the twentieth century because it was linked to a view of the radical impacts and implications of its new technologies. Breaking with the past was both predicted and desired by those who have been identified as 'utopian' planners. The prediction is typified by the assumption by Edward Bellamy in *Looking Backward* (1888) that very few of the landmarks of the late nineteenth century would survive by the year 2000 (Relph, 1987). The utopian planners believed that the past could not automatically be expected to wither away and that action was needed to get rid of it. They preferred starting again to reliance on gradual amelioration. In this respect they can be seen as allied to the decline-and-fall tradition: if left alone, cities would continue their seemingly inexorable decline into irrelevance and decrepitude. Robert Fishman describes the utopian planners as 'deeply fearful of the consequences for civilization if the old cities, with all the social conflicts and miseries they embodied, were allowed to persist'. The essential transformation of cities 'meant the extensive rebuilding and even partial abandonment of the cities of their time' (Fishman, 1977, pp. 12-13). What was distinctive about the utopians was their optimism that such a transformation could be effected.

Typical of these attitudes were both the title and the message of Frank Lloyd Wright's *The Disappearing City* (1932) which depicted a time when the great urban centres had ceased to exist. He believed this was inevitable. In the age of the automobile and the telephone the great cities were doomed: they were 'no longer modern'. Wright was convinced that the big city was as obsolete as the horse-and-carriage and therefore destined to be replaced by something better (Fishman, 1977, pp. 122-123). So in a way the enormous decaying and largely abandoned centres which so many American cities had acquired by mid-century had not only been predicted but even hoped for. The utopians either had little to say about what would happen to the legacy of the old city or, like Le Corbusier, they planned for drastic action to replace it altogether. It could not be learned from, as Geddes believed, because of its irrelevance.

Garden cities, one of the most influential planning concepts of the century, were based on the idea of starting from scratch. The idea was that garden cities would reduce the concentration and overcrowding in the old cities and ultimately make them more habitable as well. But some advocates of the idea came close to foreseeing the complete abandonment of old cities. Garden city enthusiasts do not appear to have given much thought to heritage. Ebenezer Howard, the pioneer advocate of garden cities, wrote that 'these crowded cities have done their work'. He believed that 'each generation should build to suit its own needs; and it is no more in the nature of things that men should continue to live in old areas because their ancestors lived in them, than it is that they should cherish the old beliefs which a wider faith and a more enlarged understanding have outgrown' (Howard, 1965, p. 146). From this point of view, heritage was no more than a curiosity, certainly not something from which anything could now be learned. It is ironic that, at the start of the new century, garden city projects are themselves becoming the focus of intense preservationist concern.

The Reconstruction of Historic City Centres by Dictators

The reconstruction of historic city centres by many of the century's dictatorships and other forms of autocratic regime was another cause of loss of heritage. Schemes implemented by dictators almost always involved the modernization of historic centres by means reminiscent of Haussmannization, namely cutting wide avenues through densely built-up quarters, often the districts inhabited by working-class opponents. Ideology tended to predominate at the expense of urban design considerations in dictatorial approaches to urban reconstruction (Cavalcanti, 1997). It might be argued, of course, that dictatorships simply had the power and the ideological motivation to carry out more ruthlessly what was attempted by more democratic regimes via urban renewal and a host of other urban planning programmes in the name of modernization and progress.

There were variations in dictators' approach to the retention of heritage. For example, Mussolini in reconstructing central Rome did not want to get rid of all heritage. He wanted to enhance the focus on the ancient monuments of imperial Rome in order to provide an ideological underpinning for his new Roman empire (figure 11.1). So there was 'a large-scale demolition of building fabric, consisting of medieval, baroque and later architectural developments, regarded as worthless and decadent by the Fascist regime' (Cavalcanti, 1997, pp. 79-80; Kostof, 1992). Some dictators effected a deliberate rupture of continuity and tradition in order to emphasize the revolutionary newness of the order which they were seeking to impose. This is true, for instance, of the Third Reich in Germany (1933-45) and the Ceausescu regime in Bucharest (1965-89).

The Impacts of the Second World War and the Triumph of Modernism

Modernism, which prevailed in town planning and architecture by mid-century, drew on utopian assumptions and was profoundly hostile to conservation. To modernist architects and planners, the existing urban environment was fit only to be discarded. Le Corbusier wrote that 'the increasing disorder in our towns is offensive: their decay damages our self-esteem and injures our dignity. They are not worthy of the age. They are no longer worthy of us' (quoted in Ravetz, 1980,

Figure 11.1. The opening of the Via dell'Impero, Rome by the Italian Fascist regime. This 900-metre long route was designed to link the Piazza Venexia and the Colosseum. Its contruction involved the demolition of vast swathes of old Rome, including 5500 housing units. (*Source*: Istituto Centrale per il Catalogo e la Documentazione, Rome)

p. 44). Le Corbusier was an admirer of Haussmann. His Plan Voisin maintained the tradition Haussmann had established by envisaging the sweeping away of the narrow streets of central Paris and the complete reconstruction of the city, with the monuments of the past preserved like museum pieces in parks surrounding skyscrapers (Fishman, 1977). By mid-century, the desire to sweep away the past had become predominant among planners. Planning was seen by many practitioners and especially theorists as scientifically devised improvement. It was believed that cities should and could be made better. In particular, great faith was reposed in the new technologies which were believed to have made the old city completely obsolescent and irrelevant. The past came to be regarded as an impediment to progress (figure 11.2).

The triumph of modernism was greatly promoted by another major transforming influence on the relationship between planning and heritage: war. Not only did the Second World War cause loss of heritage on a vast scale, its ideological nature, and the enthusiasm for a better way of life after the war that the democracies encouraged in order to maintain morale, resulted in enormous additional momentum behind the impulse to make a new start (Cherry, 1988). The hatred of large, overcrowded industrial cities that had been seen, for instance, in the garden city movement, now could be given full rein.

Throughout Europe after the Second World War the main desire was to rebuild the cities (Diefendorf, 1989, 1993; Kostof, 1992). British planners 'regarded the blitz as a providential and a cleansing process'. 'Because it was loathed, much of the existing urban fabric was seen as wholly expendable' (Ravetz, 1980, pp. 21-24). It was not just bomb-damaged areas that were made the target of reconstruction planning. Cities that had been spared devastation during the war seemed willing to inflict on themselves in peacetime equivalent forms of damage to secure the apparent benefits that they could observe accruing from these opportunities for wholesale reorganization.

Nevertheless, much heritage did survive, even where the intention was to get rid of it. This was a common theme of the twentieth century: master plans were many, but they were seldom fully realized and planning had to proceed in a more piecemeal way. This helped the survival of heritage because continuing use of the inherited structures had to be part of the planning. It seldom turned out to be possible to start all over again, even in cities as devastated as those of Germany (Ladd, 1997).

In this reconstruction of city centres there resurfaced one of the key issues in the twentieth century debate on the relationship between heritage and planning: what should happen to old, narrow, winding city streets and to what extent should they be straightened out or even removed to accommodate modern traffic flows? Modernists such as Le Corbusier hated old streets. There was a concerted attack on them in Germany after the Second World War (Diefendorf, 1989). Some historians have seen what happened to old streets in mid-century as going a long way beyond the Haussmannite legacy. In the view of Donald Olsen, for instance, the town planners of the 1960s were far more ruthless than Haussmann; he, at least, designed streets, whereas a central principle of modern town planning seemed to have become the destruction of the street (Olsen, 1986).

The flaw in the new urban planning was the assumption that older urban cores could be opened up to accommodate modern traffic and communications flows on an indefinite basis. That did not prove possible. The intensifying congestion that resulted required radically different solutions, notably diverting traffic away from historic city cores. In the later part of the century, after immense damage had been done to heritage, this approach began to prevail. One reason for this was the rise of heritage tourism which placed a new value on preserving city centres in a high degree of historical integrity. Planners have developed the concept of entertainment, theatre, and arts districts in which historic ambience and the protection of historic structures play a big part. In many European cities, the historic cores are now largely given over to these functions, and the protection and enhancement of heritage are top priorities.

THE REACTION AGAINST MODERNISM

Gradually, after the early 1960s, a change began to emerge in the attitudes of

Figure 11.2. Clean-sweep planning, New Zealand-style. A project of 60 students on the 1947-48 Summer School of Design organized by the Architectural Centre in Wellington, New Zealand (Wilson, 1996) focused on the redesigning of the Te Aro Flat area in central Wellington, one of the earliest settled parts of the city (post-1840). Finding 'the surface indications of the rapid disintegration of a city' (squalid housing, congested streets), they came up with a model for the complete reorganization of the area. This model, which caused a sensation and was widely exhibited and discussed, proposed the obliteration of almost the entire existing physical fabric of the area. 'It was in the best Corbusian manner – abstract slab blocks disposed in parkland.' It typified the worldwide reaction against 'slums', linked with the Depression. The plan was not implemented. (*Source*: Allan Wild)

Figure 11.2 (*a*) The students at work surveying conditions in Te Aro.

Figure 11.2 (*b*) A photograph of existing buildings (including the War Memorial Carillon, the Museum, and the High School) taken by the students whose model envisaged the almost complete disappearance of this heritage.

Figure 11.2 (*c*) The model from the harbour side. The site shown in figure 11.1 (*a*) can be seen in the middle rear of the model.

planners to heritage. There was an increasing convergence between planning and preservation, although the timing of this varied from country to country. One reason is that some planners, Thomas Sharp in Britain for example, became involved in creating plans for historic towns which involved having to reconcile the protection of their historic cores with the management of modern problems such as traffic flows (Cherry, 1981). While this was initially a specialized line of business, lessons learned from it tended to spill over into general urban design and into the protection of heritage features in townscapes less obviously 'historic' than those of places such as Exeter, Oxford, and Durham on which planners such as Sharp reported. An important new influence was urban design, an approach to planning which aimed to discover features of inherited townscapes that went beyond simply the architectural styles of individual buildings – to street layouts, for instance – and then incorporate these into the redesigning of the urban environment. The turn to 'townscape' owed much to the urban morphological studies pioneered in particular by M.R.G. Conzen (Slater, 1990).

Changing attitudes to the Victorian legacy were especially significant. This was now an era several steps removed from the present time and becoming bathed in a certain nostalgia typified by the political call for a return to 'Victorian values'. There were even changed attitudes towards the most notorious Victorian legacy: the slums. Slum clearance was continued vigorously in Britain during the first half of the century (Yelling, 1992). But there was a strong reaction to the devastation caused to decayed inner-city areas in the cities of the United States in the 1950s and 1960s by urban renewal programmes. Intellectually, this reaction was spearheaded by the critiques of Jane Jacobs and also the sociologist Herbert Gans who discovered communities – which Gans called 'urban villages' – in the districts that had been written off (Gans, 1962; Jacobs, 1961). The 'slum' began to be reinterpreted as a construct imposed on working-class districts from outside and used as a justification for high-handed, undemocratic bulldozing of communities. At this time, there was renewed interest in Engels' 1842 analysis of the 'slums' of Manchester (Marcus, 1974). An official of the Ministère de l'Equipement central office in Paris described the evolution of French urban redevelopment policies:

> In 1958, we thought of slums, and it was a national priority to get rid of them. In 1962 with the *Loi Malraux*, we said that we demolished too much and should preserve the interesting neighborhoods – only the interesting ones . . . Now we try to conserve more, not just the historically interesting. (Webman, 1982, pp. 38-43)

Rehabilitation increasingly replaced demolition and removal of the population as the preferred response to perceived slum conditions. This is one area in which conservation began to appeal as an alternative planning strategy. Indeed, it became for some an alternative planning ethic, linked with new values such as sustainability and use of existing resources, in this case housing stock, as against the wasteful duplication of infrastructure. In other words, what appeared to be a growing sensitivity of planners towards heritage was not necessarily always the result of enthusiasm for heritage *per se*. Often they were moved by much more practical considerations to do with economics and the failure or unaffordability of 'clean sweep' planning.

The process by which planning and conservation became integrated was a slow one. For some time in Britain, they actually went on side by side without much effort to make connections (Ravetz, 1986). Some planners manifested an intense dislike for preservationists and what they perceived to be their detrimental impacts on urban progress. Well known in this regard is the attitude to 'preservationism' of Dame Evelyn Sharp, Permanent Secretary at the Ministry of Housing and Local Government in Britain in the 1960s. To her it was, according to Richard Crossman, her Minister, 'a word of abuse'. She saw 'a clear-cut conflict between "modern" planning and "reactionary" preservation' (Delafons, 1997, p. 76). Even into the 1990s, there were still pleas at conferences on historic preservation in the United States for a greater integration of heritage protection into planning and land use regulation. Detailed suggestions were offered as to how this could be done, for example model zoning ordinances incorporating historic preservation (Lee, 1992).

A major reason why it came to seem important to take account of the urban historical environment in planning was a political one: the connections between built heritage and urban identity. In numerous towns and cities in the United States, the drive to create 'historic districts' has come from concerted city initiatives where restoration of pride in the community is often seen as linked to the restoration of buildings and districts whose features exemplify eras when there was optimism about the city's prospects (Hamer, 1998). This is one of the areas where the interests of preservationists and planners coincided or overlapped. The strategy was to use the past to help towns, cities, and neighbourhoods cope with the challenges of the present and future. An example of this is the 'Main Street' movement which spread from the United States to numerous other countries. Making the business district an historic district, and taking advantage of tax credits, were major components in many Main Street rehabilitation strategies in the United States.

The Historic District as the Focus for the Convergence of Interests

Preservationists learned that coalitions with city planners were one of the most effective and necessary strategies for the achievement of preservation objectives. One result of this was that, because planners worked at the level of districts and zones rather than individual structures, historic preservationists found themselves being drawn into the same spheres of action and concentrating on the creation of historic districts.

During the 1970s and 1980s, historic preservation in the United States and many other countries moved from its traditional concern with individual structures to an emphasis on the contexts of structures. The focus was increasingly on preserving areas in which there may have been no individual buildings that would have been considered worthy of preservation in accordance with traditional criteria. The major problem with concentrating on individual structures had been that it ignored what was happening to their contexts and settings. Many preservationists were converted to the district approach as it became clear that a focus on individual buildings was too narrow and that it would be in the interests of the survival of individual structures themselves that an effort be made to protect their contexts. Buildings can lose a great deal of their meaning once they deteriorate or become surrounded by incompatible buildings and activities. The modern historic district approach to historic preservation has involved a rejection of the Haussmannite approach to the contexts of individual structures, even 'monuments'. Haussmann isolated individual buildings to enhance their meaning. But, as Geddes and others saw early on, the reverse was often the result. Context was meaning, especially if interpreted in relation to change over time: the dwellings of the poor that came to surround charity-dispensing cathedrals, for instance, or the rows of lawyers' offices adjacent to courthouses in the United States.

Once preservation moved beyond the individual building, planning had to come into the story. It might have been possible to focus on a building as a house museum in total detachment from its surroundings and from what was being planned for and done to that environment. This often did happen with deleterious effects on the message that the building was supposed to convey about its place in history. But a district, unless it is walled off, has to involve a co-existence of history with many other aspects of urban life – with some, like traffic, often destructive of heritage values – requiring planning intervention. Preservation may itself contribute to planning objectives, through, for example, the rehabilitation of a district so that people will want once more to live in it and a city's tax base will be reinforced.

The creation of historic districts – termed 'conservation areas' in Britain and other countries – is one of the major planning developments in the second half of the century (Larkham, 1996; Hamer, 1998). It has been a principal area of convergence between preservationists and planners. In the United States, significant precedents for the historic district included colonial Williamsburg, Charleston, and New Orleans. But the large-scale district orientation of modern historic preservation is principally a consequence of reactions to urban renewal in the 1950s and 1960s. Blitzed downtowns and large-scale loss of inner-city heritage became a legacy of that era. Often as a result of public reaction to what had been wrought, small districts left over after urban renewal and freeway construction were preserved explicitly as fragments and made to fulfil a symbolic function as representative of all else that has been removed. Planners were not necessarily involved in this, and some such exercises might be regarded as cynical exercises in appeasing opposition and public indignation. Urban renewal programmes had some more positive spin-offs of relevance to the development of a symbiosis between planning and preservation, especially federal government funding which provided assistance for the retention, protection, and restoration of existing housing and neighbourhoods. The major area of convergence, however, was zoning. It is through the development and extension of this major tool of twentieth century urban planning in the creation of locally designated historic districts that the most effective management of the protection of built heritage has been achieved in the United States.

Historic districts have created a whole new form of usable past. Insofar as planners had been interested in the past, that interest had derived more from the pure ideals of the past than from surviving remnants which represented that heritage only in a degraded and compromised form. Built heritage, if it has survived into the present, has more often than not done so in a corrupted form, particularly if, as is the case with the sort of heritage that influences town planning, it involves entire districts. So there was not, and there never has been, a necessary connection between drawing inspiration for planning and urban design from the past and historic preservation. Such heritage as has influenced planning has not always consisted of actual, physically available structures. The 'village' is an example of the idealized heritage that has influenced much design. It is an ideal that has seldom related to actual extant villages. It became a vision of an ideal community, in the case of Britain at least, reaching deep back into folk memories of a pre-Industrial Revolution or even (in the case of the Norman Yoke myth) pre-feudal past (Ravetz, 1986, p. 53).

In the late twentieth century, with the emergence of strictly regulated and restored historic districts, the situation changed considerably. These areas became purged of the decay and detritus of the centuries and were restored to as close to their original intended state as increasingly sophisticated preservationist strategies and techniques could contrive. The ability of such purified districts to furnish inspiration for innovations in urban design was manifested in such movements as the New Urbanism in the United States which was based in substantial part on intensive study of ideal form as it appeared to be revealed in restored historic districts.

Third World and Post-Colonial Issues

So far we have been considering only Europe and the United States. In relation to much of the rest of the world, this apparent bias has relevance insofar as the twentieth century was also an era both of imperialism and of post-colonialism. The ideas of Europeans concerning the relationship between heritage and planning were exported to many other countries and then given particular reorientations according to the strategies of the colonialists. The influence of these ideas was still being felt at the end of the century, even in countries that had long since emancipated themselves from colonial rule.

Colonial planners treated the heritage of colonies in diverse ways, although local urban traditions were frequently ignored. In this respect Geddes was again a notable, although not very effective, exception (Home, 1997; Meller, 1990). A characteristic model of planning was the 'dual city', one reason for which has been identified as the enthusiasm of European architect-planners for the colonies as virgin fields on which to implement their ideas. This meant that they left the old cities alone and built new ones elsewhere. Another motive was the segregation of the races, as in Morocco where the French policy of conserving the 'native quarters' had the effect of preserving some heritage (King, 1990; Kostof, 1992; Wright, 1991).

The development of 'dual cities' has revived in the management of many Third World cities. In Latin and South America, for instance, one frequently finds an 'Old' and a 'New' city. Examples are Quito in Ecuador and Havana in Cuba. The 'Old City' may be a World Heritage Site, carefully preserved as a tourist attraction, while modern development is concentrated in the 'New City'. A distinction has also developed between 'upper' and 'lower' cities. The former developed initially as a defensible site. Now that function is no longer required, a distinction has emerged between the 'upper' tourist historic city and the 'lower' modern commercial city (Ashworth and Tunbridge, 1990). This type of dualism appears in the New World in Quebec City and is common in Europe.

By the end of the century, almost all colonies had been granted their freedom, a culmination to the century symbolized for many by the handing over of Hong Kong to China in 1997. Many former colonies had achieved independent nationhood only after prolonged and often bloody struggle against the imperial power. Revolutionary movements which had spearheaded campaigns of liberation then came to power themselves, and their revolutionary ideologies formed the basis of the national identities propagated by the new regimes. In addition, there was the collapse of the Soviet empire in the late 1980s and early 1990s, bringing new post-Cold War regimes to many countries which had already experienced German occupation during the Second World War. Post-revolutionary and post-colonial countries are thus nations whose heritage is multi-layered, both temporally and culturally. And it remains intensely political. Which heritage will give legitimacy to the current regime? Will any, or should there be a clean slate? The heritage of the colonial era will often have many bad associations which lead to a desire to see it downplayed (Slater, 1990). Planning history has a critical role to play in all this because imperial powers often manifested their control through the design of cities and the architecture and positioning of monumental buildings. New nations have frequently to determine what is to be done with such legacies. The wholesale rejection of inherited planning by revolutionary regimes can have seriously damaging impacts on sensitivity to urbanistic values. This has been noted, for instance, in regard to the post-1959 adoption of modernistic principles in Cuba (Prestamo, 1993).

CONCLUSION

The management and protection of heritage within the immense complexity of a modern urban environment requires a high level of planning skills. To some, what has happened in the process is an inevitable subordination of heritage values to the agendas which planners are principally responsible for determining and implementing. For example, aspects of the survived past which receive protection tend to be those which lend themselves to exploitation by planners as a form of usable history. Others that do not qualify in this way have a much harder time surviving.

However, once the 'clean sweep' approach to urban planning, and master-planning in general, fell out of favour, the view did gradually emerge that knowledge of urban history could be a guide to planners by enabling them to understand better why a city has developed in the way it has. This was, for instance, the position that had been taken by Geddes and later by his friend and correspondent Lewis Mumford (Miller, 1992). But this influence cannot derive just from surviving built heritage. It has to involve what might be called 'deep' heritage, i.e. not just buildings but hidden features of the inherited townscape and indeed of the culture itself that affect how we live today and constrain what can be done to change our cities. The impact has been reduced by the slowness of urban historians to become involved in the public history and historic preservation areas (Hamer, 1998). However, notable contributions were made by the American planner Kevin Lynch, who pioneered the application of cognitive mapping to understanding urban form 'from below' – how the urban environment is actually perceived and experienced by the people who live in it (Lynch, 1960). Late in the century, the 'Power of Place' strategy applied by Dolores Hayden and her colleagues in Los Angeles suggested new ways in which, through evocations of memory, continuity could be retrieved from urban settings in which physical heritage has survived only in highly fragmented forms (Hayden, 1995). These developments held out some promise for the emergence of a new, more democratic understanding of the role of heritage in the planning of the urban environment.

REFERENCES

Ashworth, G.J. and Tunbridge, J.E. (1990) *The Tourist-Historic City*. London: Belhaven Press.

Bellamy, E. (1888) *Looking Backward: 2000–1887*. Boston: Ticknor.

Cavalcanti, M. (1997) Urban reconstruction and autocratic regimes: Ceausescu's Bucharest in its historic context. *Planning Perspectives*, **12**, pp. 71-110.

Cherry, G.E. (ed.) (1981) *Pioneers in British Planning*. London: The Architectural Press.

Cherry, G.E. (1988) *Cities and Plans: The Shaping of Urban Britain in the Nineteenth and Twentieth Centuries*. London: Edward Arnold.

Collins, G.R., and Collins, C.C. (1965) *Camillo Sitte and the Birth of Modern City Planning*. London: Phaidon Press.

Creese, W.L. (1966) *The Search for Environment: The Garden City. Before and After*. New Haven: Yale University Press.

Delafons, J. (1997) *Politics and Preservation: A Policy History of the Built Heritage, 1882-1996*. London: E & FN Spon.

Diefendorf, J.M. (1989) Artery: Urban reconstruction and traffic planning in postwar Germany. *Journal of Urban History*, **15**, pp. 131-58.

Diefendorf, J.M. (1993) Planning postwar Vienna. *Planning Perspectives*, **8**, pp. 1-19.

Fishman, R. (1977) *Urban Utopias in the Twentieth Century: Ebenezer Howard, Frank Lloyd Wright, and Le Corbusier*. New York: Basic Books.

Gans, H. (1962) *The Urban Villagers*. New York: The Free Press.

Geddes, P. (1915) *Cities in Evolution: An Introduction to the Town Planning Movement and to the Study of Civics*. London: Williams & Norgate.

Hamer, D.A. (1990) *New Towns in the New World: Images and Perceptions of the Nineteenth-Century Urban Frontier*. New York: Columbia University Press.

Hamer, D.A. (1998) *History in Urban Places: The Historic Districts of the United States*. Columbus, Ohio: Ohio State University Press.

Hayden, D. (1995) *The Power of Place: Urban Landscapes as Public History*. Cambridge, Mass: MIT Press.

Hill, D.R. (1993) A case for teleological urban form history and ideas: Lewis Mumford. F.L. Wright, Jane Jacobs and Victor Gruen. *Planning Perspectives*, **8**, pp. 53-71.

Home, R. (1997) *Of Planting and Planning: The Making of British Colonial Cities.* London: E & FN Spon.

Howard, E. (1965) *Garden Cities of To-Morrow*. London: Faber and Faber.

Jacobs, J. (1961) *The Death and Life of Great American Cities*. New York: Random House.

Jordan, D.P. (1995) *Transforming Paris: The Life and Labors of Baron Haussmann*. New York: The Free Press.

King, A.D. (1990) *Urbanism, Colonialism, and the World-Economy: Cultural and Spatial Foundations of the World Urban System*. London: Routledge.

Koshar, R. (1993) Against the 'Frightful Leveler': Historic preservation and German cities, 1890-1914. *Journal of Urban History*, **19**, pp. 7-29.

Kostof, S. (1992) *The City Assembled: The Elements of Urban Form Through History*. London: Thames and Hudson.

Ladd, P. (1997) *The Ghosts of Berlin: Confronting German History in the Urban Landscape*. Chicago: The University of Chicago Press.

Larkham, P.J. (1996) *Conservation and the City*. London: Routledge.

Le Corbusier, L. (1973) *The Athens Charter*. New York: Grossman Publishers.

Lee, A. (1992) *Past Meets Future: Saving America's Historic Environments*. Washington DC: The Preservation Press.

Lynch, K. (1960) *The Image of the City*. Cambridge, Mass: MIT Press.

Marcus, S. (1974) *Engels, Manchester and the Working Class*. New York: Random House.

Meller, H. (1990) *Patrick Geddes: Social Evolutionist and City Planner*. London and New York: Routledge.

Meller, H. (1997) *Towns, Plans and Society in Modern Britain*. Cambridge: Cambridge University Press.

Miller, D.L. (1992) Lewis Mumford: Urban historian, urban visionary. *Journal of Urban History*, **18**, pp. 280-307.

Mumford, L. (1961) *The City in History*. New York: Harcourt Brace.

Olsen, D.J. (1986) *The City as a Work of Art*. New Haven: Yale University Press.

Prestamo, F.J. (1993) City planning in a revolution: Cuba, 1959-61. *Planning Perspectives*, **8**, pp. 188-212.

Ravetz, A. (1980) *Remaking Cities: Contradictions of the Recent Urban Environment*. London: Croom Helm.

Ravetz, A. (1986) *The Government of Space: Town Planning in Modern Society*. London: Faber and Faber.

Relph, E. (1987) *The Modern Urban Landscape*. Baltimore: Johns Hopkins University Press.

Slater, T.R. (ed.) (1990) *The Built Form of Western Cities: Essays for M.R.G. Conzen on the Occasion of his Eightieth Birthday*. Leicester: Leicester University Press.

Sutcliffe, A. (1970) *The Autumn of Central Paris: The Defeat of Town Planning 1850-1970*. London: Edward Arnold.

Webman, J.A. (1982) *Reviving the Industrial City: The Politics of Urban Renewal in Lyon and Birmingham*. London: Croom Helm.

White, M. and White, L. (1977) *The Intellectual Versus the City: From Thomas Jefferson to Frank Lloyd Wright*. Oxford: Oxford University Press.

Wiener, M.J. (1981) *English Culture and the Decline of the Industrial Spirit, 1850-1980*. Cambridge: Cambridge University Press.

Williams, R. (1973) *The Country and the City*. London: Chatto and Windus.

Wilson, J. (ed.) (1996) *Zeal and Crusade: The Modern Movement in Wellington*. Christchurch: Te Waihora Press.

Wright, F.L. (1932) *The Disappearing City*. New York: Payson.

Wright, G. (1991) *The Politics of Design in French Colonial Urbanism*. Chicago: University of Chicago Press.

Yelling, J.A. (1992) *Slums and Redevelopment: Policy and Practice in England, 1918-1945, with Particular Reference to London*. London: University College London Press.

CHAPTER 12

Open Space, Urban Planning and the Evolution of the Green City

Maurits Van Rooijen

The conscious provision of open space is an integral part of modern town planning. By and large, it was toward the very end of the nineteenth century that green open space became an important, though not obligatory, element in urban extensions. The provision of green open space only became an obvious integral part of town planning – at least in the industrialized world – during the second half of the twentieth century. No hard statistical information is available, but it is a safe guess that in many large cities in the Western world, the amount of green open space reached its highest levels after the 1960s. The average amount of green open space per inhabitant in urbanized areas went up from just a few square metres per inhabitant in the first half of the century, to some 20-30 square metres by the 1950s-1960s. Subsequently, in many European cities, the figure is likely to have reached averages of some 40 square metres per capita in the 1980s (Van Asperen, 1983).

The presence of this amount of green open space in the urban environment is the result of a long process. Understanding that process is important, not just to gain some insight into town planning itself, but also to gain a greater understanding of the meaning of green open space in cities today. The purpose of this chapter is broadly to analyse the provision of substantial green open areas in modern cities against the historical development of society set on an inexorable evolution toward the multi-faceted concept of the 'green city'. Though the research on which this chapter is based originally had a focus on Dutch town planning history, it was enriched with results of more or less comparable research in other European countries such as the United Kingdom, Germany, Austria and France. Since then the study has evolved to more theoretical concepts that might now be less accurate as empirical description but should have gained in usefulness at a more global level.

Categorizing Urban Open Spaces Historically

There are many ways of analysing urban open space. Green open areas in cities, whether free-standing or as part of a metropolitan system, have been mostly categorized in terms of form and/or function. In these terms, Cranz (1982) identifies four historical stages in America: the Pleasure Ground (1850-1900), the Reform Park (1900-1930), the Recreational Facility (1930-1965), and the Open-Space System (1965 and after). Clearly there has been a direct influence between Old World and New World; first mainly from Old to New, and through the twentieth century, more from New to Old:

> The original North American urban park prototype drew on European examples such as Birkenhead Park and the Bois de Boulogne. These were in turn developments of earlier prototypes – the royal parks and the private country estate – which were expressions of an elite group's engagement with nature, art and landscape painting. Downing, Olmsted and their contemporaries developed the concept of the North American urban park as scenic antithesis to the urban condition and tool for promoting democracy and social welfare. (Thompson, 1998, p. 906)

But mutual influence was not a matter of copying. The American parks derived not just from European antecedents but 'from an anti-urban ideal that dwelt on the traditional prescription for relief from the evils of the city – to escape to the country. The new American parks were thus conceived as great pleasure grounds meant to be pieces of the country, with fresh air, meadows, lakes, and sunshine right in the city' (Cranz, 1982, pp. 3-5). It would be incorrect to presume that the American stages as identified by Cranz apply without modification to Europe or other continents.

In order to formulate a framework general enough to be used as a reference for research of the development of green town planning in different parts of the world, this study takes into consideration four elementary factors: political/legal; social/functional; economic/financial; and design/environmental. As in Cranz's (1982) study, the focus is on the original intention, design and usage, but added to these are financial parameters and economic context. Furthermore, from an 'Old World' perspective a longer time-scale needs to be taken into consideration, appreciating that current ideas and attitudes are the results of a long-term process. Also, developments in Europe have been more diverse than in the United States, with park-specific historical stages less useful. The categorization presented here therefore refers to spatial types in a societal context rather than to stages in *the* history of green town planning. The elementary categorization acknowledges the importance of the historical dimension, the evolution of concepts, changes in ideas and practice and, at the same time, long-term continuity, recognizing that societies develop at different speed and along various routes.

The framework presented is based on a typology of the basic or original concepts behind specific forms and functions of man-made nature areas in cities:

- Common open space;
- The royal public garden;
- The bourgeois park;
- The people's park;
- The garden city/new town.

This typology recognizes the time dimension that is crucial to the planning process, by indicating an evolution of planning ideas and intentions spanning several centuries. It considers, one might say, the length of the historical roots of specific forms of green open space. Importantly, however, the main purpose is to understand the present rather than describing the past. Thus this approach identifies the evolutionary process underlying contemporary green city planning and is even able to look toward the future. It is equally important to recognize that the green spaces which one would consider typical for a specific category still exist, still function, and in fact are often still created. Alternatively, specific open spaces might have (in fact, often have) evolved according to changing fashions, concepts and public demands. It should also be appreciated that, when identifying for the sake of clarity only a very limited number of categories, many areas will show a mixture of features or might be considered as transitional between categories.

COMMON OPEN SPACE

One could define the *common* as open space in its purest form: the natural, wild, unspoiled, and uncultivated state of the environment, without clearly defined ownership. In its very origin, planning of such open space is an irrelevant concept. The original commons were typified by a state of total freedom, few rules and controls, and an absence of orderly maintenance. It would be difficult to give clear examples of such open space in the modern context, but it is relevant as a concept – admittedly a rather romantic concept – and it has remained the foundation of the so-called *commons* which survived in some societies, for instance Britain (Conway, 1991; Forshaw and Bergstroem, 1986). The original common allowed for any form of usage by any individual, yet such anarchy has had to be modified for common land in or near cities. Also, leaving maintenance to nature becomes impossible once the impact of people is no longer marginal. The *common* might be considered as a sanctuary from planning and regulation, but paradoxically it can only survive as open space due to legal protection.

THE ROYAL PUBLIC GARDEN

The first truly planned cities tended to be capital cities created by emperors, kings, princes, governors and other omnipotent rulers. Strong central authority facilitated spacious layouts and monumentalism in line with renaissance ideas about ideal towns. As early as 1484, the Italian Alberti wrote that a town should not be built purely for the sake of protection, but that public places should be created with gardens for entertainment and recreation. '*Rus in urbe*' was an acknowledged classical desire. The works of Haussmann in Paris and the creation of the famous Ringstrasse in Vienna (Reining, 1976) might be considered as late examples of such 'capital planning'.

In contrast to the *common,* which by definition is open, the park and garden typically are enclosed areas. Etymologically, both words explicitly refer to fences. Strictly speaking, a park or garden is private property and *public garden* or *park* should be considered a contradiction in terms. Historically, the park was an enclosed area where the lord could pursue his hunting privilege. Later it came to indicate the large-scale garden around the master's residence. The owner was the king, prince, or those noblemen who once had a clear place in the feudal order. In major European cities, gardens of sovereigns were opened to the public from the early seventeenth century onwards, starting in London (Brace, 1986; Lasdun, 1991), Paris (Leveque, 1982), Berlin (Ritter, 1982), and Vienna. This was partly due to the fact that royal families moved away from the city to the safety and sanctity of the countryside. But more importantly, the gradual opening of royal gardens to the public should be interpreted as a political phenomenon, illustrating changes in the structure of power in society. It shows the recognition by the ancient regime of the need for concessions to maintain or recover some social harmony.

The public park could be interpreted as an example of enlightened governance. One might also consider it as a bid by the aristocratic establishment to retain stability by recognizing the changing times. Yet the physical form of the public park retained the features of the private garden and the design was still very much a high culture-based art form. Ownership remained with the king, the sovereign, or the State. The public park was a gift to the people, or more precisely to the appreciative upper and middle classes. The visitors were guests. They were expected to adjust to and learn from the *mores* of the higher classes. The public park offered a respectable decorum for civilized socialization, for conspicuous leisure. It explicitly excluded the lowest, uncivilized, ill-behaved classes of society (Van Rooijen, 1990*a*)

The idea of opening royal gardens to a (select) public was given a new dimension at the end of the eighteenth and during the nineteenth century in European cities when the sovereign would actually create a garden specifically for their subjects. The concept of the *Volkspark* or 'people's park' was described by Christopher Hirschfield in 1785. But the name was misleading since the intended visitors to the park were still select strata of society. Education by erecting statues for heroes, to be presented as role models, was given as one of the explicit aims. But possibly the most convincing argument for creating such open space was that a park would give the sovereign popularity amongst his subjects. In the first Viennese gardens opened to the public – the Augarten and the Prater – the benevolent Kaiser 'often goes amongst his people, without guards, just surrounded by the love of his subjects' (Hirschfeld, 1785, p. 68). Certainly at the end of the eighteenth century this was an extremely powerful argument. The sovereigns were all too aware of the literally lethal alternative to such popularity. The first example of a purpose-built public garden was probably the Theodorspark in Munich, Germany. It was the result of an initiative of the sovereign of Bayer and was opened appropriately in the year of the French Revolution, 1789.

The resulting form and function of the public garden could be described as a concession. It was a synthesis between higher and lower cultures, compromising the design of a private garden with the needs and requirements of usage by a larger public. It was the start of a process which would continue during the following centuries and which would make the modern park a typical *gesunkenes kulturgut*: an element of elitist culture which has been adopted by (and also adjusted to the taste of) lower levels in society. A very gradual democratization of parks in political-economic terms as well as social-cultural terms took place, allowing for a wider penetration of the higher cultural concept of the 'royal park' by relatively less affluent or less privileged social groups.

THE BOURGEOIS PARK

In European cities the creation of public parks entered a decisive new stage in the nineteenth century. Citizens would not just benefit from a park donated to them; they would actually create such public parks and similar spaces for themselves. Behind the creation of such parks was the idea that those who pay for the provision are basically their owners. In other words, the more one paid, directly or indirectly, for the creation of green space and for the maintenance of such provisions, the more one owned it. During most of the nineteenth century, this utilitarian idea dominated the creation of green open space in European cities.

This idea of ownership was obvious in the case of green space that was the direct result of a property developer creating a park landscape with villas. In such a project, plots of building land were sold at profit and the green environment, plus roads and further infrastructure, donated to the local authority. Those who lived in such *villa parks* had, quite literally, paid for it. The form was normally inspired by the large landscaped gardens created on the estates of wealthy landowners in 'English' style, following the works of landscape architects such as Capability Brown. Villa parks were created on former family estates near railway stations and at commuting distance from the cities. Later, they were also created as extensions to the cities, for instance on land that became available once it was clear that urban fortifications no longer had any military value (Van Rooijen, 1990*a*).

In Europe the creation of 'villa parks' became an important early feature of

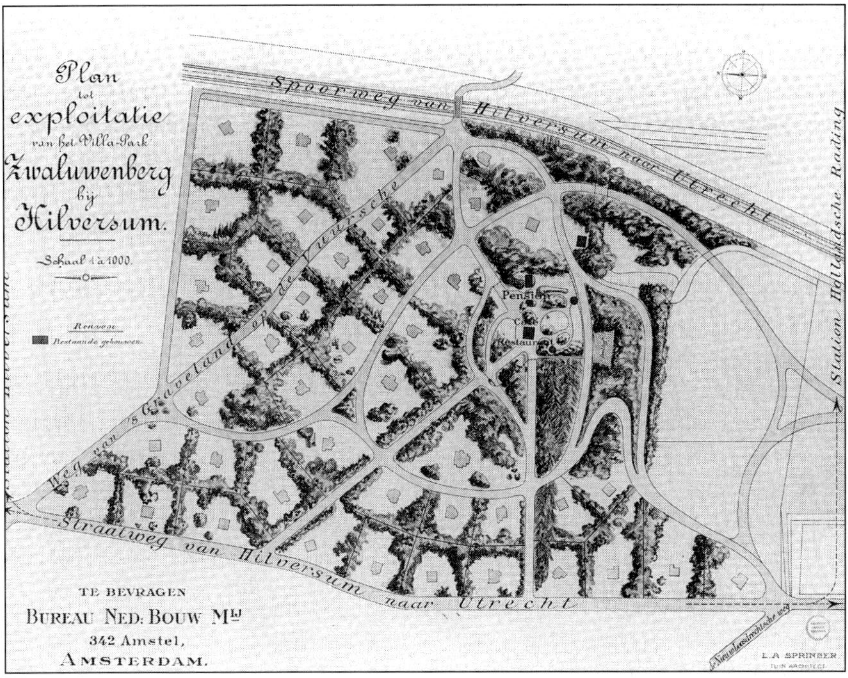

Figure 12.1. The typical design for a *villa park* by Dutch landscape architect L.A. Springer for Zwaluwenberg, located between Utrecht and Hilversum. Characteristic for late nineteenth and early twentieth century town design were green residential parks, with villas for the prosperous middle classes placed on plots scattered in a landscaped plan with winding roads. In Europe these areas were often created on country estates on the outskirts of town with good rail connection for commuting to the city. The original manor house was either demolished or re-used as a hotel, hospital, office building. (*Source*: University Wageningen, Special Collections)

modern town planning (figure 12.1). One might consider them as a triumph of the bourgeoisie. The park offered the wealthy a setting for their desirable villas and creating such a park could be profitable. Green environments were for high rate taxpayers and many were created by the direct involvement of local authorities. Squares, avenues, boulevards could all be considered as financially-sound investments. At the same time, these green forms of beautification supported civic pride, particularly of the middle classes, who considered such forms of green very much their own. They harmonized with well-designed, well-maintained private gardens and with indoor plant collections. They also suited a lifestyle with a taste for recreation in nature, including increasingly fashionable leisure pursuits such as holidays in the mountains or at the seaside, and outings to the countryside.

This form of green open space can be given the label *bourgeois park*, typifying design, intention, finance and function. A spacious green environment gives a wealthy impression and offers an appealing area for the well-to-do. In The Netherlands – historically the most urbanized part of Europe – the idea that investment in a more spacious design for town extensions could have an obvious economic benefit was argued as early as the seventeenth century. The military requirement for strong fortifications around towns made it difficult, if not impossible, to bring this idea into practice at that time (Taverne, 1978). At the end of the eighteenth century warfare had changed, making such fortifications redundant. In the following decades cities were willing even to take out substantial loans in order to create town extensions – often on the derelict former fortifications around the city – which would attract wealthy families. The costs of infrastructural works were to be recovered (ideally at a premium) from the sales of valuable plots of building land for villas. Even more importantly, wealthy families would boost the local economy and were welcome taxpayers. Many municipal parks were thus created at very low or nil cost through the sale of land around the city periphery. Maintenance costs were covered, again ideally at a premium, by the rates charged to the new inhabitants. Thus the creation of green space was perceived by many as an effective initiative to stimulate the local economy, real estate market, and employment (Van Rooijen, 1993).

Ideally, public parks, gardens, promenades, and squares would attract permanent settlers, but for some towns the same design tools were used to appeal to holidaymakers. All this was greatly helped by the arrival of the railways. They made the population more mobile. The emerging market for both longer holidays and day-trips stimulated inter-city competition. Destinations had to position themselves by targeting either the selective, exclusive end of the market or a mass market of less well-off visitors (Ward, 1998). This was not just a matter of appropriate promotional materials. A suitable physical presentation of the town was essential: strategically-placed displays of flowers in full bloom for the season and tree lined boulevards stretching from the railway station to the town centre or beach. British examples such as the seaside towns of Brighton and Blackpool, and spa-towns such as Harrogate and Bath, are famous (Conway, 1991), but the phenomenon was worldwide.

In classical liberal logic, those who financed the creation and maintenance of public space in cities were also to be the decision-makers and primary beneficiaries, although others were welcome as long as they were willing to adopt the recreational ideas of those who had paid for the provision. In terms of design, such open space respected the tastes, requirements and desires of the wealthy urban inhabitants. It was likely to include provisions for elitist leisure activities such as horse riding or cycling, in other words for sophistication 'bourgeois style'. The decision to create such municipal bourgeois parks was a matter for the city fathers, 'elected' because of their wealth and representing the interests of the higher classes in town (though at times these interests might very well coincide with the interests of the town as a whole).

In addition to hard economic motives and political realities, 'softer' drives were evident, namely those of civic pride, local chauvinism, and even charity (Conway, 1991). Donations of land and gardens have played an important role in the creation of green open space in European, especially English cities. The motives have been diverse, from an individual seeking a degree of appreciation, even some form of immortality, through indirect financial gain, to a genuine wish to contribute to a better society. Such donations have been of obvious practical importance and in some cases also of specific political importance, since they illustrated the functioning of the good-citizen concept within the liberal *laissez-faire* ideology as well as within a social-democratic concept of a more harmonious society.

People's Parks

Before the turn of the eighteenth century, the Danish lawyer J.P. Willebrand wrote about the wisdom of city authorities in creating parks, freely accessible to all, for the sake of public health and recreation (Willebrand, 1775). Yet green open space as represented by the bourgeois park was in principle created in the wealthiest residential parts of town, a rational location from a purely economic point of view. However, high density areas of low rent housing were deprived of green space. Gradually, awareness grew of the undesirability of this situation. Without questioning the social or religious motives of those who sought to improve this situation, there were self-interested arguments why change was needed. Fundamental was the acknowledgement of the interdependence of urban communities. Epidemics such as cholera had put public health high on the political agenda. Parks were explicitly recognized as the 'lungs' for the city and in the interest of the wider community no parts of town could be safely excluded from this cure.

Furthermore, it was in the 'public' interest that all classes, especially the working classes, had the opportunity to strengthen their physical condition with healthy exercise in a green environment (figure 12.2). They would then be in a better position to contribute to the development of the nation through their labour or to the defence of their nation as soldiers. But it was not just a matter of physical health. Equally important was the securing of healthy spirits in healthy bodies away from drinking and gambling. Parks and outdoor recreational facilities offered a better environment to ensure stable family life than pubs and street life. And all this also contributed to political harmony. In previous centuries the ruling élite had begun to appreciate this rationale; now capitalist enterprise started to understand it. The implicit political motive behind the creation of green open space made available to the masses was the desire for stable society. Parks for the people would contribute to a social reform agenda which offered a clear alternative to the violent revolution already witnessed in some parts of the world (notably the Russian Revolution of 1917).

Figure 12.2. H.P. Berlage's *zuiderlijk uitbreidingsplan* (Plan South) for Amsterdam (1915). In the beginning of the nineteenth century the Dutch capital Amsterdam acquired the reputation as 'mecca for social housing'. The famous extension plan for the southern part of the city was designed by H.P. Berlage. His first plan (1904) showed large amounts of green open space and upmarket houses. The second version (1915) reproduced here was aimed at the middle and upper lower classes. A large recreational area, the Amsterdamse Bos, was created on the outskirts. (*Source*: Municipal Archives of Amsterdam 'De Atlas')

As before, the widening of access did have consequences for the design of open spaces which acknowledged the specific needs and wishes of the lower classes. The social provision of green space had become a political entitlement due to changed economic circumstances. Higher income caused a demand for a more prosperous, greener living environment. Civic authorities had to take into consideration the needs and demands of an increasing number of voters. The twentieth century process of democratization has also had a major impact on the ownership concept of urban open space. In the early 1900s, the idea of the people's park as a green open space *for* the people started to become an accepted idea. But through the twentieth century this would develop further in the idea that such an amenity was *of* the people, in other words a right rather than a charitable provision.

Green open space was no longer to be created randomly, but systematically, scientifically as part of the city planning process and as part of the overall city structure. In the scientific planning approach of the twentieth century, artistic value, so clearly present in the public garden, was considered of less importance. Form followed function, but of course the form also defined the function. Locational guidelines and normative standards of provision were codified. The concept of 'people' was translated to 'communities' as hierarchies were identified and combined with hard norms about per capita space standards.

The German planners Koch (1913) and Wagner (1915) made major contributions to open space planning, setting clear quantitative and systematic guidelines. The idea that more open space was needed where there was a higher density of population presented a radically different approach from the economic principle which underpins the bourgeois park. Wagner set some simple guidelines for distances: a maximum of ten minutes to reach a local park, twenty minutes to reach a city park, and thirty minutes for sporting fields and woods. According to Wagner, districts with 10,000 inhabitants should have 13 hectares of woods, 2.4 hectares of playing fields, 1.6 hectares of sports grounds, and 0.5 hectares of walk ways. The hierarchy of green open space had to be related to categories of users, as defined by age and gender. But most importantly, the key was not the fact that this space was available in the right amounts at the right sites but that it was actually used. In the terminology of Wagner, the real value was not the *daseinswert* (just being there) but the *nutzungswert* (the actual usage). This approach identifies the success of planned open space by measuring its popularity and utilization.

Early guidelines, formulae and rules-of-thumb were often inspired by American studies. The systematic approach to open space planning was also enriched by the American concept of the parkway – linking separate green open spaces by continuous green corridors. Another idea, very popular in Scandinavian town planning, was the creation of 'green fingers' to bring the surrounding countryside into the city (figure 12.3). The British green belt idea was also influential. But most importantly and regardless of spatial form, green open space became a well-defined element within urban planning processes. It was consolidated as an integral part of a total land use package, to be rationally created and conserved in the proper proportion to housing, work, and traffic. In Europe, the planner's task was defined by the famous Athens Charter of 1933. In this manifesto, town planning was seen as a way to control urban chaos and green open space had an essential function (figure 12.4). Achieving the right balance between built forms and nature areas, and between enclosed and open spaces, was the assignment of the city planner. During most of the twentieth century, contemporary science gave seemingly unambiguous answers to what was the right balance.

Improved transport allowed cities to plan on a larger scale. No longer did the worker need to live within walking distance of the factory. Low-density town planning with more generous levels of both public and private open space could help eliminate slum areas. Paradoxically, improved transport also enabled higher salary earners to move out of the city, but simultaneously assisted towns to plan green extensions in an attempt to retain them. The creation of green open space for urban inhabitants had to stay in step with technological developments, especially with increasing mobility. As a result, the hierarchy of green open

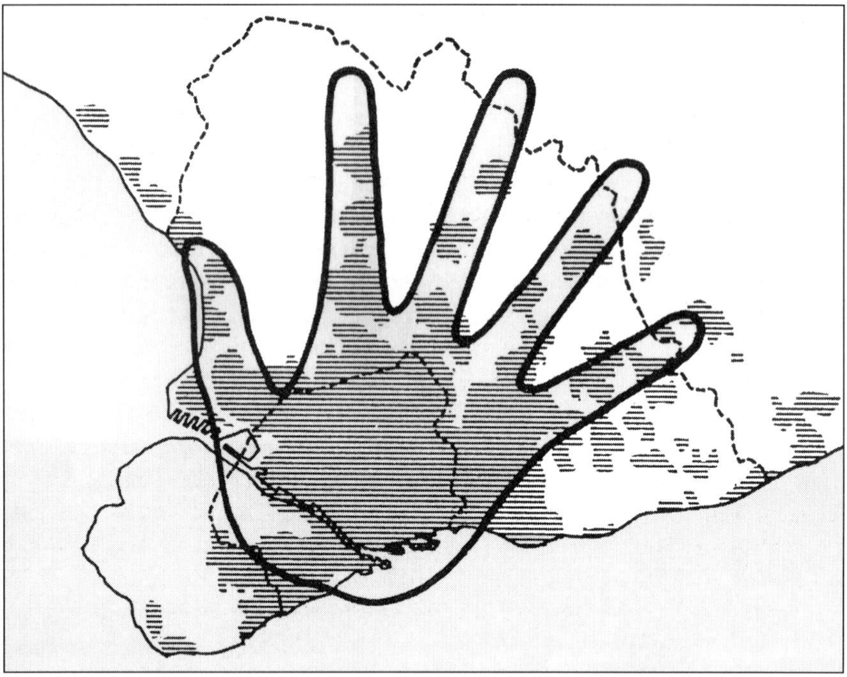

Figure 12.3. The 'Finger Plan' of Copenhagen. One of the most famous city plans developed for the Danish capital Copenhagen shortly after 1945, it proposed concentrated urban growth along narrow 'fingers' allowing for 'green wedges' (areas for recreational purpose) to stretch into town. Over the years, however, the fingers have swollen and the wedges have come under attack. (*Source*: Greater Copenhagen Council (1984) *The Green Wedges of the Capital; The Story of the Green Areas of Greater Copenhagen*)

space for cities could not remain limited to the city itself nor its immediate surroundings. Modern forms of transport not only allowed for low-density extension of cities, but also led to a 'recreational urbanization' of the countryside. A substantial range of large-scale provisions for mobile city dwellers was created to accommodate a wide variety of outdoor recreational activity: country clubs, marinas, caravan parks, holiday camps, golf courses, etc. to service rising recreation and leisure demands. This pushed the open space planning process from the city to the regional and even the national level.

GARDEN CITY/NEW TOWN

Modern planning policies toward urban green space were pragmatic and developed within large cities in reaction to economic opportunities, social issues (especially the perceived threats of imminent crises in society), and political demands. Yet it is also possible to identify the related creation of garden cities and new towns as a second 'track' in modern town planning. Lewis Mumford (1965, pp. 39-40) wrote: 'The Garden City, as [Ebenezer] Howard defined it, is not a suburb but the antithesis of a suburb; not a more rural retreat, but a more integrated foundation for an effective urban life.' Yet paradoxically the garden city movement has also had a major ideological and practical influence on the planning of many major cities during the twentieth century.

When identifying the main types of green open space as they emerged over the centuries one should not ignore that in addition to the four elementary factors identified at the beginning, ideological vision has also been a key motivation. The planning of green open space in twentieth century cities was part of a belief in a better society. The most influential ideological initiative was the garden city, the famous tree-lined path to real reform proposed by Howard (1902). The idea of creating satellite cities for reasons of health had been introduced centuries earlier by Leonardo da Vinci, who proposed to create around Milan ten new towns each of 5000 houses and 30,000 people maximum. But the garden city concept went a step further, offering a blueprint for self-contained settlements, set within the system of the new social city which would bankrupt the old city. The twentieth century new towns also differed from the new towns of earlier centuries in the function of green open space. In planned capital cities, green

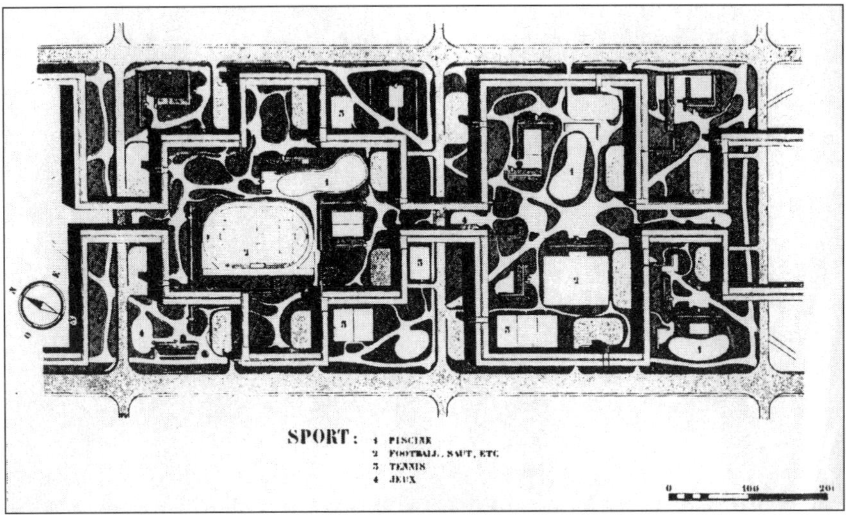

Figure 12.4. Le Corbusier's '*Ville verte*' (1930). The Swiss architect proposed a geometrical distribution of high-rise accommodation in a park-like landscape. The concentration of people allowed for large areas for sport and play at ground floor level and also allowed for sunshine to enter into the apartments. This vision had a major influence on residential planning in the second half of the twentieth century. (*Source*: Franco Migliorini (1990) *Verde Urbano: Parichi, Giardini, Paesaggio Urbano. Lo Spazio Aperto nella Construzione della Citta Moderna.* Milano: Franco Angeli)

space was used primarily as an ornament to glorify the ruler-founder and to impress the visitor. In the garden city-inspired new town, green space was an integral part of a human scale-design intended to improve living conditions, physically and mentally.

Yet it would be incorrect to assume that the emergence of the garden city was a sudden or revolutionary development. In fact, it was the result of an evolution that spanned many centuries but intensified and reached a climax during the twentieth century. In previous centuries, the urban elite would leave the town in order to settle in the more attractive countryside. This process developed in stages. First, summer residences were built; later these residences offered permanent retirement away from the city; finally they offered a permanent escape from city life. This process can be understood conceptually in terms of 'push' and 'pull' factors as well as 'stimulating' and 'restrictive' conditions. An example of a pull factor was the low cost of living in the countryside. An example of a push factor was the perceived high risk of city life in terms of criminality, fire hazards, and unhealthy conditions. A crucial stimulant was the technical development of mass transportation. Restrictive 'anti-skimming' policies (i.e. policies intended to keep those who form the economic elite of the city in town rather than allowing them to move out to rural or less heavily urbanized areas elsewhere) sought to address push and pull factors by creating quiet, spacious, green areas for settlement within the city boundaries for those with higher incomes. Cities would not just create such environments to retain their own inhabitants, but also to strengthen their appeal to residents with higher incomes from other, larger cities.

In the twentieth century, when 'skimming' caused a substantial population and wealth loss to large cities, retention sometimes became a sufficient aim in itself. The garden city, in its original form, was an attempt to extend the benefits of living in a peaceful, healthy, spacious, green environment, away from, yet near to, the big city, to the lower classes instead of just to the happy few. As such, it posed an even larger threat in that ultimately the old big city might have been left with a substantial underclass. The creation of attractive cities, with substantial green areas and garden suburbs, was one response. This answer was possible because of better transport, economic growth, and the political power of cities. The latter allowed for changes in city boundaries, changes in tax systems, and modified funding models, all of which have helped large cities to survive. In the twentieth century, many cities were able to redeem themselves as green environments, retaining their positions as centres of culture and economic growth, while at the same time offering large residential areas with good living conditions (Van Rooijen, 1990*b*).

GREEN TOWN PLANNING OF TOMORROW

The 'green city' emerging from the environmental imperatives of the twentieth century has now embraced a new meaning of the word 'green': ecological. The issue is no longer just the provision of green open space as a counterpoint to bricks, concrete and tarmac, or as an escape from city life. Green defines a more holistic approach, with ecosystems and bio-diversity as the buzzwords, and with a less clear or even absent distinction between buildings and roads on the one hand and green space on the other. 'If we can create ecology parks to compensate for areas with a deficiency of nature, why should we not create more varied habitats on and around buildings?', asks Goode (1990, p. 5).

Several general planning issues are raised by the role of planned open space in the latter-day green city. Eternal economic growth as the basis for planning can no longer be taken for granted and a greater awareness of the factor 'time' seems to be emerging. Infrastructural provisions are by definition a long-term story, certainly in terms of capital investment. This might cause tension with the 'democratic principles' that make us see open space as a consumption good within a short-term horizon.

Another problem is that it is far from easy to modify the design of open spaces. If the form is a response to acknowledged needs or requirements of a certain population/community at a certain point of time, one has to find ways to adjust that form effectively in the future. Obviously, future users will have different requirements. Hence the form of new spaces may well need to be so unspecific that they accommodate a wide range of possible uses, or they need to be designed with low-cost future changes in mind.

But form is only part of the story. The abundant provision of green open space in some cities has made maintenance a major issue. Though parks have been created with low maintenance costs in mind, this only applied to the maintenance costs at the time of formation. The awareness has slowly dawned that cheap maintenance at a specific moment is likely to become relatively expensive maintenance in the future. The circumstances determining the costs of maintenance will have changed. Hence, for example, the maintenance of the bourgeois park was originally very low because labour costs were low, but now labour is expensive and so is the maintenance of such nineteenth century greenery.

We also need to consider that old green space will require renovation in addition to maintenance. Combined with this is the political pressure to limit expenditure and therefore maintenance, the consequences of which will not show up immediately, but will result in increased future costs for renovation. Permanent low-maintenance open space is an economic imperative in modern town planning. So the trend has emerged to design urban open space which gives an impression of green space but contains a high level of low maintenance hard surfaces. Alternatively, relatively large nature areas are being created that up to a certain extent are self-sufficient in terms of maintenance. Of course, the formal argument for the creation of such open space is not cost cutting but ecological conservation. Thus the motive of *dasein* (just being there) for the presence of green open space in the urban environment, rather than the actual usage, is being reintroduced. At the other extreme, recognizing old open space (historical parks and gardens) as elements of our cultural heritage politically justifies high expenditure on careful maintenance, renovation and even restoration (Hennebo, 1985).

The green town planning of tomorrow might lead to an even more radical change, whereby ownership is handed over completely to the inhabitants, not just in terms of the ownership of land, but the planning process itself (Van Rooijen, 1987). Ecology parks, city farms, allotment gardens, neighbourhood wildspace, and local community gardens are grassroots matters, creating green jewels often in run-down urban districts:

> A vital aspect of this grassroots movement in cities is that people are able to take control of something for themselves. Success in establishing a green city must depend on local action and local responsibility, whether it is fish ponds in Calcutta or a paper recycling center in downtown Detroit. Self-reliance is the key. Bureaucracies may facilitate, but at the end of the day it is local people who make things work. (Goode, 1990, p.7)

This view on green provisions reintroduces the concept of common open space in a modern context. A similar revaluation in more ecological and grassroots-political terms might happen to the garden city concept.

CONCLUSION

The green city of the late twentieth century was not just a physical reality but also a way of living and a manner of thinking (Hough, 1990). The ideas that open space should exist in an urban environment, should be created with a variety of forms and functions, and should be planned and maintained carefully, were the result of a long, slow development. Though the birth of the green city is a twentieth century affair, its incubation has been a process reaching far back in history. At the most general level, this development has been in step with the broader evolution of urban society; that is, over time the creation of different types of green open space has reflected broader cultural, political and economic currents.

But as thinking about the provision of green open space in cities has evolved in concert with further knowledge, fashions, and innovations, simple ideas have become more complex, resulting in new and transitional forms, each rooted in history. Some are still clearly present and identifiable; others might be perceived as dated. All have become incorporated in planning principles that often will be taken as self-evident, whilst old ideas occasionally re-emerge as new, or are redefined, revalued or enriched in response to emergent issues. Cranz (1982, p. 253) notes that contemporary attitudes toward parks are threefold. Some assume that 'parks are just plots of land preserved in their original state'. Others recognize that parks are 'aesthetic objects'. The third, more sophisticated view, is that the elements of the urban park represent part of an evolving planning strategy 'so that today, as in the past, the citizen visiting a park is subject to an accumulated set of intended moral and social lessons'. History explains why these different perceptions are all correct in their own way.

The origin of the idea of urban open space being 'wasteland', saved from building and preserved for unrestricted usage, can be found in the common open land. But of greater importance has been the origin of the public park as a private property, an exclusive refuge with artistic design, slowly made accessible to select groups of visitors. This public garden developed into the bourgeois park, offering higher levels of accessibility but retaining many of the essential design features of the elite garden. Finally, this evolved into the more genuinely democratic form of the people's park. Mass usage rather than exclusivity became the main intention, though the form continued to define appropriate functions. Perceptions of ownership evolved accordingly. Yesterday's guests became hosts when wider layers of society acquired their politically rightful place in the park. Thus urban green open spaces have illustrated and simultaneously confirmed changes in society, accumulating in their diversity the richness of history.

One could interpret parks as didactic tools to maintain harmony in society, yet at the same time parks gave to new groups of people that which they had become entitled, in political, social and economic terms. From the early seventeenth century onwards in European cities, the creation of parks was part of a peaceful path to real reform, intentionally and unintentionally. They were elements in the transition of power from the ancient regime, with its enlightened rulers fearing a violent alternative, to the bourgeoisie, and later, from the upper middle classes, with its enlightened entrepreneurs and professionals, to the lower middle and respectable working classes.

In this social-cultural context, usage has remained the dominant feature of green urban space, combined with its aesthetic value, even though the latter gradually lost its high culture status with advancing democratization. In the cities that obtained wealth through entrepreneurialism, the economic value of green open space has been a major motive. It was the foundation for the bourgeois park, and 'hard' economic arguments have remained prominent until today. At the same time, the 'softer' drive for green open space as an ecological antidote to pollution caused by high levels of urbanization and industrialization is rapidly gaining force with growing concern over global-ecological issues.

Twentieth century town planning was enabled by technological progress and economic growth to use scientific methods to attempt to control urban chaos, with green open space as a major structuring element. Society at all levels placed great trust in the self-confidence of town planners, though gradually it has become difficult for planning to live up to expectations. Up to a certain extent the same applies to the garden city movement. New towns offered an attractive alternative to city life and have been highly influential during the twentieth century, but did not achieve the dramatic results originally proposed. Equally, green open space has never fully bestowed an arcadian quality on cities. But it should be acknowledged that the quality of urban life has still been substantially enhanced by the planned provision of green open space.

REFERENCES

Brace, M. (1986) *London Parks and Gardens*. Cambridge: The Pevensey Press.

Conway, H. (1991) *People's Parks: The Design and Development of Victorian Parks in Britain*. Cambridge: Cambridge University Press.

Cranz, G. (1982) *The Politics of Park Design: A History of Urban Parks in America*. Cambridge, Mass: MIT Press.

Forshaw, A, and Bergstroem, T. (1986) *The Open Spaces of London*. London: Allison and Busby.

Goode, D. (1990) A green renaissance, in Gordon, D (ed.), *Green Cities: Ecologically Sound Approaches to Urban Space*, pp. 1-8.

Gordon, D. (ed.) (1990) *Green Cities: Ecologically Sound Approaches to Urban Space*. Quebec: Black Rose Books.

Hennebo, D. (ed.) (1985) *Gartendenkmalpflege*. Stuttgart: Ulmer Verlag.

Hirschfeld, Ch.C.L. (1785) *Theorie der Gartenkunst*, part V. Leipzig: Weidmann (reprinted 1985, Frankfurt: Olms).

Hough, M. (1990) Formed by nature: A definition of the green city, Gordon, D. (ed.), *Green Cities: Ecologically Sound Approaches to Urban Space*, pp. 15-20.

Howard, E. (1902) *Garden Cities of To-Morrow: A Peaceful Path to Real Reform*. London: Faber and Faber.

Koch, H. (1913) *Gartenkunst im Staedtebau*. Berlin: Verlag Ernst Wasmuth.

Lasdun, S. (1991) *The English Park: Royal, Private and Public*. London: Andre Deutsch.

Leveque, J-J. (1982) *Jardins de Paris*. Paris: Hachette.

Mumford, L. (1965) The garden city idea and modern planning, in Howard, E., *Garden Cities of To-Morrow: A Peaceful Path to Real Reform*. London: Faber and Faber, p. 35.

Reining, H.R. (1976) Die Entwicklung der oeffentlich zogaenglichen Gruenflaechen im Bereich der Wiener Ringstrasse. PhD thesis, University of Vienna.

Ritter, N. (1982) *Das Gruene Berlin*. Berlin: Stapp Verlag.

Taverne, E. (1978) *In 't Land van Belofte: in de Nieue Stadt*. Maarssen: G. Schwartz uitgeverij.

Thompson, C.W. (1998) Urban parks: The historic legacy and the contemporary American experience, in Freestone, R. (ed.), *The Twentieth Century Urban Planning Experience: Proceedings of the 8th International Planning History Society Conference and 4th Australian Planning/Urban History Conference*. Sydney: University of New South Wales, pp. 906-911.

Van Asperen, H.S. (1983) Samenhang, ontwerp en uitvoering bij het scheppen en instandhouden van groenvoorzieningen. PhD thesis, University of Wageningen, The Netherlands.

Van Rooijen, M. (1987) Wohnungorientierte Parkanlagen in den Niederlanden. *Das Gartenamt*, **11**, pp. 697-701.

Van Rooijen, M. (1990*a*) De Wortels van het Stedelijk Groen: een studie naar ontstaan en voortbestaan van de Nederlandse groene stad. PhD thesis, University of Utrecht, The Netherlands.

Van Rooijen, M. (1990*b*) Garden city versus green town: The case of Amsterdam 1910-1935. *Planning Perspectives*, **5**, pp. 285-293.

Van Rooijen, M. (1993) The green towns of Holland. *Dutch Crossing*, **49**, pp. 107-123.

Wagner, M. (1915) *Staedtische Freiflaechenpolitik.* Berlin.
Ward, S.V. (1998) *Selling Places: Marketing and Promotion of Towns and Cities 1850-2000.* London: E & FN Spon.
Willebrand, J.P. (1775) *Grundriss einer schoenen Stadt in Absicht ihrer Anlage und Einrichtung.* Hamburg, Leipzig, publisher unknown.

Chapter 13

Diversity, Difference and the Multi-Layered City

Susan Thompson

Planners and urban theorists are currently engaged in a flurry of excitement about the notion of the postmodern, or what I term the 'multi-layered' city. There is a sense that this is a new area of study and policy interest, but the debates about the issue go back some time. The ways in which the city is viewed and understood may have changed. As Fincher and Jacobs (1998, pp. 1-2) cogently state, '"The city" as an object of analysis has been irredeemably unsettled, and many of the more resilient ways urban processes are understood have been rendered problematic'. But to imagine that this is a novel area of scholarship is fanciful. While the concept of difference has been opened up, the difficulties of serving multiple publics and defining the 'common good' have long been contemplated (see for example, Wilson, 1966).

Nevertheless we are in the midst of a renewal of interest and understanding of difference and the ways in which the 'multiple' city (Sandercock, 1998*a*, p. 3) is constructed. There is also a challenge to orthodox research methodologies and an acknowledgement that the so-called 'expert' has to be open to hearing the individual stories of those who inhabit, and ultimately create, the multi-layered city. In our efforts to understand this environment, it is easy to forget that it is made up of different voices of struggle and loss, leaving and resettling, hardship and reward. And while the macro processes which shape the contemporary metropolis must be studied and comprehended, the lives of ordinary people also command close attention. We have to be open to hearing their voices in a multiplicity of empowering and affirming ways.

It is a truism to say that any account of the multi-layered city, no matter how extensive, can only be partial. This is related to the author's interpretation of the 'texts' she or he inevitably selectively chooses to consult, as well as the experiences upon which she or he draws. Accordingly, it is important to state from where one comes and the context within which one writes. This is particularly so given the subjectivities and representations that I discuss in this chapter as well as an acknowledgement that 'identities [including mine] are not neutral' (Weeks, 1990, p. 89). I write from both an academic and practice position, given that I have occupied both stages at one time or another. My own subjectivity is also related to my middle-class, white, Anglo-Celtic background and the fact that I live and work in the country of my birth.

The partiality of my account is also inexorably linked to the way in which I have used the term 'multi-layered' and interpret the related terrain of migration, multiplicity and multiculturalism. In describing the rich tapestry of the multi-layered city, my definition is necessarily broad. It encompasses both traditional 'axes of difference' – gender, race and class – as well as aspects of physical ability, religious affiliation, sexual orientation and life stage (Fincher and Jacobs, 1998, p. 3). I have deliberately not used the word 'multiculturalism' to describe the multi-layered landscape of the postmodern city. Lippard's (1997, p. 5) term of 'multicenteredness' as an extension of the frequently abused expression of multiculturalism was influential in this respect. So too was the notion that 'multiculturalism is best understood as one aspect of pluralism, and . . . that in the defence of pluralism' it should not be permitted to dominate (Kukathas, 1993, p. 19).

The chapter is organized in four main sections. In the first I discuss the issue of global migration and the emergence of the multi-layered city, indicating that it has been in existence for some time, but not consciously acknowledged as such. Using this backdrop, I then turn to an examination of the multi-layered city as a contemporary phenomenon. I am particularly interested in how the traditional frameworks used to understand the modernist city have been challenged and largely usurped by urban theorists validating the reality of the postmodern city. I then outline some of the ways that planning is responding to this emerging urban condition, initially focusing on how, in the past, it disregarded the voices of marginality. My empirical concern is an Australian one, with an emphasis on local government. The chapter concludes with some ideas and challenges for a more sensitive planning practice which affirms individual voices of difference, as well as being responsive to the macro urban processes evident in the city of today.

GLOBAL MIGRATION AND THE EMERGENCE OF THE MULTI-LAYERED CITY

Historically, the multi-layered city has always been evident, but there has been a reluctance consciously to engage with its complexities. Urban theorists have worked from a presumption of homogeneity, attributing difference and diversity to the geographic and social 'margins' of the metropolis. There has, however, been an awareness of the contribution of global migration in creating a multi-layered city. For centuries, people have moved from country to country seeking a better life, often fleeing intolerable situations in their homeland. The impact of such migration has been variously commented upon in the urban planning literature over time. For example, Ward (1971) documented the emergence of ethnic 'ghettoes' in American cities between 1840 and 1920. He described how new settlers crowded into districts which provided poor quality but cheap accommodation, with and accessibility to growing employment opportunities. The advantages of living in a familiar community setting were recognized in this account. Gans (1971) exposed a similar form of low rent neighbourhood, accommodating either ethnic groups or 'single men, pathological families, people in hiding from themselves or society, and individuals who provide the more disreputable of illegal-but-demanded services to the rest of the community' (Gans, 1971, p. 301). This description, while pejorative, shows an appreciation of social difference in the city.

There was no such appreciation in the postwar urban renewal programmes which were aimed at clearing out the ethnic poor from inner city neighbourhoods. There may well have been an altruistic, although paternalistic, motive in the public sector's relocation schemes, but the poor were problematized and their needs for community misunderstood and disregarded (Gans, 1966). Grief reactions similar to the loss of a loved family member or close friend were documented (Fried, 1963), even though there was little critique of the process and the underlying assimilationist agenda.

Discussing the social and spatial character of world cities in 1969, Thomlinson included the phenomenon of 'minority-group enclaves', declaring that for existing residents 'resentment is common but rarely achieves high intensity' provided the newcomers are white and do not arrive in large numbers (Thomlinson, 1969, p. 190). The issue of 'exotic diversity' was also addressed in this text, but the reference related to the range of activities and degree of mobility that a 'real city man' can find in a large metropolis. Class and cultural conflicts are documented in Laslett's (1996) historical account of migration to Los Angeles between 1900 and 1970. Issues of race, colour and religion were at the forefront of prejudices that emerged between those coming from outside the United States and those settling in the city from other parts of the country. These accounts demonstrate that an understanding of any contemporary city's patterns of diversity can only be enhanced by looking at the history of migration and the subsequent growth of disparate communities.

This is also the case when one is examining national perspectives on migration and the emergence of diversity. Countries such as Australia, Canada and America are often referred to in the context of 'having been built on immigration'. Burnley, Murphy and Fagan (1997) contend that for Australia immigration has played a major role in population growth since white colonization in 1788. Some nine million people have come to the country in the intervening period, largely drawn to the urban centres along the coast. Other accounts document a long history of migration, increasing population diversity and the evolution of related government programmes and policies (e.g., Freeman and Jupp, 1992; Jupp, 1991; Lack and Templeton, 1995).

Even notions of multiculturalism have been put into an historical context. Qadeer (1997) distinguishes between an 'old' and 'new' multiculturalism as part of current ethnic studies. The 'old' was a 'private affair' for the immigrant. He or she was to be assimilated into a host society which conferred no rights. Migrants were generally working class and confined to immigrant districts. The 'new' multiculturalism acknowledges that public policy has a role in ensuring that post-World War Two concepts of human rights and equality between citizens are honoured. Contemporary cultural diversity has also been transformed into an economic asset given the context of a post-industrial economy and increasing internationalism and globalization.

Given this historical overview, it is evident that the multi-layered city has been with us for a while. Urban theorists have focused their attention on migration and its impact on the physical and social fabric of the metropolis. The emphasis has often been negative: on attempts to reduce diversity and return the landscape to a state of homogeneous order. In the next section I consider the qualities of the phenomenon of the multi-layered, multiple, postmodern city. This site has emerged in the last decade and now represents a vibrant area of academic scholarship in urban studies and planning.

THE MULTI-LAYERED CITY

The contemporary multi-layered city is one of composite images. There are displays of power and wealth, epitomized in corporate, downtown areas. There are examples of poverty and decay, evident in dying inner-city industrial precincts. There are images of the 'immigrant city' (Sassen, 1996, p. 23) – showing a new found wealth and status in some localities, and, in others, a sense of struggle and displacement. The multi-layered city is both a social and spatial 'coming together' of difference and diversity, chaos and order, fascination and intrigue – a sensual delight, at the same time challenging notions of tolerance and feelings of belonging. The multi-layered city is imagined and real, a creation of our own subjective experiences of the urban landscape as well as a response to the personal – our gender, age, ethnicity, class, physical ability, religious beliefs and sexual orientation. People not only make up the multi-layered city in all its guises, they formulate different understandings of what it is and how it operates. The opening up of spaces to hear individual stories has advanced our collective understanding. But these narratives have also challenged traditional preconceptions about the city, created conflict, and unsettled those endeavouring to establish order and predictability within its borders.

Urban theorists, planners, philosophers, art historians and social scientists are amongst those researching and writing about the multi-layered city. There has been an 'intertextual implosion of representations' (King, 1996, p. 3) breaking down traditional boundaries of academic disciplines, as well as their notions of theorizing and analysis. Totalizing discourses are no longer adequate to describe the city, and unless we consider both socio-cultural forces and the impacts of global economic restructuring, we will fail to understand the complexities of this urban phenomenon. Sandercock (1998*a*, pp. 2-3) argues that the processes of economic change have been privileged for too long and that more attention must be given to contemporary socio-cultural factors. These are apparent on a global scale and encompass the following themes:

- the age of migration and multicultural citizenship;
- the age of post-colonialism and the reclaiming of land by indigenous people; and
- the rise of civil society and the age of minorities.

These processes are fundamental to an understanding of the textures and qualities of the multi-layered city, which are considered below.

A City of Groups and Individuals

The culture of the multi-layered city encompasses 'different genders, ethnicities, ideologies, races, classes, sexual orientation, nationalities, theoretical difference of every shape and form' (King, 1996, p. 2). This landscape is also about 'hierarchized grids of difference' (Pratt, 1998, p. 44). Although it encompasses much more than multiculturalism, definitions of this term can assist our understanding of citizens' rights and obligations in communities which have varying and sometimes conflicting allegiances to each other and society as a whole. *The National Agenda for a Multicultural Australia* (Office of Multicultural Affairs, 1989) defines the term multiculturalism as embodying the following three principles:

- Cultural identity: the right to express and share cultural heritage, language and religion;
- Social justice: the right of all citizens to equal treatment in all spheres of life;
- Economic efficiency: the need to capitalize on skills of all citizens, regardless of background.

The Report of the Standing Committee on Multiculturalism of the Canadian Parliament acknowledges that the recognition of diversity in a plural society needs to be officially supported. It assumes that everyone has an ethnic origin and that all cultures deserve respect (Hindess, 1993). This philosophy sees the community as a 'mosaic of beliefs, practices and customs' rather than a 'melting pot' assimilating different groups into one homogenous mass (Qadeer, 1997, p. 482).

These definitions reinforce the acceptance of difference and diversity within and between ethnic groups. They can also be used to argue for respect across other expressions of difference. The latter may incorporate a multiplicity of identities, with varying intersections depending on the context and identification of an individual or group. Rather than simply considering the issue of ethnicity for example, it is more pertinent to examine its juxtaposition with gender and class. It is also important to consider who is behind the creation of meaning. Until recently, there has been an assumption that migrants have similar experiences regardless of gender. Men have traditionally spoken on behalf of their communities, providing a misleading sense of unity (de Lepervanche, 1992). Women were commonly viewed as the 'appendages of either protective males or the patriarchal state' (Fincher *et al.*, 1994, p. 150). Even under the rubric of feminism, migrant women have been marginalized by their white sisters too quickly to stereotype experiences and responses. One such example is the home. Frequently discounted as a site which demeans women, it can be a place of power, helping migrant women to atone for the losses sustained in the leaving and settling process. It also represents achievement in the face of hardship and provides a means of maintaining cultural, religious and personal links with the past (Thompson, 1994; Thompson, 1996).

Researchers are engaging with the complex interplay of social categories as they map the different groups which inhabit the city. Power structures and degrees of oppression and harassment are charted. Knopp (1998) argues that this approach is a reaction against the theory of class relations which dominated Marxist interpretations of the city. Class must be considered in relation to gender, race and sexual orientation, as well as how it fits in terms of the interests served by a 'social totality' which is 'racist, patriarchal and heterosexist' (Knopp, 1998, p. 151). Mapping sexuality and urban space reveals diffuse notions of power and oppression across the city. Different responses to the homosexual community's demand to have a place free of victimization and violence have been uncovered. For example, Sydney's gay district along Oxford Street, just east of the Central Business District, displays the community's economic power and hosts the annual Gay and Lesbian Mardi Gras expressing homosexual pride. Outside this inner city area, gays do not have a space and 'lacking a space they lack visibility' (Murphy and Watson, 1997, p. 80). Other cities, such as Edinburgh in Scotland, respond differently. The social and religious conservatism of its people has 'produced strong resistances to gay male visibility and power' (Knopp, 1998, p. 159). These patterns of resistance to, and the acceptance of difference, ebb and flow across geographic space and time, and are a key quality of the multi-layered city.

A City of Conflict and Social Polarization

The multi-layered city is characterized by varying manifestations of conflict. In some instances there is open hostility to those who are perceived as different or marginal. In other cases, there is a vacillation between an acceptance of the opportunities which difference affords, and a rejection of it as destructive to community norms (Fincher and Jacobs, 1998). As 'multi-ethnic, multi-racial and multi-national populations' come to dominate the city, there are rising levels of fear which can undermine fundamental values and expectations (Sandercock, 1998*a*, p. 164). The threat cuts across economic, cultural, religious and psychological layers, fuelling a deep-seated apprehension about profound change and its consequences. These fears are producing a new world disorder – rising levels of violence and racism against those who are different and do not belong, and increasingly repressive responses to groups hitherto perceived as marginal and who are asserting rights and claims over space (Sandercock, 1998*a*).

Holston (1998) uses the term 'spaces of insurgent citizenship' to describe places where 'new residents with histories, cultures, and demands that disrupt the normative and assumed categories of social life' (p. 50) oppose the 'modernist spaces that physically dominate' (p. 39) contemporary cities. These conflicting sites challenge the state as the 'only legitimate source of citizenship rights, meanings and practices', representing an 'assertion of legitimacy' by hitherto marginalized groups (p. 39). Such an 'assertion of legitimacy' is to be found throughout the multi-layered city (see, for example, Fincher and Jacobs, 1998; King, 1996; Sandercock, 1998*b*). And these stories are not confined to academia. One only has to open a newspaper or turn on the television to find demonstrations of citizen opposition to state sponsored and supported development proposals. Struggle and conflict are at the heart of these narratives. The push for change and the manner in which rights are asserted vary, and in some cases can be seen as contradictory. Different opinions and activist positions collide within and between groups as they struggle to be heard against those in power (Anderson, 1998). Such a plurality of responses is common but can be characterized by the state as indicative of a disorganized battle, destined to fail.

Fear of the other is most evident in a community's response to multiculturalism and migrant presence. Currently there is a global movement towards more stringent regulations for migrants. This can be seen in different world cities as legislation is enacted to control and subjugate the migrant's influence and his/her level of success. Paris is an extreme example; foreigners are only wanted if they do not openly practice cultural difference (Sandercock, 1998*a*). Political rhetoric can also be the catalyst for questioning policy which supports marginal groups. This occurred in Australia when an independent member for the Federal Parliament ignited a 'race debate' (Hanson, 1996). The episode received considerable media attention both nationally and internationally. The citizenship of indigenous and Asian-Australians was publicly questioned and opinion polls used to 'demonstrate' high levels of support for the racist views being expounded. The electoral demise of this politician has seen a reduction in attention given to the matter, but the episode exposes the fragility of enacted legislation which supports expressions of difference.

In Miami the immigrant is blamed for a general moral decline as well as a host of other problems such as job losses and an overburdening of the welfare system. These difficulties are situated within a 'broader discourse' that perceives immigration as a 'foreign threat' and immigrants 'unwelcome invaders' (Croucher, 1997, p. 21). The different cultural practices which emerge, including language, challenge 'the hegemony of the status quo' (Croucher, 1997, p. 62).

Fear of the other is not only confined to ethnic expressions of difference. Certain forms of urbanization that seek physically to protect those who fear are increasingly apparent in the multi-layered city. Home owners in America are fortifying their residences, ironically exploiting the democratic process to 'exclude and discriminate' (Holston, 1998, p. 53). Walled estates express a personal and group identity that communicates privilege and a particular social order, polarizing the 'haves' and the 'have nots'. Everyday public places are also becoming exclusive and separatist as security guards request those without a clear purpose to move on and out. This is especially the case in shopping malls which have become pseudo public spaces. Young people 'hanging about' are seen as a potential threat. These practices reinforce the sense of a divided and polarized city. 'Safe', patrolled and ordered spaces contrast with 'unsafe' and unsupervised sites.

A City of Celebration

The multi-layered city is not, however, all negative. Difference and diversity are celebrated across the metropolis in a variety of guises and often as a response to an ongoing insistence to be seen and heard. This is the case in relation to Sydney's Gay and Lesbian Mardi Gras. In 1999, a crowd of more than 600,000 people cheered on 6000 marchers and 200 floats (Dennis, 1999). The procession was the twenty-second, marking an acceptance of homosexual culture. Even the State Police Service is represented, traditional Australian sporting stereotypes of masculinity challenged, and multicultural displays of homosexuality given a place. However, it is important to note that pockets of resistance still remain. One of the more farcical is led by the Reverend Fred Nile, a member of the State Parliament, who with a band of supporters, prays for rain each year to disrupt the event!

Local government in Sydney is also celebrating diversity. This is related to multiculturalism and the use of municipal slogans. Fairfield, in Sydney's outer west, and the most culturally diverse local government area in the country, 'Celebrates Diversity'. Canterbury, also in the west, promotes itself as a 'City of Diversity'. Nevertheless, it cannot be assumed that these slogans permeate council policies and practices. There are varying levels of commitment at both the political and professional officer levels (Thompson *et al.*, 1998).

Examples of local pride in difference and diversity underscore the search for a 'geographic belonging' as an antidote to a 'prevailing alienation' (Lippard, 1997, p. 7). This opens up possibilities for planning to engage with the complexities of the multi-layered city. I discuss some examples below showing that planning is capable of responding to the needs of disparate groups and the opportunities this presents. The picture is however, somewhat mixed and begs the question as to how we might institute a more systematic response so that planners can better develop equitably and culturally sensitive policies and practices.

PLANNING AND THE MULTI-LAYERED CITY

In this section, I explore planning's overwhelming failure to grapple with the reality of spatial and social expressions of difference and diversity. There has been, and still is, a reluctance to hear the voices at the margins. This can be seen in the historical overview which follows, but is encouragingly less evident in contemporary practices which attempt to acknowledge, empower and enliven the multiple layers of today's metropolis.

Uncovering the Voices of Marginality

Mainstream planning history, like other historiographies, has presented a particular interpretation of events that reinforces a sense of professional continuity and celebrates it as a modernist project serving the 'public interest'. But just who is the public being served and how can conflicting desires, demands and dreams be addressed? In critically evaluating planning history, Sandercock (1998*b*) opens up possibilities for other stories to be told that challenge the 'official story'. The latter emanates from within the profession and while certain individuals are singled out for positive assessment, 'planning itself is the real hero . . . slaying the dragons of greed and irrationality' (Sandercock, 1998*b*, p. 4). In this context, the discipline is an assumed set of knowledges and practices which are never critically scrutinized: 'There is little soul searching about planning's failures' nor any 'application of theories of power/knowledge/control' (Sandercock, 1998*b*, p. 4). This has resulted in a tendency to ignore ideologies of 'class, gender or ethnic origins' as well as planners' biases and the effects of their practice on issues of gender and ethnicity.

Sandercock further asserts that mainstream planning histories do not acknowledge the '*noir* side', or the hidden and dark aspects, of the profession. A critical evaluation may well reveal how planning can contribute to 'social repression, economic retardation or environmental degradation' (Yiftachel, 1995, p. 217). Historical accounts rarely discuss the women, indigenous and ethnic people who actively created communities and spaces for belonging (Sandercock, 1998*b*). It is assumed that because these people were not classified as 'planners' their participation in urban affairs was of little consequence when compared to those sanctioned by the power and privilege of professional status – in the main, white, western men. The failure of mainstream planning history to acknowledge 'non-planners' is also related to the discipline's boundaries. There has been a tendency to privilege the physical and technical, marginalizing the social and community side of the profession. The latter is often seen as a 'soft' activity, associated with nurturing and femininity – 'an add-on function to mainstream planning' with its emphasis on development control and policies firmly grounded in 'objective' scientific methodology (Watson, 1993, p. 144).

The 'new' planning histories show how the voices of marginality have largely been, until recently, silenced by those with the institutional power to define both their practice and its domain. Hayden's (1981) chronicle of early material feminists' attempts to collectivize home life reveals pioneering designs for commercial kitchens, child care centres and cooperative housing in the years between the American Civil War and the Great Depression. Not only does this account disclose significant women who were active community planners, it suggests that the political ideologies which they espoused were branded subversive and extremist.

Other feminist analyses of urban planning further reinforce this discomfort with the moral condition of woman – a force to be restrained and kept in her 'right' urban place. The suburban home – where she could nurture children and spouse – was where she belonged, and planning policy needed to ensure that she desired nothing more (Wilson, 1991). Women's access to the public space of the city had to be controlled as it was perceived to be a site of sexual promiscuity. In this context, contemporary planning projects aimed at making cities safer for women (and so the argument goes, safer for everyone) may not be all they seem (see for example Trench, Oc and Tiesdell, 1992; Wekerle and Whitzman, 1995). Even though woman's occupation of public space today is legitimate, in some cases her right to be there continues to be questioned. Witness, for example, media accounts of female rape and murder victims whose standard of dress is described. And what of the outcomes of these 'safer city programmes'? Are the problems simply 'moved on' to an area where there is less community pressure for change or back into the private sphere where women, children and the elderly are particularly vulnerable in the face of abusive relationships? If it is the latter, the planner can walk away, falsely believing that it is now 'safer' to negotiate the urban landscape.

Planning history has also ignored women of colour, especially those from working class backgrounds (Hayden, 1995). With greater sensibility towards stories of difference, there is a recognition that those 'who have known well the structuring of difference and the importance of identity politics' have a significant role to play in undermining the 'singular positions of authority' which have dominated understandings of the city (Fincher and Jacobs, 1998, p. 1). Gloria Anzaldùa is one such writer/activist. She speaks about being a 'border woman', growing up between two cultures and being uncomfortable in this position (Soja, 1996, p. 127). Her voice and those of fellow activists not only present a challenge to hegemonic practices and values, they offer empowering possibilities for the communities which have hitherto been marginalized.

Planning's Response to the Multi-Layered City

So how is contemporary planning theory and practice responding to the challenges created by the multi-layered city? Historically, the issue had been largely ignored (Watson and McGillivray, 1995). Planning's 'modernist' agenda can be seen as a major inhibiting factor. So too the discipline's failure to acknowledge contradiction and conflict, which is at the heart of social life. Such indeterminancy cannot be incorporated into totalizing plans which arrogantly assume a 'rational domination of the future' (Holston, 1998, p. 46).

Qadeer (1997) argues that planning has to deal with two key issues which have a relationship with multiplicity, multiculturalism and pluralistic societies. The first is the equity of the planning process. Planning policies must stand up to scrutiny in terms of justice and how they meet the disparate needs, wants and demands of citizens. This cuts across gender, age, ethnicity, ability, religious affiliation, life stage and sexual orientation. Equity must be evident in both procedural and substantive aspects of policy. Planners need to be able to assess cultural bias in their work, as well as suggest ways to balance competing mainstream and minority demands. The second aspect which planning must tackle is the enclave. This may be defined as an area which has a high proportion of people from a particular ethnic background living and working in the locality. There is also a balancing act between the 'advantages of neighbourhood homogeneity' and the 'goal of openness and equal access by all' (Qadeer, 1997, p. 483).

City planners are increasingly being forced to grapple with these matters. There are various examples displaying differing degrees of success, but underlying them all is an imperative to negotiate spatial expressions of conflict 'over who can live where and with what social, economic and cultural rights' (Sandercock, 1998*a*, p.7).

One interesting area where progress is apparent is that of conservation planning. The failure of traditional heritage control and protection to recognize anything but Eurocentric norms has been exposed by Armstrong (1994) and Dubrow (1998). Rather than working on the basis of long held assumptions about what is 'worth preserving', migrant communities have to be asked about their priorities (Armstrong, 1994, p. 105). Lived experiences must be heard and different cultural ideologies valued. Dubrow (1998, p. 57) argues for the 'potential for using historic preservation as an instrument of a democratic and inclusive approach to planning'. She examines the ways in which different sites with historical significance for minority groups such as women, people of colour and homosexuals have been preserved and celebrated. Long term discrimination has acted against preservation, but grass roots activism continues to create positive change and a recognition of sites previously ignored. Morris (1997) addresses the 'invisibility of women's heritage' and develops a methodology to identify places of significance. This is no easy task, especially when overlays of class and ethnicity are considered (Hayden, 1995)

Planning legislation, the foundation of practice, is another area warranting attention. Its complexity is a major barrier for those whose mother tongue is not the same as their country of resettlement. This not only has implications for understanding what is going on, but limits participation in the planning process. Information in appropriate languages is essential, as is the education of culturally-sensitive planners (Antonios, 1994). The Eurocentric assumptions that have driven planning in the West need to be acknowledged and current regulatory practices questioned in order to determine an appropriate response for contemporary communities.

The history of urban governance records a chronic vacillation 'between celebrating and enhancing . . . diversity, on the one hand, and regulating and repressing it, on the other' (Fincher and Jacobs, 1998, p. 1). Examples of responsiveness to diversity are gradually finding their way into the planning literature, government reports and community stories of success and failure. I consider one such case below. This study shows the potential for a holistic response to difference which encompasses planning and its many related practices.

RESPONDING TO THE MULTI-LAYERED CITY: GOVERNANCE IN AUSTRALIA

Local government in Australia is in a good position to respond to the challenge of the multi-layered city. Although the third tier of government in a country which has federal and state jurisdictions, the local level is closely linked to its state counterpart. Local government is administered by councils under the direction of locally-elected councillors. In New South Wales (NSW), the most populous state in the nation, significant legislation is in place to encourage local government to act responsively to its diverse constituents (Thompson *et al.*, 1998).

In 1987, the NSW Department of Local Government combined with the Ethnic Affairs Commission of NSW to help coordinate a *Local Ethnic Affairs Policy Statement* (LEAPS) project. LEAPS initiatives focused upon improving access to local government services, translation provision, and communication between government and different ethnic groups at the local level. This programme was later reinforced with the legislative backing of the NSW *Charter of Principles for a Culturally Diverse Society* (Ethnic Affairs Commission NSW, 1993). NSW has incorporated this *Charter* into its local government legislation, requiring councils to 'exercise their functions in a manner that is consistent with and actively promotes the principles of cultural diversity' (Ethnic Affairs Commission NSW, 1993, p. 1). Councils must also report annually on programmes undertaken to promote services, as well as ensuring that they are accessible to people from diverse cultural and linguistic backgrounds. These institutional requirements have compelled many organizations to confront the need for reform, and to embrace, rather than shrink from cultural difference.

Other state legislation acknowledges local government as a significant player in service provision for diverse cultural groups. It also recognizes the important role of encouraging participation in the processes of governance, especially consultation about urban planning matters. Discrimination on the grounds of race is outlawed and the need has been established for special provisions to ensure equitable access to services and facilities. In 1998, a new regulation was added to the NSW *Local Government Act* mandating councils to develop a detailed social plan integrating issues relating to indigenous communities and people from non-English speaking backgrounds.

In the first extensive national survey of council policies and practices, respondents were asked to identify the ways in which they endeavour to meet the needs of diverse cultural groups (Thompson *et al.*, 1998). Many councils were found to be working in a variety of innovative ways to reach both immigrant people and traditionally marginalized groups such as the poor, aged, disabled, youth, gays and women. One Sydney municipality acknowledged that cultural diversity 'presents a challenge to deliver services in ways that are equitable and easily accessible to all residents'. It saw it had an ethical responsibility to go beyond employing specialist officers and supporting cultural events and grants – an 'holistic workplace change' was necessary:

> It is about a new way of thinking and acting . . . it is about working from a social justice framework . . . Access is about providing all our citizens with the same opportunities to maximize quality of life objectives. Equity . . . is about recognizing that although all citizens are meant to be equal in their rights and in the way they are treated by government, they are in fact very different in their needs, interests and values. Treating everybody the same simply perpetuates existing inequalities . . . intervention must provide appropriate rules, programs and service provision to combat barriers . . . [such as] language, culture, race or religion, [and] gender, if we are to provide a socially just customer service commitment. (Liverpool City Council, 1992, p. 2)

The research uncovered a growing commitment to implementing policies and programmes of inclusion to ensure that people from different cultural backgrounds are afforded the opportunity to participate in the processes of local governance. A wealth of innovative policy and practice was evident across all council departments including planning, engineering and health. As residents, all members of the community have a right of service and participation. To refuse to acknowledge this is effectively to deny citizenship.

Practices of inclusion are particularly noticeable in the big cities of Sydney and Melbourne where there are significant levels of diversity (Thompson *et al.*, 1998). Nevertheless, even in these urban regions a systematic response to the needs of different groups is a long way off. Some councils are surprisingly ignorant of the nature of their local diversity, and have failed to identify and abandon discriminatory and iniquitous practices in the provision of services and facilities. Research of this kind is important in uncovering significant problems in both policy and practice. This type of work can also be used to raise awareness, initiate change and bring about more equitable practices at the grass roots level.

A WAY FORWARD

The multi-faceted nature of the multiple, multi-layered city necessitates a new way of thinking and acting in urban governance and planning. For some time now, theorists have been urging change in practice, but the response has been patchy. Resistance ranges from denials that any adjustment is warranted, to personal prejudices against expression of difference and arguments that it is too costly to effect reform. A more optimistic reading of the situation suggests a readiness to acknowledge past biases and culturally-insensitive practices. Both positions are evident in some of the examples presented in this chapter. However there is a need continually to reiterate the urgency of transformative processes across the many faces of planning and related governance. In this final section I consider what these might be. Interspersed with the suggestions however, are many perplexing questions which need to be articulated and debated. It is pointless to pretend that the problems facing contemporary cities will be easily solved. But the old ways will no longer suffice and by engaging with these difficult issues, the prospects of moving towards a more equitable and responsive planning practice in the twenty-first century are markedly enhanced.

Educating Planners

Planning education has to challenge students to assess critically their discipline's theoretical base and past practices – both successes and failures. Educators have to be open to the changing notions of what constitutes planning and enable students to see beyond the dominance of economic analyses and solutions. This can be done by exposing the socio-cultural processes of change which are fundamental to an understanding of the multi-layered city. It is imperative to leave the lecture hall behind and meet the people who live and work in the city – not just the wealthy and powerful, but the aged, the homeless, ethnic and indigenous communities, school children, gays and lesbians. Communication skills, including negotiation and conflict resolution, can also be practised in the field. Students can further enrich their learning in this context by sharing their own subjectivities and life experiences.

Sandercock (1998*a*) posits a comprehensive framework for 'preparing planners for the twenty first century'. She argues that this encapsulates specifying a particular domain for the discipline, as well as opening it up to environmental and design programmes, a greater emphasis on ethics, and the teaching of 'key literacies'. The latter includes 'traditional' skills in addition to those which will enhance the planner's ability to be 'more humble, more collaborative, more respectful of the value of difference and cultural diversity' (Sandercock, 1998*a*, p. 228). Teaching qualitative methods is an important component here.

Research Methods

Planners, students and practitioners alike, have to accept that the old ways of analysis are problematic (Fincher and Jacobs, 1998; Thompson, 1996). There needs to be widespread and rigorous use of qualitative methods, alongside the more traditional approaches. Qualitative methods challenge planning orthodoxies and recognize that the dominant mode of scientific rationalism, used in isolation, is hindering the delivery of people-centred planning. Qualitative methods allow us to consider the emotional and spiritual significance of place. They enable the acknowledgement of human emotions, spirituality, aspirations and dreams as significant concerns of planning theory and practice. This is central to unravelling the complexities of the multi-layered city.

Examining Our Own Subjectivity

Planning practitioners and urban theorists have to position themselves in relation to their work. This entails an acknowledgement of one's subjectivities, which may be personally challenging, especially in the context of racism (Hage, 1998; Sandercock, 1998*a*). Affirming one's position does not mean being apologetic for who one is, but rather it is an exercise in awareness of biases and prejudices. It is also an important recognition that one does not possess all the answers and that there is a willingness to hear and heed other voices. Stating one's position is further complicated because it is not simply an intellectual action. It requires an honest examination of long held beliefs and prejudices, and an acknowledgement of their potential to influence one's thinking, theorizing and practice.

Equity and Empowerment

Planning for the multi-layered city involves transforming 'structures of power and privilege, the ways oppressed groups can, through a politics of identity and a politics of place, reclaim rights, resist, and subvert' (Fincher and Jacobs, 1998, p. 2). This poses a serious dilemma for the practising planner if she or he represents the State. How can this practitioner work with and encourage groups to undermine those who support the planner's position? Do we set up a two tiered system: one planner working within the traditional institutional framework using negotiation and mediation, collaboration and consensus building, and the other a 'grass roots community based' professional helping marginalized communities to find their voice? (Sandercock, 1998*a*, p. 7). This is a difficult issue which requires continued debate and thought. Academics and researchers must also confront their power and privilege. The use of elitist language is one area of considerable concern (Massey, 1991). So too are the narrow and conservative modes of communication used to disseminate ideas and research findings. More innovative and creative means need to be trialed if we are to address effectively the realities of the 'multiple' city.

CONCLUSION

The multi-layered city is an exciting concept with which to engage. It has captured the imagination of policy makers, planners, researchers, artists, philosophers and those who inhabit and create this space. The textures, qualities and complexities of the multi-layered city are slowly being unfurled in a plethora of exciting and innovative scholarship, some of which has been touched on in this chapter. Researching this site, a history-in-the making, is a challenge and demands a holistic approach. We cannot simply intellectualize its character, nor can we rely solely on the scientific, rational approach to bring about understanding. To be an effective planner in this challenging environment necessitates moving beyond the inherited professional wisdom – reinventing modernism's 'activist commitment' (Holston, 1998, p. 39) – towards a humanistic and sensitive practice where 'the other' is viewed with respect and valued in the planning process. It also means genuinely contemplating the possibilities for a shared common culture (Sennett, 1994).

REFERENCES

Anderson, K. (1998) Sites of difference: Beyond a cultural politics of race polarity, in Fincher, R. and Jacobs, J. (eds.) *Cities of Difference*. New York: The Guilford Press, pp. 201-225.

Antonios, Z. (1994) *State of the Nation 1994: A Report on People from Non-English Speaking Backgrounds*. Canberra: Australian Government Publishing Service.

Armstrong, H. (1994) Cultural continuity in multicultural sub/urban places, in Gibson, K. and Watson, S. (eds.) *Metropolis Now: Planning and the Urban in Contemporary Australia*. Sydney: Pluto Press, pp. 102-113.

Bourne, L. (ed.) (1971) *Internal Structure of the City: Readings of Space and Environment*. New York: Oxford University Press.

Burnley, I., Murphy, P. and Fagan, R. (1997) *Immigration and Australian Cities*. Sydney: The Federation Press.

Croucher, S. (1997) *Imagining Miami: Ethnic Politics in a Postmodern World*. Charlottesville: University Press of Virginia.

de Lepervanche, M. (1992) Working for the man: Migrant women and multiculturalism, in Saunders, K. and Evans, R. (eds.) *Gender Relations in Australia: Domination and Negotiation*. Sydney: Harcourt Brace Jovanovich, pp. 82-96.

Dennis, A. (1999) When too much just isn't enough. *The Sydney Morning Herald*, 1 March, p. 8.

Dubrow, G.L. (1998) Feminist and multicultural perspectives on preservation planning, in Sandercock, L. (ed.) *Making the Invisible Visible: A Multicultural Planning History*. Los Angeles: University of California Press, pp. 57-77.

Ethnic Affairs Commission of New South Wales (1993) *New South Wales Charter of Principles for a Culturally Diverse Society*. Sydney: Ethnic Affairs Commission of New South Wales.

Fincher, R. and Jacobs, J. (eds.) (1998) *Cities of Difference*. New York: The Guilford Press.

Fincher, R., Foster, L., Giles, W. and Preston, V. (1994) Gender and migration policy, in Adelman, H., Borowski, A., Bustein, M. and Foster, L. (eds.) *Immigration and Refugee Policy: Australia and Canada Compared*, Vol. 1. Melbourne: Melbourne University Press, pp. 149-184.

Freeman, G. P. and Jupp, J. (1992) *Nations of Immigrants: Australia, the United States, and International Migration*. Melbourne: Oxford University Press.

Fried, M. (1963) Grieving for a lost home, in Duhl, L. (ed.) *The Urban Condition: People and Policy in the Metropolis*. New York: Basic Books, pp. 151-171.

Gans, H. (1966) The failure of urban renewal, in Wilson, J. Q. (ed.) *Urban Renewal: The Record and the Controversy*. Cambridge, Mass: MIT Press, pp. 537-557.

Gans, H. (1971) The West End: An urban village, in Bourne, L. (ed.) *Internal Structure of the City: Readings of Space and Environment*. New York: Oxford University Press, pp. 300-308.

Gibson, K. and Watson, S. (eds.) (1994) *Metropolis Now: Planning and the Urban in Contemporary Australia*. Sydney: Pluto Press.

Hage, G. (1998) *White Nation: Fantasies of White Supremacy in a Multicultural Society*. Sydney: Pluto Press.

Hanson, P. (1996) Appropriation Bill (No. 1) 1996-97, Second Reading, Member for Oxley. *Commonwealth Parliamentary Debates*, 10 September, pp. 3860-3863. Canberra: House of Representatives.

Hayden, D. (1981) *The Grand Domestic Revolution: A History of Feminist Designs for American Homes, Neighborhoods, and Cities*. Cambridge, Mass: MIT Press.

Hayden, D. (1995) *The Power of Place: Urban Landscapes as Public Policy*. Cambridge, Mass: MIT Press.

Hindess, B. (1993) Multiculturalism and citizenship, in Kukathas, C. (ed.) *Multicultural Citizens: The Philosophy and Politics of Identity*. Sydney: The Centre for Independent Studies, pp. 31-45.

Holston, J. (1998) Spaces of insurgent citizenship, in Sandercock, L. (ed.) *Making the Invisible Visible: A Multicultural Planning History*. Los Angeles: University of California Press, pp. 37-56.

Jupp, J. (1991) *Immigration*. Sydney: Sydney University Press.

King, A.D. (ed.) (1996) *Re-Presenting the City: Ethnicity, Capital and Culture in the Twenty-First Metropolis*. Basingstoke: Macmillan.

Knopp, L. (1998) Sexuality and urban space: Gay male identity politics in the United States, the United Kingdom, and Australia, in Fincher R. and Jacobs, J. (eds.) *Cities of Difference*. New York: The Guilford Press, pp. 149-176.

Kukathas, C. (1993) The idea of a multicultural society, in Kukathas, C. (ed.) *Multicultural Citizens: The Philosophy and Politics of Identity*. Sydney: The Centre for Independent Studies, pp. 17-30.

Kukathas, C. (ed.) (1993) *Multicultural Citizens: The Philosophy and Politics of Identity*. Sydney: The Centre for Independent Studies.

Lack, J. and Templeton, J. (1995) *Bold Experiment: A Documentary History of Australian Immigration since 1945*. Melbourne: Oxford University Press.

Laslett, J.H.M. (1996) Historical perspectives: Immigration and the rise of a distinctive urban region, 1900-1970, in Waldinger, R. and Bozorgmehr, M. (eds.) *Ethnic Los Angeles*. New York: Russell Sage Foundation, pp. 39-75.

Lippard, L. R. (1997) *The Lure of the Local: Senses of Place in a Multicentered Society*. New York: The New Press.

Liverpool City Council (1992) *Local Ethnic Affairs Policy Statement (LEAPS) Annual Report*. Liverpool: Liverpool City Council LEAPS Committee.

Massey, D. (1991) Flexible sexism. *Environment and Planning D: Society and Space*, **9**, pp. 31-57.

Morris, M. (1997) *Placing Women: A Methodology for the Identification, Interpretation and Promotion of the Heritage of Women in Tasmania*. Hobart: Office of the Status of Women.

Murphy, P. and Watson, S. (1997) *Surface City: Sydney at the Millennium*. Sydney: Pluto Press.

Office of Multicultural Affairs (1989) *National Agenda for a Multicultural Australia . . . Sharing Our Future*. Canberra: Australian Government Publishing Service.

Pratt, G. (1998) Grids of difference: Place and identity formation, in Fincher, R. and Jacobs, J. (eds.) *Cities of Difference*. New York: The Guilford Press, pp. 26-48.

Qadeer, M. (1997) Pluralistic planning for multicultural cities: The Canadian practice. *Journal of the American Planning Association*, **63**, pp. 481-494.

Sandercock, L. (1998*a*) *Towards Cosmopolis: Planning for Multicultural Cities*. Chichester: John Wiley.

Sandercock, L. (ed.) (1998*b*) *Making the Invisible Visible: A Multicultural Planning History*. Los Angeles: University of California Press.

Sassen, S. (1996) Rebuilding the global city: Economy, ethnicity and space, in King, A.D. (ed.) *Re-Presenting the City: Ethnicity, Capital and Culture in the Twenty-First Metropolis*. Basingstoke: Macmillan, pp. 23-41.

Sennett, R. (1994) *Flesh and Stone: The Body and the City in Western Civilization*. London: Faber and Faber.

Soja, E. W. (1996) *Thirdspace: Journeys to Los Angeles and Other Real-and-Imagined Places*. Cambridge: Blackwell.

Thomlinson, R. (1969) *Urban Structure: The Social and Spatial Character of Cities*. New York: Random House.

Thompson, S. (1994) Suburbs of opportunity: The power of home for migrant women, in Gibson, K. and Watson, S. (eds.) *Metropolis Now: Planning and the Urban in Contemporary Australia*. Sydney: Pluto Press, pp. 33-45.

Thompson, S. (1996) Women's Stories of Home: Meanings of Home for Ethnic Women living in Established Migrant Communities. PhD thesis, University of Sydney.

Thompson, S., Dunn, K., Burnley, I., Murphy, P. and Hanna, B. (1998) *Multiculturalism and Local Governance: A National Perspective*. Sydney: New South Wales Department of Local Government/Ethnic Affairs Commission of New South Wales/ University of New South Wales.

Trench, S., Oc, T. and Tiesdell, S. (1992) Safer cities for women. *Town Planning Review*, **63**, pp. 279-296.

Ward, D. (1971) The emergence of central immigrant ghettoes in American cities: 1840-1920, in Bourne, L. (ed.) *Internal Structure of the City: Readings of Space and Environment*. New York: Oxford University Press, pp. 291-299.

Watson, S. (1993) Cities of dreams and fantasy: Social planning in a postmodern era, in Freestone, R. (ed.) *Spirited Cities: Urban Planning, Traffic and Environmental Management in the Nineties*. Sydney: Federation Press, pp. 140-149.

Watson, S. and McGillivray, A. (1995) Planning in a multicultural environment: A challenge for the nineties, in Troy, P. (ed.) *Australian Cities: Issues, Strategies and Policies for Urban Australia in the 1990s*. Melbourne: Cambridge University Press, pp. 164-178.

Weeks, J. (1990) The value of difference, in Rutherford, J. (ed.) *Identity: Community, Culture, Difference*. London: Lawrence and Wishart, pp. 88-100.

Wekerle, G. and Whitzman, C. (1995) *Safe Cities: Guidelines for Planning, Design and Management*. New York: Van Nostrand Reinhold.

Wilson, E. (1991) *The Sphinx in the City: Urban Life, the Control of Disorder, and Women*. Berkeley: University of California Press.

Wilson, J. Q. (ed.) (1966) *Urban Renewal: The Record and the Controversy*. Cambridge, Mass: MIT Press.

Yiftachel, O. (1995) The dark side of modernism: Planning as control of an ethnic minority, in Watson, S. and Gibson, K. (eds.) *Postmodern Cities and Spaces*. Oxford: Blackwell, pp. 216-242.

CHAPTER 14

Global City Regions: Planning and Policy Dilemmas in a Neo-Liberal World

Allen J. Scott

Contrary to many recent predictions, the ever increasing openness of national economic and political systems is not leading to the evanescence of geographical distinctions but rather to a reassertion of the role of selected cities and regions as privileged sites of production, employment, consumption, and social life. The persistent vigorous growth of huge cosmopolitan agglomerations of capital and labour, that I identify later in this chapter as *global city regions*, represents perhaps the most striking illustration of this proposition.

Much has been written in recent years about globalization and its relations to urban and regional development, though there is also much debate about the nature of the actual processes at work in these spheres of social life. There is, in particular, a great deal of controversy about the substantive meaning of the term 'globalization', and a number of scholars have been forthright in expressing their scepticism about some of the more extreme visions that have been proclaimed under the banner of this term (e.g. Hirst and Thompson, 1996).

In the present account, I shall take it that an identifiable process of globalization is indeed under way, at least in regard to certain critical economic variables, though with the stipulation that it has so far not moved much beyond an early phase and that it remains remote from any final state of fulfillment. Possibly the best way of identifying this process overall is to say that it seems to involve a progressive interpenetration of national capitalisms one with another in such a way that basic economic relationships (input-output chains, investment flows, rates of profit, prices, and so on) are all evolving in a direction whose end point, if it were ever achieved, would eventually be representable as a single worldwide (but not geographically homogeneous) capitalist system. Any trend in this direction, however, is also subject to political regulation on the part of sovereign states, which in some observable instances have helped to accelerate it, and in others have patently impeded its operation. Certainly, if the sovereign state is itself undergoing various changes as a result of these developments, it remains a potent and in many respects an unpredictable element of the new economic and political geography that is coming into existence as we enter the twenty-first century, and in opposition to some pronouncements (e.g. Ohmae, 1995), it shows no clear signs of withering away in the near future. The long-run process of globalization, as defined, is thus likely to be characterized by many wayward trends, and it may well be subject at some point to significant retardation or even reversal as a result of political pressures.

One of the important direct impacts that globalization is having as it runs its course is its tendency to induce much detailed restructuring of the geography of capitalism. Among the more noteworthy of these impacts is the emergence of an inter-continental mosaic of large city regions constituting the basic economic motors of the world economy. But why, in a period of space-time compression (Harvey, 1985), marked as it is by the continual reduction of geographic barriers and hence by increasing mobility of capital and labour, are such regions coming to be so dominating an element of the world map? Why is it, in particular, that so much modern economic activity is concentrating and reconcentrating in them instead of dispersing far and wide? What concomitant changes are observable in the internal structure and functions of global city regions? What new urban and regional policy problems are posed by this situation, and how are these problems to be dealt with in view of the spreading ethos of neo-liberalism in the world at large? What implications, if any, do our answers to these questions have for conceptions and practices of citizenship and democracy in the emerging global system of city regions?

The bulk of this chapter is devoted to a search for ways of addressing these questions. As we proceed in this task, however, one basic methodological proviso needs to be established at the outset. In any really viable analysis of the issues at hand, it is not sufficient simply to juxtapose the phenomenon of globalization and the phenomenon of city regions and to proclaim that they are connected. If there is indeed a defensible problematic of the global and the local to be articulated, then we must be able to point unambiguously to their reflexive interactions. In its most accomplished form, this would entail a clear demonstration that globalization has definite and transformative impacts on regional development and urbanization, and that these latter phenomena in turn exert some decisive influence over the trajectory of globalization.

SETTING THE SCENE: CITY AND REGION IN THE POSTWAR ERA

The Space Economy of Fordism

In the advanced capitalist societies over the immediate postwar decades, a tight functional relationship between the national economy and the sovereign state over a definite territorial expanse was the invariable order of the day. In each case, this relationship was consolidated by the central role of Fordist mass-production industries as the champion propulsive sectors of the national economy, and by the enactment of national Keynesian welfare-statist policies designed to provide the stabilizing mechanisms for these industries and to maintain their essential social bases. To be sure, international trade was expanding greatly in absolute terms during the classical Fordist mass-production period over the 1950s and 1960s, though it was only toward the end of the 1960s that it began to grow at a pace that significantly outstripped the rate of growth of gross domestic product (Dicken, 1992; Scott, 1998). The period of the long postwar boom in North America and Western Europe can thus be seen as a sort of apogee of national capitalisms, even as it was also a period in which an international economic system was coming once again (as in the late nineteenth century) to be a major but nevertheless subservient element of the worldwide economic order. Much of this order was presided over by the United States in a *Pax Americana* opposed on virtually all fronts to the communist bloc.

This was also a period marked by distinctive intra-national patterns of urban and regional development in North America and Western Europe. In the great manufacturing belts, swarms of overgrown industrial cities flourished apace, their economies based on growth pole industries such as car production, machinery, and domestic appliances. With few exceptions, the major cities of the Fordist mass production era were remarkably successful in economic terms, and all the more so as their industrial underpinnings were well protected from internal malfunctions and from foreign competition by activist national policies and administrative arrangements. The central problems of the industrial cities of this era took the form of a succession of locational upheavals and associated forms of social contestation generated by continual dramatic urban growth and internal spatial re-organization (Harvey, 1973; Scott, 1980).

In response to these problems, batteries of remedial counter-measures were routinely set in motion in large metropolitan areas. Urban planners thus went to work on such tasks as the zoning of land uses, the charting of outward suburban expansion, the design and construction of intra-urban transportation networks, urban renewal programmes, and the provision of public housing. This wholesale intervention of planners in guiding urban development had many beneficial impacts on the workings of the Fordist mass production economy, not simply by helping to make large cities – where the main elements of the mass-production system were concentrated – function more efficiently, but also by promoting high levels of consumption. In particular, rates of home-ownership and use of the private car were increasing rapidly as a consequence of planned changes in the urban environment. In sympathy with the modernist precepts of the times, many urban theorists conceptualized all of this planning activity in terms of a highly schematized, rational-comprehensive approach to social engineering (cf. Scott, 1980).

In the more advanced capitalist countries, the main industrial cities, together with the areas immediately surrounding them, also came to operate as thriving core regions relative to the remaining parts of each national territory. The latter, by contrast, functioned as less developed and less prosperous peripheries. A typical dynamic of core-periphery interaction was accordingly initiated such that (*a*) growing core regions with high levels of economic opportunity attracted streams of migrants from peripheral areas, while (*b*) low-wage, low-skill branch plants flowed selectively into relatively stagnant peripheries in order to tap their pools of cheap surplus labour (Hirschman, 1958; Myrdal, 1957). In this dynamic, peripheries became caught in a developmental bind in which they seemed doomed permanently to lag behind core regions.

Regional planners invested considerable effort in attempts to counter the problem of the relative economic backwardness of national peripheries (Keating, 1997). Regional planning in the postwar decades was focused principally on attempts to re-direct income and growth from core to peripheral areas, and it was orchestrated above all from central governmental agencies such as the Economic Development Administration in the United States, the Department of Regional Economic Expansion in Canada, the Department of Economic Affairs in Britain, the Délégation à l'Aménagement du Territoire in France, and the Cassa per il Mezzogiorno in Italy. Attempts were also made to create important new growth centres in some parts of the periphery (e.g. southern Italy or Fos-sur-Mer in France) by means of lavish investments in physical infrastructure and the offer of major financial investments to plants in growth pole industries to locate in the vicinity. Most of these initiatives, however, failed to achieve anything even close to their initial goals.

The Crisis of Fordism: The Rise of Neo-Liberalism

By the early to mid-1970s, the Fordist mass-production system was entering an extended period of crisis, and in parallel with this trend, much of the old Keynesian welfare-statist policy system was becoming at best unworkable and at worst counter-productive. The crisis was signalled by, among other things, an epidemic of plant closures and unemployment in formerly prosperous manufacturing regions and greatly accelerated dispersal of production to peripheral regions (Bluestone and Harrison, 1982). As the 1970s wore on, it was the core rather than the periphery that now increasingly seemed to present the most pressing problems, and indeed some formerly peripheral or semi-peripheral areas such as the United States Sunbelt, the Third Italy, or parts of southern Germany were at this stage starting to show signs of vigorous new industrial growth. The crisis of core regions was further aggravated by a greatly expanded exodus of branch plants, not just to locations in national peripheries, but now also to many different areas in the Third World. This flood of foreign direct investment helped to divert yet more capital and employment away from the economies of the main Fordist mass production regions, while simultaneously pushing the international system toward a climacteric that Fröbel *et al.* (1980) labelled the *new international division of labour*, and that in a sense foreshadowed the shift into globalization as such.

The symbolic and practical manifestation of the end of Keynesian welfare-statism is expressed in the late 1970s and early 1980s by the accession to political power of the Thatcher government in Britain and the Reagan administration in the United States. In retrospect, the historic mission of these two political regimes was perhaps less to inaugurate a new positive system of regulation than it was to dismantle the old policy arrangements and to open up economic and social life to ever-intensifying market forces. While Britain and the United States are still today probably the most advanced cases of the official neo-liberalism that was thus unleashed in the contemporary world, this general shift in the political winds has tended steadily to spread to other areas, and most recently has figured prominently in the International Monetary Fund's programme of fiscal reform for East and South-East Asia.

One of the earliest casualties of the dramatic economic and political changes that were occurring toward the end of the 1970s was urban and regional planning. The combination of economic crisis and drastic national policy reform undermined in significant ways both the range and the intensity of much previous planning activity. Redistributive regional planning operations were fairly quickly cut back. Former peripheries in most of the advanced capitalist countries were in any case no longer showing strong signs of stress relative to core areas, just the contrary was now becoming a more common state of affairs. Urban planning apparatuses were also curtailed in one metropolitan area after another, with large capital-intensive development projects in particular being severely cut. Not only was the tax base in many of these metropolitan areas shrinking rapidly – especially those areas most directly impacted by the crisis of Fordist mass-production – but intergovernmental fiscal transfers from higher levels of government were starting to be reduced as well. In the United States, federal and state subsidies to county and city governments declined sharply after 1975, even though a shift of responsibility for many kinds of social services from upper to lower tiers of government was simultaneously forcing local governments into higher levels of expenditure. Cutbacks in intergovernmental transfers were especially drastic in the country's three largest cities, namely, New York, Los Angeles, and Chicago.

In a first wave of reaction to the critical situation that developed after the second half of the 1970s, many local governments in different parts of the world began to shift away from planning interventions that yielded little or no immediate financial return and toward more narrowly focused entrepreneurial approaches directed to enlarging the municipal business base. Urged on by local growth coalitions and booster groups, urban officials turned to the tasks of persuading new investors to locate within their jurisdictions and of building up good local business climates (Eisinger, 1988; Mayer, 1994). New urban and regional economic development programmes thus expanded apace in the guise of public relations efforts, incentive packages for in-coming firms (tax reductions, financial subsidies, special land deals), municipal brokering of labour relations arrangements in favour of new investors, and attempts at urban re-imaging (Amin and Thrift, 1995; Bartik, 1991; Clarke and Gaile, 1998; Fainstein, 1991; Gittell *et al.*, 1996; Hall and Hubbard, 1996). Much was made, too, of the alleged growth potential of high-technology industry, and many localities poured enormous resources and energy into efforts to emulate the experience of Silicon Valley (Miller and Côte, 1987) though with almost universally unsatisfactory results. With the increasingly widespread and aggressive adoption of local economic development programmes of these sorts over the 1980s, the predictable result was that regions became ever more embroiled in wasteful bidding wars in the race for scarce investments. These unregulated and predatory tournaments have often tended in practice to provide more benefits to distant corporate shareholders than they have to local residents, and when they result in the diversion of firms away from locations where they would have operated more efficiently (in terms of joint private and public benefits), they actually have a negative-sum effect.

In many respects, the current situation of deepening globalization, combined with a prevailing neo-liberal policy environment, is reinforcing these trends at the local level, and this state of affairs is liable to prove destabilizing. Certainly the current conjuncture does not appear to be one in which a viable long-run system of regulation has yet made a decisive historical appearance, most notably at the local level. We might say, along with Peck and Tickell (1994) that the currently widespread influence of neo-liberal ideals is actually representative of a transitional phase between Keynesian welfare-statism and some yet to be determined after-Fordist system of regulation, rather than a durable response to the problems of economic coordination (including urban and regional management) in the current phase of capitalist development. My argument is centred on the proposition that the minimum condition of stability and growth in local economic systems is dependent on the formulation of a policy agenda committed to institution-building in pursuit of trans-market efficiency gains and orderly long-run change in structures of production at both the intra- and inter-regional levels.

GLOBAL CITY REGIONS: FORM AND FUNCTION

A New Geographic Phenomenon?

The emerging system of global city regions can be viewed in geographic terms as a complex mosaic spread out across the entire world (Amin and Thrift, 1992; Scott, 1998; Storper, 1992; Veltz, 1997). Each individual city region comprises a central urbanized mass encircled by an indefinite stretch of dependent hinterland, itself invariably punctuated by a network of discrete urban centres. The mosaic is linked up across national boundaries to form a dense global grid of economic, social, and cultural exchanges. Over the last decade or so, this burgeoning spatial structure has begun to override (though not to supplant entirely) the older pattern of core-periphery interdependency that was such a salient feature of the geography of Fordism.

Although an explicitly spatial definition of global city regions has been offered here, there can in practice be no *a priori* specification of the boundaries and territorial extent of any given case. We can, however (taking a cue from Marxian theories of social class determination), identify two factors that must usually be present for a determinate outcome to be achieved. First, there must be a set of *objective conditions*, meaning in this context the presence of a large and multi-faceted complex of economic activities generating forceful polarization effects at the regional level. Second, a set of *political practices* must also be in evidence such that definite alliances are established between adjacent territorial units resulting in their joint aggregation into a single constituency which acquires powers of collective decision-making and action. In this latter sense, and in view of the balkanization that often characterizes local government, the world's city regions are fairly clearly still in an extremely fluid and preliminary form, though a number of prominent metropolitan and megalopolitan areas (e.g. the New York-New Jersey region, Montreal, South-East England, the Ile de France, Tokyo, Mexico City, Hong Kong, Singapore, Seoul) would seem to be well on the way to global city region status. More complex cases can be cited in the attempts at territorial coalition-building now going on in northern Italy around the region of 'Padania,' or in the Seattle-Vancouver area to construct the region of 'Cascadia,' or in the formation of a 'Trans-Manche Region' consisting of Nord-Pas-de-Calais in France, the western extremities of Belgium, and parts of Kent in Britain. Some of these examples consist of trans-border units, and these types of city regions, too, in the dawning global era will in all likelihood become familiar elements of the mosaic (Ohmae, 1995).

The Economic Base of Global City Regions

This evolving situation is in large degree being propelled forward by the massive growth and innovation effects unleashed by new rounds of post-Fordist industrialization in the world economy together with the revitalization that has occurred in more traditional industries by means of neo-Fordist reforms. There can be no clear-cut demarcation between these two types of economic activity, and the terms post-Fordist and neo-Fordist are in any case extremely unsatisfactory. Roughly, the first type is made up of sectors like much high-technology manufacturing (e.g. advanced semiconductor devices, robotics, biotechnology), neo-artisanal and cultural-products industries (e.g. fashion clothing, designer furniture, film, multimedia), and a wide assortment of business, financial and other services. The second type includes a notable variety of industries in assembly, process, and packaging sectors (along with dependent networks of subcontractors) that were very much part of the landscape of Fordism, but which have restructured over the last couple of decades in ways that allow for considerably more flexibility, discretionary performance on the part of workers, and product variety. Post-Fordist and neo-Fordist industries are typically made up of large phalanxes of small- and medium-sized producers, but they almost always comprise large producers too, often branch plants of multinational corporations.

The significance of the historic shift beyond a dominant Fordist paradigm of industrialization and economic growth is that the crisis of the great classic industrial cities over much of the 1970s and part of the 1980s has now been succeeded by strong new rounds of concentrated regional expansion (Saxenian, 1994; Scott, 1993; Storper and Walker, 1989). Such expansion has been especially pronounced in the ascending system of global city regions, some of which are located in parts of the world that were once sometimes thought of as development-resistant Third World peripheries. The resurgence of city regions in the contemporary period is due above all to the renewed proclivity of many types of productive activities to agglomerate intensely together in geographic space, while the spatial clusters thus formed also come to function as staging posts and consumption points in worldwide commodity chains.

As I have tried to show at length elsewhere (Scott, 1993; 1998), these developments are closely connected to the advancing tide of post-Fordist and neo-Fordist industrialization with its peculiar tendency to demassification and externalization of production systems and their resynthesis into value-adding networks. Of special note are the broad locational forces set in motion by the changing character of economic production in present-day capitalism. In contradistinction to the centrifugal spatial drift that was so much in evidence as Fordist mass production matured over the 1960s and 1970s, these forces have reinvigorated processes of centripetal growth in different parts of the world. Their roots lie in the transformation of technologies, of patterns of industrial organization, and of labour markets, that has been occurring over the last couple of decades, and they mutually reinforce one another in ways that keep pushing units of capital and labour together into large spatial aggregates. Once these aggregates have formed, they generate potent increasing returns effects so that their growth leads onward to yet more growth.

Of all the many different sources of these increasing returns effects, four are of special interest here, though detailed explanations of their inner logic must be sought elsewhere (e.g. Scott, 1988; Storper and Walker, 1989). First, since much contemporary industry is subject to a dynamic of intra-sectoral specialization and complementarity, selected sets of producers find it to their advantage to cluster together in order to optimize access to one another for the purposes of exchange, collaboration, and risk reduction. Second, growth-inducing learning and innovation effects are apt to be generated in abundance within the interactive and multifaceted agglomerations formed in this manner. Third, local labour markets encourage locational clustering because of the myriad advantages that accrue from the formation and pooling of agglomeration-specific skills, aptitudes, habits, and sensitivities. Fourth, agglomeration is intensified by the emergence of distinctively local business cultures, conventions, institutions, and reputation effects, leading in many instances to the streamlining of economic production and exchange processes. The operation of these four sets of variables is reinforced yet again by the rich infrastructural goods and services that are regularly supplied by local governments wherever dense concentrations of productive activity and human settlement occur.

Cities and regions with these kinds of (socially-constructed) endowments dramatically exemplify the maxim that the whole is greater than the sum of the individual parts, and they function as vortexes of agglomeration economies generating long-lasting virtuous circles of local economic growth. Obviously, various bottlenecks (such as traffic congestion, pollution, land shortages, or proliferating mismatches in the locations of jobs and residences) may periodically disrupt such growth, but concerted remedial public action is almost always forthcoming in the effort to unleash new rounds of expansion.

City Regions and Globalization

If the production complexes that form the economic base of major city regions are tightly anchored in geographic space, recent changes in the technologies of transportation and communication also make it possible for rising portions of their output to be sold on world markets (Lipsey, 1997). Specialized multinational companies often play a decisive role in this process of global distribution (Scott, 1997). Moreover, city regions that are successful in the international commercialization of their products are liable to experience a strengthening of their intra-regional scale economies, and a widening of their economies of scope as the local division of labour responds to concomitant extensions of the market (Cooke, 1997). The net result will be the formation of a yet more robust system of local competitive advantages, and a propensity for the economic bases of global city regions (though not necessarily their non-basic activities) to become more specialized relative to one another (Krugman and Venables, 1995; Leamer, 1995).

These dynamics have profoundly transformative effects on the economic and social character of global city regions. They induce significant re-structuring of their economies, and via this process they also exert powerful impacts on labour markets. The economies of global city regions often exhibit, as well, a pronounced tendency to bifurcation in their employment structures. Thus, on the one hand, global city regions are places where highly qualified professional, managerial, technical, and creative workers find ready employment in the upper echelons of the production system. On the other hand, they are also places where low-wage, low-skill employment frequently abounds. While this bifurcation phenomenon is probably not uniquely due to globalization, it is certainly exaggerated and reshaped by it, and especially by the impacts of international trade (which helps to put a premium on rewards to skill while exerting a dampening effect on the wages of unskilled labour) and by the downward pressure that is further put on wages at the bottom end of the income ladder as a consequence of massive immigration of workers from less developed countries (Leamer, 1995; Sassen, 1997). This latter syndrome is greatly accentuated by the symbiotic relationship that is often observable in global city regions between a proliferation of sweatshop factories and low-grade service activities on the one hand, and a ready supply of compliant, politically-marginalized immigrant labour on the other.

In a study of the Los Angeles household furniture industry, I have described a salient empirical example of this syndrome (Scott, 1996). Over the 1970s and 1980s, employers in this sector steadily substituted low-skill immigrant (mainly Latino) workers for more skilled native-born workers, a trend that was evidently driven by manufacturers' short-term search for reductions in operating costs. But while short-term costs were certainly reduced, product quality and innovativeness were also seriously compromised, and over the longer term this brought the industry into devastating head-to-head competition with cheap producers located in Latin America and East and Southeast Asia. As a consequence, the Los Angeles household furniture industry today is facing a condition of widespread structural failure, and in the absence of some decisive remedy to this situation will almost certainly continue to give ground before the onslaught of foreign competition. Caught in this vicious circle, employers in this and other sweatshop sectors in Los Angeles continue to push down on wages and working conditions, thereby opening wider the social gap that has become so strong a feature of the city's socio-economic structure in the 1980s and 1990s.

Thus, even though global city regions represent some of the most dynamic and prosperous economic communities in the world today, they are also prone to various kinds of failure, and in numerous cases their aggregate wealth masks a rampant underside of poverty and exploitation.

PROSPECTIVE PLANNING AND POLITICAL AGENDAS FOR GLOBAL CITY REGIONS

The Neo-Liberal Dilemma

These remarks alone are surely adequate to prompt some scepticism about the neo-liberal model despite the degree of legitimacy that is conferred on it by reason of its concurrence with the recent extended period of aggregate economic growth in the United States and Britain. Further drift in this political direction is likely to exacerbate the problems of global city regions, not to mention those of other spheres of economic activity, and to curtail prospects for more orderly and progressive forms of development (Allen *et al.*, 1998).

In a nutshell, neo-liberalism eventually undermines the possibilities of its own long-run success by steadily eating away at the buttresses of collective order in capitalism. It clears the way for the extension of competitive market processes into the far corners of society, but the very same action is liable to bring a series of mounting breakdowns, which are not simply expressions of technical market failures, but more importantly are a function of the steady dissolution that is also brought about of the necessary and complex complementarities in capitalism between privatization and competition on the one side, and communality and cooperation on the other. Even when markets themselves operate efficiently over a specified social domain, they can never work independently of an infrastructure of extra-market social and political arrangements.

Above and beyond these observations, it is possible to devise a not too far-fetched scenario suggesting that the encounter of neo-liberalism with its own secular limits may not be far off. Thus, while the taxpayers' revolt that is one of the hallmarks of neo-liberalism has forced cutbacks in many essential governmental services, its impacts have actually been mitigated by recent vigorous growth in the economy. This growth has helped to fill out the public coffers despite declining tax rates, and thus to maintain a measure of effective regulatory order and positive governmental action. Economic growth in capitalism, however, can never proceed uninterruptedly upward, and the coming economic downturn, once it arrives, is likely to be characterized by a cumulatively negative effect. In the first place, it will tend to accentuate governmental efforts to cut already strained public services still further because of falling GDP. In the second place, it will diminish yet further the capacity of governments at all hierarchical levels to raise taxes on incomes. The net outcome is likely to be either a magnification of the economic slow-down and its negative social consequences, or a severe fiscal crisis for the state should it attempt to engage in counteractive measures.

Viable alternatives to neo-liberalism are not only conceivable but are also embryonically present in certain non-Anglo-American forms of capitalism, notwithstanding the current macro-economic disarray that is detectable in some of them (Albert, 1991; Wade, 1990). Here, I shall propose some outlines of a prospectively social democratic approach to economic order in the contemporary world, with particular emphasis on the specific problems and predicaments of global city regions. This alternative approach is likely to prove at least partially capable of bending the economic trends currently sweeping across the world's principal city regions into a more benign shape than they might otherwise assume.

Tasks for Planners and Urban Policy Makers

Many novel questions about urban planning and policy have come to the fore as global city regions have become more self-assertive and as nation states have found themselves less capable of providing a protective umbrella for all the geographic and sectional interests within their dominion. This is not to say that traditional urban planning issues have disappeared – on the contrary, they are as alive as ever – but that an entirely new and unexpected set of problems must now be dealt with, and nowhere more so than in the economic structures of large city regions. Moreover, the remedial tasks that call ever more urgently for completion as these problems multiply entail vastly more complex forms of collective action than the stop-gap entrepreneurial approaches that so many urban administrations have favoured in the recent past.

Global city regions represent constellations of interdependent local interests locked increasingly into competitive (but also collaborative) relationships with one another across national boundaries. They are made up of unique and immobile agglomerations of industrial and service activities that jointly generate increasing returns for all as an effect of the social division of labour, learning and innovation mechanisms, labour market processes, and the cultural norms that often materialize in local production complexes and that help to guide business behaviour and interaction. The economies of these regions, then, are structured in such a way that the destiny of any single producer is almost always intertwined with the destiny of all other producers located in the same region. Because of this inherently communal nature of production, together with the circumstance that markets alone cannot guarantee the optimal production and allocation of all the complex externalities that run through these agglomerations, there is of necessity a positive role to be played by local institutions of collective order able to provide critical coordination and temporal steering services. Such institutions may assume a variety of forms, ranging from agencies of local government, through hybrid arrangements such as quangos or private-public partnerships, to purely civil organizations such as manufacturers' associations, chambers of commerce, labour unions, and community groups (Clarke and Gaile, 1997, 1998; Scott, 1998).

What, we may ask, should be the specific operating features and objectives of such institutions? I shall largely confine my answers to this question to matters of economic interest, while nonetheless acknowledging that many complex social issues are simultaneously very much at issue (Friedmann and Wolff, 1982; Sassen, 1997). Above all, I shall focus on the central problem of how strategically to build and sustain local competitive advantages in a world where national boundaries exert a steadily decreasing effect on economic interactions. In whatever way we may approach this problem, it is important to bear in mind the principle that while the appropriate lines of policy action are always describable in generic language, the nature of the issues under consideration here demands that they also be concretely formulated in *agglomeration-specific* terms, i.e. as a set of measures responding to the always idiosyncratic qualities of any given city region.

The varieties of planning and policy activity which are now to be briefly discussed can be conveniently divided into two main categories: synchronic coordination of economic relationships, and long-term planning procedures. However, there is no really clear dividing line between the two categories.

Synchronic Coordination

Various forms of coordination are called for to deal with a series of functional gaps and breakdowns that appear repeatedly in urban economic systems. These occur in a multitude of different guises, and are often most severe in the small- and medium-sized firm segment of the economy. Such firms are generally less effective than large firms in raising capital, in marketing their outputs, in acquiring information, or in any kind of activity that requires significant resources and organizational back-up. Yet they represent in many ways the lifeblood of the urban economy, not just because they usually account for the vast majority of producers in any given instance, but also because of their role as a critical mainstay of large firms via their diverse subcontracting and service operations.

Firms of all sizes, but again small- and medium-sized firms in particular, also suffer from an incapacity to provide labour training and research in socially-rational quantities, and this problem is notably acute because of free-rider problems where individual firms withhold investment in training and research in the hope that they can tap into positive spill-overs from other firms. Training and research, however, are indispensable components of local competitive advantage in both capital- and technology-intensive sectors as well as in many of the neo-artisanal and new cultural-products industries that are growing rapidly in global city regions and for which innovative design capacity is an important factor in commercial success. There is thus a major role to be played by extra-market institutions capable of providing training and research programmes relevant to local needs.

In addition, the webs of transactional relationships that are an intrinsic aspect of localized agglomerations of producers are usually susceptible to much operational improvement by means of network brokering and mediation activities. There is a rapidly increasing demand for such activities in agglomerated production systems as policy-makers and other concerned groups seek to improve local industrial performance by encouraging the formation of collaborative manufacturing networks with high levels of inter-firm synergy. Among the numerous examples of this general approach that have been described in the literature, mention may be made of the ambitious industrial extension programmes that have recently been set up in Quebec and New Hampshire (Ferland *et al.*, 1996; Gittell *et al.*, 1996).

Long-Term Planning

Long-term planning procedures are an essential ingredient of the regulatory apparatus of global city regions because the economies of these regions are deeply caught up in processes of path-dependent development, leading in turn to a corresponding dynamic of 'temporal lock-in'. Without some sort of steering mechanisms to keep the local economy on a viable long-term track, it is entirely possible for developmental pathways to be selected – because they offer economic agents attractive rewards in the short run – that become increasingly problematical over the long run. The case of the Los Angeles household furniture industry is a noteworthy example of this type of temporal trap (Scott, 1996). Given this industry's current trajectory of development, it is questionable whether any effort to redirect its future evolution into a more viable configuration could succeed, and even if it could, it would certainly be extremely painful and costly. Had there been some institutionalized means of taking stock of trends in the Los Angeles household furniture industry through the 1970s and 1980s, and of putting appropriate counteractive strategies into effect at an early stage, the problems of the industry might well have been somewhat less serious than they now are.

In the same way, appropriate steering and collective decision-making procedures can help regions to reap critical 'first mover' benefits. There is by no means any straightforward method of securing such benefits, though they can sometimes be fostered by extra-market support for infant industries that may not be viable in the short run if left to their own devices. The expectation here is that any industry selected for treatment in this manner will begin to flourish without public support once it has reached a critical stage of development, especially if and when it starts to generate localized positive externalities.

The Inter-Regional Dimension

These two principal varieties of planning and policy activity can be characterized as attempts to underpin both the defensive and offensive economic capabilities of global city regions, and they seek to achieve these goals by taking full advantage of the peculiar logic of locational agglomeration and increasing returns.

We might well expect, however, that any concerted bid by city regions actually to move vigorously forward in this way would also have important repercussions on their mutual interactions. First, the formation of strong intra-regional institutions capable of coordinated local economic action would tend to facilitate the organization of beneficial inter-local joint ventures where selected sets of city regions combine together in the pursuit of synergistic association (Cooke, 1993). Second, however, the same kinds of institutions would also certainly result in sharpened inter-regional competition as a consequence of their likely positive effects on the economies of their home bases. In either case, we might well expect rising degrees of politicization to occur in the worldwide inter-regional economic system. This putative politicization is apt to be especially acute given the increasingly zealous pursuit by multinational corporations of localized positive externalities (Dunning, 1998), while their accountability to local communities is usually quite ambiguous. Some early hints about the nature of this process of politicization can be found in the concerns that have been expressed (and in some cases, the outright retaliation that has ensued) as a result of the more predatory forms of regional economic activism that have come to be widely practised since the 1980s. This and allied problems immediately raise the difficult issue of how to build and operate workable systems of supra-regional coordination across different national jurisdictions. The need for this kind of coordination will predictably become more pressing, and will eventually no doubt evoke some sort of institutionalized response, as perhaps exemplified in rudimentary form by the European Committee of the Regions.

It goes without saying that these different planning and policy tasks represent a significant expansion of what we normally mean by urban and regional management, and as such, they pose important new challenges to professionals in the field. Any serious attempt to face up to these challenges presupposes major changes in training programmes for urban and regional managers as well as in their working environments. Among all of the other things that they must do, urban and regional managers in the global city regions of the future will need to be able to deal with complex local business issues while at the same time creatively administering relevant public goods systems and maintaining a careful eye on the promises and dangers of the changing global situation.

In Search of a Political Context

Several theorists have recently pointed out that the deepening trend to globalization seems to have brought with it a certain abridgement – though by no means a vanishing away – of the regulatory powers of the nation state. This is perhaps most evident in declining levels of national economic sovereignty (cf. Amin and Thrift, 1995; Peck and Tickell, 1994), and in the early signs of an erosion of the legitimacy of the state as the privileged fountainhead of citizenship and democratic order in the contemporary world (Holston and Appadurai, 1996; Mouffe, 1992; Sullivan, 1995).

One further symptom of this condition is the appearance of new forums of social regulation that are beginning to complement and to substitute for the traditional nation state. Some of these forums have a purely sectoral expression; others have an explicitly territorial identity. Three of these are of special concern here. The first is definable as the global level itself, in the guise of international organizations, contractual regimes, formal and informal diplomatic encounters, and so on. The second is constituted by plurinational blocs, the European Union being the most advanced specimen of the type. The third is made up of a multitude of localities, with the global city region epitomizing, in prospect, the most conspicuous case. Together with the existing – but restructuring – system of nation states, this emerging political geometry may well present a foretaste of the basic scaffolding of a new regulatory regime that seems to be coming into being. We can thus think of our political affiliations as now beginning to disperse throughout a quadripartite system of global, plurinational, national, and regional regulatory structures (cf. Albrow *et al.*, 1997). In this dawning scheme of things, global city regions will presumably play a particularly important role, for not only are they constituted as vital assemblages of productive activity, but they also represent concentrated common milieux of daily life and experience for large groups of individuals.

The city region thus has special meaning as a community of interest, even in a globalizing world, although it must be stressed that this remark has nothing to do with romantic visions of a pre-modern civic republicanism or Tocquevillian communitarianism. What is implied is a conception of the city region as a domain of citizenship distinct from other levels of political organization. This notion of citizenship stands in contrast to the usual meaning that it carries as a birthright granted by the national state. Programmatically, we might think of it in addition as a civil attribute obtainable by residence in a particular place, and bringing with it substantive rights and obligations specific to that place (Holston and Appadurai, 1996). In this sense, it would now be possible to acquire citizenships many times over as individuals move – even across national boundaries – from one city region to another in the course of their lives, thus voiding the disabling and dysfunctional effects of outsider status on participation in local political affairs. One important consequence of any real reform in this direction would be the enfranchisement of the large marginalized immigrant populations that exist in many global city regions.

A more democratic and participatory order at the city region level would also have many implications for the ways in which any practical implementation of the kinds of planning and policy measures discussed earlier might be handled, and for the benefits that they might be expected to confer on different interests and constituencies. Such measures are not just technocratic tools; they are also powerful instruments of social adjustment with potentially profound implications for issues like distribution, equity, and social justice. I would contend, further, that if these measures are to realize their full potential in enhancing the vibrancy of the economies and collective life of global city regions, it is imperative that they be harnessed to the task of moving toward a more social democratic future, in the local as in other political arenas, and away from neo-liberalism. Any shift of this sort is obviously going to require more than a series of simple administrative reforms; it depends, in addition, on our ability to carve out and to strive for a progressive vision of what is possible in the new world order.

In the absence of political mobilization around a clearly articulated and workable programme of social democracy, a different and much less benign future may well lie in store for global city regions. Any continuation of the neo-liberal agenda in the world's great city regions – quite apart from its chances of long-run failure in the economic arena – is almost certain to result in even higher levels of social exclusion and polarization than are observable today, and hence in a greatly augmented potential for serious social unrest (Autès, 1997).

CONCLUSION

Over the immediate postwar decades, the urban question came to be formulated primarily in terms of the internal socio-spatial organization and associated modernist planning systems of the large metropolis (Scott and Moulaert, 1997). In both academic and policy circles, this question was often linked to a further series of concerns pertaining to the developmental trajectories of core and peripheral regions, and the persistent inequalities between them. At the dawn of the twenty-first century, the substantive content and meaning of urban and regional analysis are shifting significantly into other registers. Today, the major issues are concerned more and more with the historical and geographical appearance of a global system of large city regions and with the ways in which the dynamics of these city regions are influenced by their deepening cross-national interdependencies. I have concentrated my attention on the economic side of this matter because what is at stake here is so fundamental to the overall shape of the future. There are, however, many pressing social and cultural issues that also call urgently for investigation, not least being the internal social inequalities that currently abound in global city regions and their concomitant potential for disorder and unrest.

Notwithstanding the triumphalism of Anglo-American neo-liberal capitalism in the present conjuncture, and its aggressive self-assertion as a model for the rest of the world, there are actually many reasons for supposing that this framework of economic and political organization may well reach exhaustion in the not too distant future as it runs up against its own intrinsic outer limits. There is, however, a real alternative to neo-liberalism in the guise of a reconstructed social democratic model, one of whose elements is to build on current trends in global city regions while seeking to mould them in politically progressive ways. In addition, barring a major global catastrophe, the quadripartite regulatory system will doubtless become a more clearly delineated feature of world political geography in future years. We face many puzzles about the emerging shape of things at each tier in this system, as well as about the interrelations between the different tiers. The tier that is primarily made up by global city regions is particularly puzzling because, as a new 'city-centric capitalism' crystallizes out on the international landscape (Brenner, 1998, p. 8), these regions come to function as the geographic foundations of the whole edifice.

REFERENCES

Albert, M. (1991) *Capitalisme contre capitalisme*. Paris: Editions du Seuil.

Albrow, M., Eade, J. , Dürrschmidt, J. and Washbourne, N. (1997) The impact of globalization on sociological concepts, in Eade, J. (ed.) *Living in the Global City: Globalization as a Local Process*. London: Routledge, pp. 20-36.

Allen, J., Massey, D., Cochrane, A., Charlesworth, J., Court, G., Henry, N. and Sarre, P. (1998) *Rethinking the Region*. London: Routledge.

Amin, A. (ed.) (1994) *Post-Fordism: A Reader*. Oxford: Blackwell.

Amin, A. and Thrift, N. (1992) Neo-Marshallian nodes in global networks. *International Journal of Urban and Regional Research*, **16**, pp. 571-587.

Amin, A. and Thrift, N. (1995) Institutional issues for the European regions: From markets and plans to socioeconomics and powers of association. *Economy and Society*, **21**, pp. 41-66.

Autès, M. (1997) Public action, local democracy and the challenge of economic globalization, in Moulaert, F. and Scott, A.J. (eds.) *Cities, Enterprises and Society on the Eve of the 21st Century*. London: Pinter, pp. 229-243.

Bartik, T. J. (1991) *Who benefits from State and Local Economic Development Policies?* Kalamazoo: W. E. Upjohn Institute.

Bluestone, B. and Harrison, B. (1982) *The Deindustrialization of America: Plant Closings, Community Abandonment, and the Dismantling of Basic Industry*. New York: Basic Books.

Brenner, N. (1998) Global cities, glocal states: Global city formation and state territorial restructuring in contemporary Europe. *Review of International Political Economy*, **5**, pp. 1-37.

Clarke, S.E. and Gaile, G.L. (1997) Local politics in a global era: Thinking locally, acting globally. *Annals of the American Academy of Political and Social Science*, no. 551, pp. 28-43.

Clarke, S.E. and Gaile, G.L. (1998) *The Work of Cities*. Minneapolis: University of Minnesota Press.

Cooke, P. (1993) Globalization of economic organization and the emergence of regional interstate partnerships, in Williams, C.H. (ed.) *The Political Geography of the New World Order*. London: Belhaven, pp. 46-58.

Cooke, P. (1997) Regions in a global market: The experiences of Wales and Baden-Württemberg. *Review of International Political Economy*, **4**, pp. 349-381.

Dicken, P. (1992) *Global Shift: The Internationalization of Economic Activity*. New York: Guilford.

Dunning, J. H. (1998) Location and the multinational enterprise: A neglected factor? *Journal of International Business Studies*, **29**, pp. 45-66.

Eisinger, P.K. (1988) *The Rise of the Entrepreneurial State: State and Local Economic Development Policy in the United States*. Madison: University of Wisconsin Press.

Fainstein, S.S. (1991) Promoting economic development: Urban planning in the United States and Great Britain. *Journal of the American Planning Association*, **57**, pp. 22-33.

Ferland, M., Montreuil, B. and Poulin, D. (1996) Quebec's strategy to foster value-adding interfirm competition, in Staber, U.H., Schaefer, N.V. and Sharma, B. (eds.) *Business Networks: Prospects for Regional Development*. Berlin: Walter de Gruyter, pp. 82-96.

Friedmann, J. and Wolff, G. (1982) World city formation: An agenda for research and action. *International Journal of Urban and Regional Research*, **6**, pp. 309-344.

Fröbel, F., Heinrichs, J. and Kreye, O. (1980) *The New International Division of Labour*. Cambridge: Cambridge University Press.

Gittell, R., Kaufman, A. and Merenda, M. (1996) Rationalizing state economic development, in Staber, U.H., Schaefer, N.V. and Sharma, B. (eds.) *Business Networks: Prospects for Regional Development*. Berlin: Walter de Gruyter, pp. 65-81.

Hall, T. and Hubbard, P. (1996) The entrepreneurial city: New urban politics, new urban geographies? *Progress in Human Geography*, **20**, pp. 53-174.

Harvey, D. (1973) *Social Justice and the City*. London: Edward Arnold.

Harvey, D. (1985) *The Urbanization of Capital: Studies in the History and Theory of Capitalist Urbanization*. Baltimore: The Johns Hopkins University Press.

Hirschman, A. (1958) *The Strategy of Economic Development*. New Haven: Yale University Press.

Hirst, P. and Thompson, G. (1996) *Globalization in Question*. Cambridge: Polity Press.

Holston, J. and Appadurai, A. (1996) Cities and citizenship. *Public Culture*, **19**, pp. 187-204.

Keating, M. (1997) The political economy of regionalism, in Keating, M. and Loughlin, J. (eds.) *The Political Economy of Regionalism*. London: Frank Cass, pp. 17-40.

Krugman, P. and Venables, A.J. (1995) *Globalization and the Inequality of Nations*. Working Paper No. 5098, Cambridge, Mass: Bureau of Economic Research.

Leamer, E.W. (1995) A Trade Economist's View of US Wages and Globalization. Unpublished paper, Anderson Graduate School of Management, University of California, Los Angeles.

Lipsey, R.G. (1997) Globalization and national government policies: an economist's view, in Dunning, J.H. (ed.) *Government, Globalization, and International Business*. Oxford: Oxford University Press, pp. 73-113.

Mayer, M. (1994) Post-fordist city politics, in Amin, A. (ed.) *Post-Fordism: A Reader*. Oxford: Blackwell pp. 316-337.

Miller, R.E. and Côte, M. (1987) *Growing the Next Silicon Valley: A Guide for Successful Regional Planning*. Lexington, Mass: Lexington Books.

Mouffe, C. (ed.) (1992) *Dimensions of Radical Democracy*. London: Verso.

Moulaert, F. and Scott, A.J. (eds.) (1997) *Cities, Enterprises and Society on the Eve of the 21st Century*. London: Pinter.

Myrdal, G. (1957) *Rich Lands and Poor*. New York: Harper and Row.

Ohmae, K. (1995) *The End of the Nation State: The Rise of Regional Economies*. New York: The Free Press.

Peck, J. and Tickell, A. (1994) Searching for a new institutional fix: The *after*-fordist crisis and the global-local disorder, in Amin, A. (ed.) *Post-Fordism: A Reader*. Oxford: Blackwell, pp. 280-315.

Sassen, S. (1997) New employment regimes in cities, in Moulaert, F. and Scott, A.J. (eds.) *Cities, Enterprises and Society on the Eve of the 21st Century*. London: Pinter, pp. 129-150.

Saxenian, A. (1994) *Regional Advantage: Culture and Competition in Silicon Valley and Route 128*. Cambridge, Mass: Harvard University Press.

Scott, A.J. (1980) *The Urban Land Nexus and the State*. London: Pion.

Scott, A.J. (1988) *Metropolis: From the Division of Labor to Urban Form*. Berkeley: University of California Press.

Scott, A.J. (1993) *Technopolis: High-Technology Industry and Regional Development in Southern California*. Berkeley: University of California Press.

Scott, A.J. (1996) Economic decline and regeneration in a regional manufacturing complex: Southern California's household furniture industry. *Entrepreneurship and Regional Development*, **8**, pp. 75-98.

Scott, A.J. (1997) The cultural economy of cities. *International Journal of Urban and Regional Research*, **21**, pp. 323-339.

Scott, A.J. (1998) *Regions and the World Economy: The Coming Shape of Global Production, Competition and Political Order.* Oxford: Oxford University Press.

Scott, A.J. and Moulaert, F. (1997) The urban question and the future of urban research, in Moulaert, F. and Scott, A.J. (eds.) *Cities, Enterprises and Society on the Eve of the 21st Century*. London: Pinter, pp. 267-278.

Staber, U.H., Schaefer, N.V. and Sharma, B. (eds.) (1966) *Business Networks: Prospects for Regional Development*. Berlin: Walter de Gruyter.

Storper, M. (1992) The limits to globalization: Technology districts and international trade. *Economic Geography*, **68**, pp. 60-93.

Storper, M. and Walker, R. (1989) *The Capitalist Imperative: Territory, Technology, and Industrial Growth*. Oxford: Blackwell.

Sullivan, W.M. (1995) Reinventing community: prospects for politics, in Crouch, C. and Marquand, D. (eds.) *Reinventing Collective Action: From the Global to the Local*. Oxford: Blackwell, pp. 20-32.

Veltz, P. (1997) The dynamics of production systems, territories, and cities, in Moulaert, F. and Scott, A.J. (eds.) *Cities, Enterprises and Society on the Eve of the 21st Century*. London: Pinter, pp. 78-96.

Wade, R. (1990) *Governing the Market: Economic Theory and the Role of Government in East Asian Industrialization*. Princeton: Princeton University Press.

CHAPTER 15

Is Planning History?

Brendan Gleeson and Nicholas Low

From the moment that purposeful, multi-targeted government planning, including systematic urban and regional policies, began to be seriously considered in the aftermath of the Great Depression and the Second World War, opposition to it also started to form. All through the 1950s and 1960s Friedrich von Hayek turned out book after book denouncing government 'collectivist' planning. Nozick developed his theory of entitlement (essentially property entitlement) in explicit refutation of Rawls's theory of justice. Public choice theorists (e.g. Buchanan, 1977) argued for the principles of the market to prevail over political deliberation. Rothbard (1973) proposed that there was no field of government activity which could not ultimately be performed better by 'the private sector'; indeed, the very existence of public and private 'sectors' in a market society was anomalous to him.

Our aims in this chapter are twofold. The first is to explore how the mutually reinforcing politics of neo-liberalism and globalization have affected planning systems in different advanced capitalist nations, drawing attention to instances where the deregulatory agenda has been contested and even reversed. Second, we examine the impact of neo-liberalism on urban governance in Australia and explain why the deregulatory agenda has been so damaging to planning in this country. Our empirical emphasis in the paper is on planning in the Anglophone world, especially Britain and its developed former colonies.

The rest of the chapter is organized in four main sections. In the following section, we set the scene by examining the rise of the conservative ideological critique of planning in capitalist societies. Next, we review the policy responses to the politics of neo-liberalism and globalization in the developed world, contrasting the experiences of Europe and Australia. We then outline some elements of the contemporary neo-liberal assault on planning in Australia which, in a sense, seeks a return to the forms of governance (including, for some, *laissez-faire*) that prevailed until the Great Depression. Our primary empirical focus here is the National Competition Policy framework agreed to by Australian governments in 1995. We pause also to consider why the neo-liberal reform assault upon planning has proved so effective in Australia in comparison with most other developed nations. In conclusion we comment on the prospects for a revaluation of planning.

THE DECAY OF PLANNING VALUES

The movement against government planning and for 'marketization' gathered force after about 1970 (Swyngedouw, 1992). We need dwell only very briefly on this recent history which has seen the globalization of the capitalist economy, the failure of alternative economic frameworks, and the rise of neo-liberal politics both at the national level and within international political institutions (e.g. OECD, G7, World Bank, IMF).

A global economic crisis began to grip advanced capitalist nations from the early 1970s. The Long Boom that had begun to ease the painful memories of the Great Depression now foundered. The world economy slowed down as the demand for mass-produced goods began to be satisfied. The slow down was punctuated by sudden economic shocks, notably the oil crises of 1974 and 1979. The Bretton Woods agreement broke down in 1971, triggering an enormous growth in international financial transactions. The new surge of footloose investment capital encouraged a huge expansion of world trade. Also, industrial capital in the form of transnational corporations began to free itself from the West and flooded into developing nations. Information technology developed to control distant industrial plants and to manage money markets. Globalization emerged as a technological and political-economic system of control, integrating far-flung cities, regions and even nations in new international production and financial exchange systems.

This crisis contributed to a series of profound political shifts. In the developed world, Keynesian interventionism seemed unable to control the parallel growth of unemployment, inflation and interest rates ('stagflation') and was increasingly discredited. Neo-liberal interests took advantage of the sense of crisis to promote radical political reform programmes that sought to dismantle much of the Welfare State's institutions and regulatory regimes. The Soviet empire, for all its political, social and environmental deficits, had provided a working model of planning. When it began to collapse finally in the late 1980s under the weight of its economic and legitimacy crisis, there were hopes that a new form of democratic planning might evolve from the ruins of centralized Soviet-style communism. This was not to be. New and shaky post-Communist governments were often seduced by the simplistic, free-market reform programmes advocated by some Western economists and the many right-wing 'think tanks' that had been set up with vast funds by big business all over the world to promote neo-liberalism (Self, 1993, pp. 64-65).

These movements and forces produced the well-known ideological trilogy of competition, deregulation, and privatization. A new hegemony came to dominance bringing together patterns of thought – conservatism, liberalism, even forms of corporatism – which, in terms of their internal logic, are not mutually compatible. The conservative critique of planning ranges, of course, from deep antipathy to any form of spatial regulation to a variety of positions that accept some form of intervention in certain contexts and within carefully circumscribed limits. Along this spectrum of right-wing views on planning, neo-liberalism is distinguished for its profound hostility to all forms of spatial regulation, including urban and regional planning, environmental policy and economic development policies.

The broad intellectual case against urban planning rests essentially on the propositions that planning both distorts land markets and raises the transaction costs of development through bureaucratization of the urban economy. It is argued that these 'diseconomies' reduce employment growth and also stifle the ability of the land market to satisfy consumers' needs for housing and transport. For some conservatives, these diseconomies are, in some circumstances, tolerable if balanced against positive social outcomes, such as the prevention of environmental degradation and the preservation of various types of amenity (Hall *et al.*, 1973). For example, the Thatcher government's attempts to dismantle planning in Britain were to some extent frustrated by the opposition of conservative rural constituents fearful of a market 'free-for-all' in the countryside that would undo the 'containment of urban England' (Allmendinger, 1998). Again, neo-liberals are distinguished within the conservative political spectrum for their belief that markets, not regulation, produce optimal living environments. The general prescription of neo-liberals is for the removal of mandatory public regulations (notably, zoning) from land markets on the assumption that private coordinating mechanisms (e.g. covenants) will produce more efficient resource allocations.

Political conservatism rooted in neo-classical economics was hardly, however, the only critical voice to register in debates about the values of planning. Other critiques of planning values emerged, especially after the 1960s, including from Marxism, various radical democratic perspectives, and environmentalism. Whilst these perspectives shared with conservatism a critical regard for established planning values, there is, however, an important distinction to be drawn between what each saw as the *value of* planning. Generally speaking, conservatives and Marxists sought to *devalue* planning as a political project within capitalist society, whilst environmentalists and those espousing a range of radical democratic critiques have sought to *revalue* planning. Mostly these critiques were based upon a certain disillusionment with the seeming failure of planning to deliver the promised world. Few envisaged a return to the past, but rather sought to take planning further, hoping to make planning more democratic and responsive to new insights. Yet they also undermined the foundational tenets of planning without reinstating a clear vision of the future. Meanwhile, neo-liberalism was waiting in the wings to justify a return to the very world which had produced the problems planning was designed to solve.

Neo-Liberalism, Globalization and Planning

Whilst the neo-liberal discourse tends to take a broadly similar form in developed countries, its impact has been conspicuously uneven. In the English-speaking world, neo-liberal politics have been influential, to the point of redefining governance, in countries such as New Zealand, Britain and, more recently, Australia. However in the non-Anglophone, developed world – notably in Europe – neo-liberalism's purchase has been much more limited and more fiercely contested (Allmendinger, 1998; Healey and Piccinato, 1995).

In particular, there is a variety of potential consequences for planning regulation arising from globalization and neo-liberalism. This is evident when the effects of globalization on planning regulation are observed at different geopolitical scales and in different places. In some countries, such as Australia, the United States, Canada and New Zealand, the implications of a 'globally competitive economy' are enlisted in arguments against spatial regulation, whilst elsewhere globalization is cited as grounds for enhanced planning controls. It cannot therefore be said that recent and contemporary shifts towards a liberalized global economy have undermined the case for, or the strength of, planning regulations in all countries and regions. In short, globalization appears to have catalysed a new geography of planning regulation at the national and regional levels, characterized by de-regulatory and re-regulatory shifts, as well as the formulation of new spatial controls where previously they did not exist. In what follows we consider the contrasting consequences of neo-liberalism for planning regulation in two Federal political systems, the European Union and Australia.

The European Response

Globalization and neo-liberalism have not undermined the case for planning in Europe. Indeed, the reverse appears to be true and, according to Healey and Piccinato (1995, p. 4), there now exists in Europe 'a new enthusiasm for planning' (see also Newman and Thornley, 1996, p. 245). At the supranational level, the European Union (EU) has begun to develop an explicit and integrated spatial planning regime for the first time (European Commission, 1997). The planning ministers of Baltic Sea states have undertaken a similar initiative (Third Conference of Ministers, 1994). As Williams (1996, p. 3) comments:

> During the 1990s, we have seen a resurgence of interest in national and supranational spatial planning, and in the preparation of spatial planning studies of Europe as a whole. Partly, *this is a response to economic integration and the single market*, improvement of transport networks and new infrastructure such as the Channel Tunnel. [emphasis added]

Within the EU, a number of environmental and social factors, including political responses to globalization, seem to be animating the new enthusiasm for spatial planning. It is worth examining these briefly as they reveal much about the different political responses to globalization that have occurred in countries where neo-liberalism has the upper hand.

First, there is the recognition in the EU that political-economic integration – the massive broadening of the common regulatory space achieved by the Maastricht (1992) and Amsterdam (1997) Treaties – raises the need for policy coordination at the continental scale. Together, the two recent EU treaties have given freedom of movement to virtually every imaginable socio-economic phenomenon across the enormous terrain of Europe. Spatial patterns of socio-economic development will be – and already are being – changed by new flows of people and capital. The desire for spatial planning by the EU reflects an anxiety that policy is reacting to, rather than anticipating, these consequences of political-economic integration. Moreover, whilst regulatory systems have been rapidly integrated, the same cannot be said for infrastructure which, by its nature as fixed capital, resists sudden re-scaling. Transport (roads, rail and ports) is an obvious case in point, but also apparent are the frictions presented by different telecommunications systems and built environments. The Union is concerned at the prospect of a Hobbesian 'war of all against all' amongst the cities. Instead of promoting inter-urban competition for all capital resources, the Union has urged cities to seek investment that builds upon existing strengths and to cooperate in social and economic development as much as possible (Expert Group on the Urban Environment, 1996).

Why should the EU be trying to impose a geographic rationality on the market through spatial planning when it is, through the new union, elevating competition to a high principle? Surely the shift reflects a real concern to address the challenges raised for the EU by the greatly increased level of global and European competition. In the EU, the principle of competition is advanced through the ideal of Europe as a single competitive space – or a common terrain of regulation – which will thus be better equipped to compete and survive in the Darwinian struggle opened up by economic globalization. Much of the political impetus behind integration derives from anxieties over the economic power of large, non-European nations and rival geopolitical blocs (e.g. NAFTA). These politically-contentious fears lurk behind public arguments that cast integration as a means to reduce old national antagonisms and thus secure lasting peace in Europe.

The second major force behind the recent move to planning in the EU emerges from within the policy spheres of the Union itself. There is an awakening realization that there must be improved, and explicit, spatial coordination of those policies and programmes which have obvious geographic consequences. This is evident in the programmes of Directorates-General XI (Environment, Nuclear Safety and Civil Protection) and XVI (Regional Policies and Cohesion) which have both been evolving towards a more strategic and integrated model of spatial planning. These institutions of the EU seem much taken now with the idea of 'urban governance', a protean and evasive term, but one which betrays the desire for a new integrated approach towards the manifold, and interconnected, social, economic and environmental problems facing cities.

Third, environmental policy has become more influential within the Union, and this has raised the sense of 'spatial awareness' within policy spheres. Issues such as sustainable development are irreducibly spatial, and their increasing significance within the EU has had a secondary flow-on effect into the new spatial regulation domain. Importantly, environmental processes – and especially environmental externalities – have played their own role in highlighting the need for regional, cross-border and continental planning. Sometimes they have done this dramatically, as in the case of the Chernobyl accident which affected the European geopolitical consciousness to an extent that is rarely obvious to non-Europeans.

Fourth, and finally, planning has received a new political energy from the increasing recognition that uneven social development has worsened within Europe in recent decades, and may well be aggravated further by the process of political-economic integration within the Union. Social cohesion is a guiding principle of the EU, and there has been an accumulation of evidence in recent years (e.g. Dunford, 1994) to show that realization of this ideal is threatened by worsening spatial inequality and dislocation at a variety of scales. Very often the effects of globalization (e.g. deindustrialization) are cited as causes of new regional disparities. This evidence, combined with mounting political pressures within the nations and regions of Europe, has influenced the EU to adopt spatial planning as a corrective against geographic disparities that are sourced, in part, in global economic shifts.

The Australian Response

In contrast to Europe, political-economic globalization has been associated with a decline in public planning in Australia. A succession of state and Federal governments since the early 1980s have used the increasingly apparent economic impacts of globalization – including deindustrialization and rising unemployment – to justify the selective deregulation of urban change (Wiseman, 1998). Major deregulation has occurred mostly at the broad scale, at the level of metropolitan strategic planning, that had in the past sought to prevent and/or correct anomalies of resource allocation and injustices in the distribution of goods and services. As a series of conservative state governments assumed power during the 1990s, these redistributive spatial frameworks were, in most instances, either dispensed with or greatly eroded through deregulatory initiatives that aimed to enhance the investment appeal of Australian cities in the new global economic environment. These neo-liberal governments claimed that in the new global economic order investment capital was intolerant of any spatial frictions which impeded its (putative) hyper-mobility. In a context of public fiscal crises at the state level, strategic urban planning joined a cast of regulatory villains in neo-liberal political rhetoric. Electorates were told that redistributive policies were sclerotic artefacts of a pre-globalization era and that their continuance could only rob Australia's urban and regional communities of much needed investment. In short, those forms of broad-scale planning concerned with social equity had apparently been redefined by globalization as a political-economic liability.

If strategic planning had been rendered an extravagance – at least politically – by globalization, the 'costs' of statutory planning had also allegedly risen, encouraging recourse to new and debased forms of development control, such as 'performance zoning', flexible building standards, and the like. Much of the dilution of statutory planning in Australian cities had in fact been initiated and carried out by state and Federal Labor governments during the 1980s and early 1990s. By the time these Labor administrations were swept aside in the 1990s by a new wave of political conservatism, the deregulatory agenda had already been adopted and partly realized.

Paradoxically, within this relentless stream of deregulation, a small but significant counterflow of re-regulation was apparent in new policies and laws that sought to privilege certain investment environments in Australia's cities for local and international élites. As in many Western countries, Australia's residual strategic and statutory planning systems have become enmeshed in public policy machines that have sought to boost the investment appeal of capital and regional cities (cf. Molotch, 1976). In the new policy orthodoxy that has come to define urban planning, state governments act as 'entrepreneurs', soliciting new capital for their cities, whilst municipalities and local communities are expected to act in concert as economic 'growth machines' (Peel, 1995). As a consequence, strategic, and even statutory, planning have been transformed into vehicles for city promotion, rather than city regulation.

Enormous sums of public money and special facilitating legislation have been employed in numerous capital city redevelopment projects that mimic the more famous progeny of urban entrepreneurialism in North American and European cities. These publicly supported and promoted urban 'spectacles' – living and recreation spaces designed primarily for élites – are frequently located by harbours (e.g. Darling Harbour, Sydney) or significant riverfronts (Southgate in Melbourne; Southbank and Dockside in Brisbane) (Minnery, 1992; Stilwell, 1993). The social opportunity costs of these public investments – the infrastructure not supplied, the hospitals, schools and social services not upgraded or provided – have been conveniently overlooked in the boosterist discourses that have overtaken what remains of strategic urban planning. Urban policy has come to focus on the redevelopment and promotion of central cities and adjoining areas, whilst the suburbs (where the bulk of the Australian population lives) have been residualized, all but ignored in strategic thinking.

At the national level, the Federal government, though not formally involved in spatial planning, nonetheless contributed to the deregulation of Australia's cities during the 1980s and early 1990s through a series of neo-liberal initiatives that included infrastructure privatization and the formulation of a National Competition Policy (NCP) in conjunction with the states. By the early 1990s, there was belated recognition by the Federal government of the failure of new urban policy regimes to increase general welfare (Bunker and Minnery, 1992). A 'Building Better Cities' (BBC) programme was instituted in 1991, promising significant investment funds – over A$800 million in the first five years – to address the problems of locational disadvantage, housing inaccessibility and declining infrastructure (Orchard, 1995). Despite the political hopes invested in the BBC, the programme failed to reduce, or even stall, the worsening problems of locational disadvantage and spatial polarization (Stilwell, 1993). In 1996, a new conservative Federal government abolished what remained of the BBC and so ended altogether the central state's hitherto desultory interest in spatial policy.

CONTEMPORARY NEO-LIBERAL PLANNING REFORM IN AUSTRALIA

Neo-liberal politics in Australia have begun to transform some state planning systems through various policy initiatives that have undermined the place of any meaningful public intervention in urban development. In the following discussion we concentrate upon one radical reform process that has in recent years guided institutional and legislative reform in Australia: the National Competition Policy framework. We argue that this reform initiative – virtually now the orthodox view of governance in Australia – constitutes an assault upon the core values that have animated planning practice in the postwar era.

National Competition Policy (NCP)

In the early 1990s, the Australian (Labor) government, in conjunction with the states, established the framework for a National Competition Policy (NCP). In 1995 members of the Council of Australian Governments enshrined the NCP in a set of written agreements. The broad thrust of the NCP reflects the familiar desire of neo-liberal interests to reduce government to a minimal role in the economy, essentially to correcting limited aspects of market failures. NCP has already had far-reaching effects in many areas of public sector activity in Australia and, as we will show, its impact is now being felt in the planning arena. The political potency of the NCP derives in large measure from the extent of its political-economic aims which include reform of both public services and public regulation. Critically, the reforms aim at 'removing legislation that restricts competition in particular markets' at state and Federal levels (Industry Commission, 1995, p. 4).

To date, implementation of the NCP has spurred the privatization of public assets, especially state instrumentalities, and also the shift to market pricing in government business enterprises (Hood, 1997). State bodies that traditionally provided urban utilities, such as electricity, water and gas, have been the particular targets of privatization policies, though many still remain in public hands in various cities and regions (Hayward, 1997). Critics have claimed that both the formulation and implementation of NCP reforms have largely been undertaken without significant public consultation. In late 1997, an independent member of the Australian Capital Territory Legislative Assembly publicly claimed that 'The Council of Australian Governments was ruling Australia by executive decree and ramming [national competition] policies down the throats of state and territory parliaments' (*The Canberra Times*, 23 October 1997, p. 5). Moreover, the NCP framework was pressed upon local government without consultation.

The Impacts of NCP on Planning

Of central interest here is that the reform process has been extended to planning. By 1998, several state governments had embarked upon NCP reviews of their development control systems. One key effect of NCP policy on New South Wales' planning system has been the introduction of competition for development approval – hitherto the preserve of local government (Mahony, 1997). Under the new system, private practitioners can be accredited as approval authorities for certain categories of development control. The idea is that an applicant seeking planning approval will be able to choose between private and public bodies for the assessment of the proposal. Presumably this will encourage competition between accredited approval bodies, leading to greater regulatory efficiencies, and, in turn, lower transaction costs for applicants. The privatization of development control follows earlier similar reforms to building approval systems in several states.

However, as Mahony (1997) points out, the effective privatization of development control in local government threatens to reduce public participation and accountability in planning, whilst also increasing the risk of decisions that contradict broader community interests. In terms of participation and accountability, several objections arise. First, the reforms pass a vital aspect of public decision-making from elected to non-elected bodies. Moreover, the changes will also free private certifiers from the public accountability procedures that bind local government activities including plan-making and development approval. They will also 'circumvent the use of freedom of information legislation or administrative law remedies' in respect of planning decisions (Mahony, 1997, p. 11). In sum, Bonyhady (1997, p. 1) concludes that the NCP reforms to the New South Wales legislation will 'reduce public rights of participation which government made so much of in the 1970s but which are now decidedly out of fashion across the country'.

On the issue of quality, Mahony (1997) points out that private certification systems in other areas of governance – notably building control – have led to a decline in regulatory standards, sometimes threatening public safety. In late 1997, the Australian Council of Professions voiced concerns about 'state reviews of competition policy [that] could lead to unqualified people passing themselves off as experts' (ABC News, 17 November 1997). The Council noted that privatization of regulatory approval processes raised the spectre of poor and incompetent servicing in key professional areas that affect public safety, such as building control and engineering.

Perhaps most importantly, Mahony observes that the NCP changes will increase regulatory uncertainty in the planning system and thereby actually contradict the aims of the reform agenda itself:

> even from a purely economic perspective, there is a strong argument that the certainty of land use and land value provided by land use planning are matters of public benefit. Such certainty is a necessary precondition for business confidence, let alone competition. (Mahony, 1997, p. 18)

Perhaps even more worrying for advocates of public planning is the current NCP review being undertaken by another state, Victoria. The consultant brief released for the first stage of this review requires a study that will, amongst other things:

> identify any restrictions on competition arising from the Victorian planning system; consider alternative means of achieving the same restrictions on competition, including non-legislative means; and estimate the costs and benefits associated with alternative options. (Department of Infrastructure, 1997, p. 4)

In this document, 'restrictions on competition' can include 'the administration of the planning scheme *and* the actual regulation of land use and development by means of . . . planning schemes' (Department of Infrastructure, 1997, p. 5) [emphasis added]. Thus the brief intimates that the NCP principles can be used to question not only the basis of *how* public planning 'services' are supplied – as in New South Wales – but also the rationale for the development controls themselves. The review might therefore conceivably recommend both the privatization of planning approval and the reduction of development controls. This view is further justified by the prescription that the consultant study will examine the 'relevance' of 'government intervention through the planning process' in the property market (Department of Infrastructure, 1997, p. 4).

Overall, the evidence suggests that NCP threatens to undermine public urban policy – including strategic and statutory planning – from two directions: first, by eroding both the legislative sphere of planning and the role of public planners generally in development control; and second, by reducing the involvement of the state in the provision of urban services, including utilities and social-welfare programmes. Governments in Australia appear to be using the NCP framework as a means of relinquishing their power to shape the direction of urban development. Of course, this does not necessarily imply the 'ungoverning' of Australia's cities: already private sector urban development and urban financing corporations and firms have moved to control the regulatory spaces vacated by governments. Hayward writes:

> It is now safe to say that we have reached a new phase in Australian urban history, with the private sector now responsible for the provision and/or management of an unprecedented proportion of our urban infrastructure, both social and economic. (Hayward, 1997, p. 55)

In recent years, entrenched anti-planning conservatism has been revitalized by the increasing political authority of neo-liberalism. As we have shown, new agendas for planning reform have emerged, and their impacts have already been felt in state and local planning systems. But how can the efficacy of the neo-liberal planning reform initiatives in Australia be explained? Why have the politics of neo-liberalism and globalization proved far more damaging for planning in Australia than they have in Europe?

Australian Planning: Two Weaknesses

We argue that there are two main reasons why Australian planning has proven so vulnerable to neo-liberalism: first, because of the specific manner in which neo-liberalism and globalization have recast the general Australian political landscape in contrast to other developed countries (especially in Europe); and second, because of the weak institutional framework that has characterized Australian planning since its inception.

It must seem curious to European observers that Australia is embarking on a radical set of neo-liberal reforms to its economy and society, at the very time when similar, earlier initiatives, especially in Britain, have been politically discredited and when all the talk now centres upon a new 'Third Way' in capitalism (see Giddens, 1994). To explain this seeming paradox, it is necessary to recall that whilst the two major Anglophone developed nations were experiencing the full force of neo-liberalism in the 1980s – Thatcherism in Britain and Reaganomics in the United States – Australia was governed not by Conservative parties but by reformist Labor administrations at the state and Federal levels. Whilst a few state Labor governments charted social democratic courses during the 1980s, the reformist Federal governments of the 1980s and early 1990s – the Hawke (1983-91) and later Keating (1991-96) administrations – were guided by the broad tenets of neo-liberalism, though these were softened in practice by policies and programmes that provided social assistance to those most affected by economic restructuring. In general, these Federal governments embraced a milder version of neo-liberalism that accepted the 'need' for a radical economic restructuring that would secure Australia's global competitiveness, whilst rejecting the extreme post-Welfare State politics that came to redefine social policy frameworks in Britain and the United States (Argy, 1998).

If it had a kinder face, this reform vector was nonetheless broadly neo-liberal in character, and its espousal and rigorous implementation by the major social democratic party in Australia effectively foreclosed on the possibility of an alternative political-economic response to globalization. Throughout the 1980s and into the early 1990s, neo-liberalism remained as an effective bi-partisan policy setting, endorsed by both major parties whose policies were largely distinguished by relatively minor programme details rather than by marked political-economic differences. Throughout this time, neo-liberalism became deeply embedded in the Australian political and institutional landscape, and the development of NCP, largely by the Keating Labor administration, crowned neo-liberalism as the new public sector orthodoxy in Australia.

In summary, the potency of neo-liberalism in Australia is largely explained by the fact that its 'natural political enemies' – the hitherto social democratic Labor Party – surrendered to, without much resistance, the broad programme of globalization and economic deregulation that was vigorously promoted from the late 1970s by a loose coalition of powerful corporate interests, global institutions (e.g. IMF, World Bank), and economic theorists (Wiseman, 1998). A similar ideological capitulation of the New Zealand Labour Party explains that country's even more fervent embrace of neo-liberalism from the early 1980s. In both nations, planning has been targeted, at different times, for deregulatory reform as part of a broader, thoroughgoing restructuring of the public sector that has emphasized the goals of minimal government and maximum economic efficiency over other social priorities.

It also needs to be explained that planning never gained a really secure institutional footing in Australia, and has been, with few exceptions, long regarded as a relatively minor regulatory function by governments. Most importantly, no national planning framework emerged, with successive postwar Federal governments disavowing the need for explicit spatial regulation. In two notable, if brief, departures from this scenario, what emerged were policies for regional and spatial development which nonetheless lacked an explicit regulatory framework. In the early 1970s, an attempt was made by the Whitlam administration (1972-75) to establish a national approach to urban problems. A range of national urban policies was set in place to manage growth, ensure infrastructure provision, and dampen land speculation. This ambitious programme was, however, fiercely resisted by the private sector and by conservative parties and was quickly dismantled after the demise of the Whitlam government. Urban policies returned to the national political agenda during the Hawke and Keating governments (1983-96). The other programme was for 'Better Cities' in the early 1990s, mentioned previously. Some urban commentators criticized the narrowness of this programme's objectives; others pointed out that the funding was less than the cuts that the Federal government had made in grants to states for basic urban services (Forster, 1995). National attention to urban policy lapsed with the election of a new Conservative coalition government in 1996.

Thus, there has been no permanent attempt at the Federal level to establish a national urban and regional planning system. Planning has been left to the states where it has thus remained a regional endeavour that has been exposed to relentless opposition and critique from minor, localized interests. In fact, most planning activity – certainly development control – has been undertaken by local governments which are themselves the weakest, least authoritative level of administration in Australia, creatures of state legislation with no constitutional status. The continual manipulation of local government, including its planning activities, by state governments has reduced both the institutional stability and the public status of planning.

Finally, unlike in Britain (cf. Allmendinger, 1998, p. 5), planning in Australia has no 'natural constituency' that is prepared to defend it politically from hostile reform agendas. Australian rural conservatives, unlike their British counterparts, have generally displayed antipathy to planning. Broader public interest in spatial regulation has remained muted, confined mostly to local development issues. The peak professional planning body – the Royal Australian Planning Institute – has been and remains an ineffective advocate for spatial regulation, and has even at times seemed to endorse various aspects of the neo-liberal agenda. These factors have combined to render Australian planning exceedingly vulnerable to reform programmes such as the NCP.

Is Planning History?

In this chapter we have painted an, at times, rather grim picture of how neo-liberalism and economic globalization have transformed planning regulations within certain countries and regions. Nonetheless, from a broader perspective, planning is certainly not history, at least not in Europe, where, as we have shown, both the EU and member states have moved recently to reinforce spatial regulation in a variety of ways.

In other countries, such as Australia and New Zealand, the future of planning seems less certain: already neo-liberal reform initiatives have made deep inroads into spatial regulatory systems and the immediate future, it seems, promises more of the same. There are, however, signs that the broader community of planning professionals and activists in Australia is starting to question, if not yet actively resist, the anti-planning cause. In June 1998, a small but representative forum was convened to consider the cause of 'Renewing Australian Planning'. Key players from government, academic and community sectors agreed that planning was imperilled by neo-liberalism and in need of political revaluation (Gleeson and Hanley, 1998). We concur with this sentiment. There must be a political and intellectual re-inscription of the case for planning in Australia if this form of governance is to survive into the new millennium in any meaningful form. Lessons can be learned from the experiences of the EU – another (if inchoate) Federal system – and its member states. In this context, as we have shown, the neo-liberal case against planning has certainly registered but has been countered by alternative social democratic analyses of globalization and institutional reform.

In the revival of planning regulation in Europe, one glimpses the real and potential autonomy that States possess when confronted by the, admittedly powerful, processes of political-economic integration. The point is that States can and sometimes do make policy responses to globalization that resist the political blueprints laid down by the neo-liberal guardians of the new global economy. In Australia, the point has hardly registered on the broad political consciousness. Interestingly, the key premises which underlie the EU's decision to adopt spatial planning – notably, the implications of economic globalization and technological change – are precisely those advanced in Australia, and in other places, in arguments against planning. This suggests that planning advocates in Australia and elsewhere should do more to confront simplistic neo-liberal reform agendas with new arguments that highlight the social, economic and ecological imperatives for planning in an era of globalization. A broader recognition of the dependencies and vulnerabilities that characterize globalization would help to counter the myth of 'unstoppable and uncontrollable' reform that has depoliticized this process.

There is a need to recruit to the cause of planning renewal the other critiques of modernist spatial regulation – most notably, environmentalism and radical democracy – that oppose the rule of the market. As explained earlier, these well-established critical impulses have sought the revaluation rather than the devaluation of planning, centring upon new ideals of ecological sustainability and democratic participation. Since the 1960s, the social movements and thinkers representing both critiques have sought to define new planning frameworks that would protect and enhance environmental and democratic ideals.

Revaluing planning has meant literally a struggle waged, at a number of political levels, to reshape and renew planning values. For example, a range of contemporary feminist, postmodern, and other radical democratic perspectives criticize planning's tendency to adopt technocratic and authoritarian forms. In the main, however, these critiques have aspired to a democratized form of planning that would recognize and cherish social diversity and an expanded range of human values (Sandercock, 1998). In a real sense, this outlook was very much anticipated (if not completely) in the democratic vision for planning articulated by Karl Mannheim in the 1940s (Mannheim, 1951). Environmentalism, whilst generally sharing the ideal of democratization, has sought an expanded ecological agenda for planning, and in particular the dethronement of positivistic, technocratic approaches to spatial regulation. Again, there are important antecedents for this claim in the thinking of early planning advocates, such as Lewis Mumford, who saw spatial governance as critical to the reconciliation of humans to their environment (Evans, 1997).

CONCLUSION

Today there are some signs that the tide of neo-liberalism has begun to turn. The

Bute Library

Items on loan 14/11/2012 09:34

XXXXX18091

Item Details	Due Date
603668661	
Property valuation : the five	27/11/2012
605474177	
Development site evaluation	27/11/2012
611810936	
Health policy in a globalising	27/11/2012
616631311	
Development economics : D	20/11/2012
609270135	
Poverty and development in	27/11/2012
611954177	
[illegible]	[illegible]2
608937632	
Property development	20/11/2012
614794402	
Property development / by A	20/11/2012

Indicates items borrowed today

Overdue items will incur charges

Renew items at: http://library.cf.ac.uk

growth of the populist One Nation party in Australia, like the rise of Fascist parties in the 1930s, owes little to the attractiveness of that party's belief system and everything to the social breakdown caused by a return to the ideology of the natural market. The same tendency is observed worldwide (Martin and Schumann, 1997). A progressive alternative to neo-liberalism demands a return to the challenge posed in the 1940s, and at the same time a capacity to learn from the experience of the form of planning which has been realized since.

As we have intimated, there is much in the (re)constructive criticism of planning offered by contemporary environmentalists and radical democratic interests that was anticipated by earlier planning advocates. The convergence of older and newer critical thought streams offers the basis for a renewal of planning values and spatial regulation. When the 'march of events' allows, and as the ecological crisis unfolds, the case for an alternative, 'revalued planning' must be ready, and carefully articulated within a broader vision for a humane, fair, ecologically responsible society. If not, those with the will to power and without such ideas will seize their moment. Planning is undeniably history – the political challenge is to make it our future.

REFERENCES

Argy, F. (1998) *Australia at the Crossroads: Radical Free Market or a Progressive Liberalism?* Sydney: Allen and Unwin.

Allmendinger, P. (1998) Planning and Deregulation: Lessons from the UK. Paper presented at the 8th International Planning History Society Conference, 15-18 July 1998, University of New South Wales, Sydney.

Bonyhady, T. (1997) Diminished democracy. *Urban Research Program Newsletter* (Australian National University), **31**, pp. 1-2.

Buchanan, J.M. (1977) *Freedom in Constitutional Contract, Perspectives of a Political Economist.* College Station: Texas A and M University Press.

Bunker, R. and Minnery, J. (1992) *Recent Commonwealth Initiatives in Urban Affairs* (National Monograph Series, 1). Adelaide: Australian Institute of Urban Studies.

Department of Infrastructure (Victoria) (1997) *Consultant Brief for Analysis of Planning System under National Competition Policy.* Melbourne: Department of Infrastructure.

Dunford, M. (1994) Winners and losers: The new map of economic inequality in the European Union. *European Urban and Regional Studies*, **1**, pp. 95-114.

European Commission (1997) *European Spatial Development Perspective.* Brussels: European Commission.

Expert Group on the Urban Environment (1996) *European Sustainable Cities.* Brussels: European Commission – Directorate General XI.

Evans, B. (1997) From town planning to environmental planning, in Blowers, A. and Evans, B. (eds.) *Town Planning into the 21st Century.* London: Routledge, pp. 15-32.

Forster, C. (1995) *Australian Cities: Continuity and Change.* Melbourne: Oxford University Press

Giddens, A. (1994) *Beyond Left and Right: The future of radical politics.* Cambridge: Polity Press/Blackwell.

Gleeson, B.J. and Hanley, P. (eds.) (1998) *Renewing Australian Planning.* Urban Research Program Monograph, Canberra: Australian National University.

Hall, P., Gracey, H., Drewett, R. and Thomas, R. (1973) *The Containment of Urban England,* 2 volumes. London: George Allen and Unwin.

Hayward, D. (1997) The privatised city: Urban infrastructure, planning and service provision in the era of privatization. *Urban Policy and Research*, **15**, pp. 55-56.

Healey, P. and Piccinato, G. (1995) *The State of Spatial Planning in Europe*. Stockholm: Association of European Schools of Planning.

Hood, A. (1997) The not-so-level playing field: Reform of significant business activities of Queensland local governments. *Local Government Law Journal*, **3**, pp. 64-72.

Industry Commission (1995) *Implementing the National Competition Policy: Access and price regulation*. Canberra: Australian Government Publishing Service.

Mahony, S. (1997) Microeconomic Reform meets Land Use Planning – National Competition Policy and the Privatization of Land Use Decisions. Unpublished paper, available from Urban Research Program, Australian National University.

Mannheim, K. (1951) *Freedom, Power and Democratic Planning*. Oxford: Oxford University Press.

Martin, H-P. and Schumann, H. (1997) *The Global Trap: Globalization and the assault on democracy and prosperity*. London/New York: Zed Books, Sydney: Pluto.

Minnery, J. (1992) Land Use conflicts: The Case of Brisbane's Dockside. Griffith University, The Centre for Australian Public Sector Management, Brisbane Project Research Paper no. 5.

Molotch, H. (1976) The city as a growth machine: toward a political economy of place. *American Journal of Sociology*, **82**, pp. 309-332.

Newman, P. and Thornley, A. (1996) *Urban Planning in Europe: International Competition, National Systems and Planning Projects*. London: Routledge.

Orchard, L. (1995) National urban policy in the 1990s, in Troy, P. (ed.) *Australian Cities: Issues, Strategies and Policies for Urban Australia in the 1990s*. Melbourne: Cambridge University Press, pp. 65-86.

Peel, M. (1995) The urban debate: From 'Los Angeles' to the urban village, in Troy, P. (ed.) *Australian Cities: Issues, Strategies and Policies for Urban Australia in the 1990s*. Melbourne: Cambridge University Press, pp. 39-64.

Rothbard M. N. (1973) *For a New Liberty*. New York: Macmillan.

Sandercock, L. (1998) *Towards Cosmopolis: Planning for Multicultural Cities*. Chichester: John Wiley.

Self, P. (1993) *Government by the Market? The Politics of Public Choice*. Basingstoke: Macmillan.

Stilwell, F. (1993) *Reshaping Australia: Urban Problems and Policies*. Sydney: Pluto.

Swyngedouw, E. (1992) The Mammon Quest. 'Glocalisation', interspatial competition and the monetary order: The construction of new scales, in Dunford, M. and Kafkalas, G. (eds.), *Cities and Regions in the New Europe*. London: Bellhaven Press, pp. 39-67.

Third Conference of Ministers for Spatial Planning and Development (1994) *Visions and Strategies Around the Baltic Sea: Towards a framework for spatial development in the Baltic Sea region*. Karlskrona, Sweden: The Baltic Institute.

Troy, P. (ed.) (1995) *Australian Cities: Issues, Strategies and Policies for Urban Australia in the 1990s*. Melbourne: Cambridge University Press.

Williams, R.H. (1996) *European Union Spatial Policy and Planning*. London: Chapman.

Wiseman, J. (1998) *Global Nation? Australia and the politics of globalisation*. Melbourne: Cambridge University Press.

Index